ADVANCES IN APPLIED LIPID RESEARCH

Volume 1 • 1992

ADVANCES IN APPLIED LIPID RESEARCH

A Research Annual

Editor: FRED B. PADLEY
Unilever Research Laboratory
Colworth House, Sharnbrook
Bedford, England

VOLUME 1 • 1992

London, England *Greenwich, Connecticut*

JAI PRESS LTD
118 Pentonville Road
London N1 9JN

JAI PRESS INC.
55 Old Post Road No. 2
Greenwich, Connecticut 06836

ISBN: 1-55938-317-8

British Library Cataloguing in Publication Data
Advances in applied lipid research: Vol 1.
(Advances in applied lipid research)
I. Padley, Fred B. II. Series
574.19

Printed in the United States of America

CONTENTS

LIST OF CONTRIBUTORS

Robert G. Ackman	Canadian Institute of Fisheries Technology Technical University of Nova Scotia Halifax, Nova Scotia, Canada
Keith D. Bartle	School of Chemistry University of Leeds, England
Tony A. Clifford	School of Chemistry University of Leeds, England
Stephen C. Cunnane	Department of Nutritional Sciences University of Toronto, Ontario, Canada
R. Julian Davies	DSIR Industrial Processing Division Lower Hutt, New Zealand
Peter Eigtved	Novo Nordisk A/S Bagsvaerd, Denmark
Peter N. Gillatt	Leatherhead Food RA Leatherhead, England
Jane E. Holdsworth	DSIR Industrial Processing Division Lower Hutt, New Zealand
J. Barry Rossell	Leatherhead Food RA Leatherhead, England

PREFACE

This is the first volume of a series aimed at the applied area of lipid research. The topics selected have been chosen intentionally not to focus on a single theme but more to offer something of interest across a broader field.

The contributors to this volume have been chosen because of their close involvement with the specific areas under review. The areas hopefully are seen as topical by the reader. These are certainly subjects of research which are starting to be applied and for which I anticipate a growing interest in the future.

My sincere thanks to the authors for the time and effort which they have put into compiling these reviews which I trust will be of value to scientists and technologists working in the lipid field.

Fred B. Padley
Series Editor

ENZYMES AND LIPID MODIFICATION

Peter Eigtved

OUTLINE

Advances in Applied Lipid Research, Volume 1, pages 1–64

ISBN: 1-55938-317-8

1. INTRODUCTION

1.1 Background and Scope

In its broadest sense, enzymes and lipid modification would include the very large number of different substances and enzymatic reactions known from the biochemistry of lipids in different organisms. In this chapter, emphasis will be on the more common lipids of industrial significance and the enzymes that are most likely to be applied. However, in the field of applied lipid research, biotechnology will enable our knowledge of biochemical pathways to be used to industrial lipid production and modification to a greater extent in the future.

A recent source of detailed information about lipids is *The Lipid Handbook* [1]. The lipids covered here will mainly be the naturally occurring fatty acid based substances and their derivatives: acylglycerols, glycerophospholipids, fatty acids and esters thereof. Products based on these are important in foods and as oleochemicals. Steroids, icosanoids and other lipid classes of pharmaceutical and physiological significance are extensively described in terms of biosynthesis and biotransformation elsewhere. The metobolism of dietary fats and synthesis of lipids in humans will not be included in any detail.

Many enzymes involved in lipid modifications are not easily separated from their natural environment and applied on a larger scale. They may be co-factor dependent, intracellular, unstable, part of complex enzyme systems and/or difficult to obtain. The major exception to this is the extracellular lipases from microorganisms, which alone offer many established and possible lipid modifications. Therefore lipases will be given primary attention.

The most recent monograph on lipases apeared in 1984 [2]; however, the discussion on microbial lipases and their applications is rather limited and a recent review on this area can be found in [3]. An excellent review on present and future applications of lipases was published in 1985 [4]. Since then, much new information has been published, especially in the fields of the molecular biology, structure and application of lipases. Although largely outside the scope of this review, it should be mentioned that the use of lipase stereoselectivity in organic synthesis is a rapid growing area.

1.2 Biotechnology of Lipids

Focusing on the applied and technical aspects of enzymes and lipid modification, the term biotechnology should cover the subject. The biotechnology of lipids includes the production of specific lipids in plants and microorganisms by cloning and mutation techniques. These technologies are expected to compete in the future with *in vitro* enzymatic lipid modifications. State-of-the-art and references will be given but no detailed description. Modifications in the fatty acid chain is a field of considerable interest, but with difficult tasks to solve if reactions are to be applied on a large scale, e.g. biohydrogenation of vegetable oils.

The world-wide annual production of vegetable oils in 1990 was about 56 million tons [5]; the addition of animal fats and marine oils brings the total above 70 million tons. Most of this is processed by physical and chemical methods to make it suitable as foods, ingredients and chemicals. With the growing concern to use less energy, avoid certain chemicals in factories and the environment, and produce pure, natural and healthy products, we may expect biocatalytic processes to become much more common in the oils and fats industry.

It is important to view the research within the lipid biotechnology as a whole, aiming at the optimal solution based on our current understanding and level of technology. A number of recent publications with a broad coverage of the subject are available [6–12].

2. THE ENZYMES

2.1 Lipases

2.1.1 Definitions and Basic Characteristics

Lipases, or glycerol ester hydrolases (EC 3.1.1.3), are enzymes which occur widely in microorganisms, plants and animals. They catalyse the hydrolysis of triglycerides to form di- and monoglycerides, free fatty acids, and glycerol. In this way, triglycerides from natural oils and fats provide energy and essential fatty acids for living organisms. Lipases differ from other enzymes because their natural substrates are insoluble in water. Their catalytic activity is expressed at the interface between the enzyme/water phase and the lipid/organic phase, e.g. when triglycerides are in the form of emulsified droplets in water containing lipase in solution.

In a famous experiment by Sarda and Desnuelle [13] it was shown that the activity of pancreatic lipase was greatly increased when the solubility limit of the partial water soluble substrate triacetin was passed and interface thus formed. Traditionally this has been one of the ways to distinguish between esterase and lipase activity; however, the division is not clear-cut.

Some enzymes are referred to as esterases because of a high activity towards short chain fatty acid esters or their broad substrate specificity. Others are named after a particular ester substrate, like cholesterol esterase or cutinase. Serine proteases may have a significant esterase activity. One of the most interesting lipase features is the possibility of using the reverse reaction, synthesis, or a combination of hydrolysis and synthesis, interesterification, in very low water content systems.

2.1.2 Sources

Although widespread in nature, lipases have only recently become available in larger quantities for industrial purposes. Some of the major application possibilities have only been established within the last 10–15 years, and this is short compared with the developments on production of and now large markets for proteases and carbohydrases. In addition, yield improvement, purification and immobilization of extracellular microbial lipases have demanded new technology, partly because of the special interfacial properties of lipases.

Microbial sources have been the most attractive as thermostable lipases without co-factor requirements and of different specifications have been extensively described. However, traditional applications like enzyme-modified milk fat has also been based on animal lipases such as pre-gastric esterase and pancreatic lipase. In addition, the mammalian lipases (and phospholipases) have been studied in detail over a long period of time, and our understanding of the structure and function of these enzymes are in many ways more advanced than for the microbial lipases.

With genetic engineering techniques now available, recombinant or mutant enzymes may be produced in a suitable microorganism with a more directed approach towards desired properties and efficient expression. Table 1 lists some selected mammalian and microbial lipases, their glyceride specificity and some further characteristics.

Plant lipases. Plant lipases [30–37] have been intensively studied, especially in connection with the germination of oil seeds, where their function is to mobilize energy and provide carbon skeletons from the stored triglyceride [14]. Lipase associated with the oil palm mesocarp may be endogenous and/or microbial. It is heat inactivated before milling to avoid oil loss by hydrolysis [32]. Mustard and rape lipases have been characterized as possible enzymes for biotechnological purposes [33]. Rape lipase specificity studies show low activity towards 4-*cis* and 6-*trans* unsaturated fatty acids [34] and this has been used in the enrichment of GLA [35].

Animal lipases. Animal lipases, especially the human enzymes, have been described in great detail due to their biological significance [38]. Moreover, many lipases of animal origin have interesting specificity properties and have

Table 1. Selected lipases and characteristics.

Source	Specificity[a]	Other characteristics	References
Mammalian			
porcine pancreatic	1,3-	co-lipase dependent	[14]
human gastric	1,3-(non-)	acid stable and active, essential –SH	[15]
human bile salt activated	non-	bile salt dependent	[16]
human lipoprotein	1,3-	apo C-II dependent	[17]
Microbial			
Aspergillus niger	1,3-	enzymes with variable fatty acid specificity	[18]
Candida cylindracea	non-	high hydrolytic activity	[19]
Candida antarctica			[20]
A-lipase	non-	very thermostable	
B-lipase	non-(1,3)-	high esterification activity	
Chromobacterium viscosum	1,3-		[21]
Geotrichum candidum	non-	*cis*-Δ-9 fatty acids, isoenzymes	[22]
Humicola lanuginosa	1,3-	alkali stable and active	[23]
Mucor miehei	1,3-	A- and B-isoenzymes	[24]
Penicillium cyclopium	non-	mono/diglyceride specific, other lipase components	[25]
Pseudomonas fluorescens	1,3-		[26]
Rhizopus arrhizus	1,3-	some phospholipase A_1 activity	[27]
Rhizopus delemar	1,3-	polyester activity	[28]
Staphylococcus hyiscus	non-	lipase and phospholipase activity	[29]

[a] Positional specificity on glycerides.

been applied in areas such as flavour enhancement, digestive preparations, organic chemistry, analytical and diagnostic systems, etc. Bovine and porcine pancreatic lipase have been available from pancreatic powder (pancreatin) and widely used [14]. Pre-gastric esterases are produced in the salivary glands and can be obtained from the abomasum of milk-fed calves, lambs and kids. They are important in the manufacture of certain (Italian) cheeses due to their short-chain fatty acid specificity [39]. Pig liver esterase (and other enzymes) has been used in synthetic organic chemistry because of its stereospecifity in various reactions [40].

Digestive lipases. Among the mammalian digestive lipases, a number are used or being considered as digestive aids or pharmaceutical preparations. People with exocrine pancreatic insufficiency, such as cystic fibrosis and chronic pancreatitis patients, are dependent on enzyme supplementation; lipase being the most important and obtained from pancreatin [41, 42].

Gastric lipases are natural acid stable enzymes and could be prepared as oral preparations. The rabbit gastric lipase has been isolated for this purpose, and the human enzyme was recently cloned and expressed in yeast [15]. The name lingual lipase is often used, but is only appropriate in the rat and mouse where it is mainly secreted from the tongue tissue. In polygastric species, like those mentioned above, secretion of pre-gastric lipase occurs in other tissue, but in man and most other mammals acid lipase production is associated with fundic cells of the stomach [43, 44]. Gastric lipase levels are lowered as a function of age (above about 50 years of age) but apparently not due to common diseases [15].

Bile salt activated lipase (BSAL) from human milk is a third major digestive lipase. In the breast-fed newborn, pancreatic lipase production is low and BSAL improves the digestion of milk fat [16, 45, 46]. A recombinant enzyme might be used as a digestive preparation, especially in connection with infant formula fed babies. Pancreatic carboxyl ester lipase (cholesterol esterase) is homologous to BSAL [47]. Lipoprotein lipase (LPL) [17] plays a major role in the regulation of lipid metabolism and digestion of parenteral lipid emulsions [48]. Inherited LPL deficiency due to inactive mutant enzymes has been reported, but occurs only in about 1 in 1 000 000 individuals [49].

Lipases from fish. Digestive lipases from cod have been studied in terms of fatty acid specificity [50]. A lipase with high activity, preference or positional specificity towards long-chain polyunsaturated fatty acids, e.g. EPA and DHA, would be of interest, but has not been reported so far.

Microbial lipases. Some recent surveys detailing the many sources, characteristics and uses of microbial lipases have been performed by Stead [51] and Godtfredsen [3]. They reflect the wide distribution of enzymes with different properties, screening and research activities, and industrial interest. Occurrence in nature is associated with microorganisms growing on lipid substrates and induction of lipase has also been achieved this way in the production of lipase by fermentation. Lipases of pathogenic bacteria have been studied [52], especially in relation to acne [53]. Psychrotrophic bacterial lipases are of importance in milk fat spoilage [54].

Extracellular lipases are common and most easily recovered, although binding to cells or mycelia may be stronger than for other enzymes due to their tendency to attach to interfaces. A microbial classification of lipases from bacteria, fungi and yeasts does not lead to corresponding groups of property-related enzymes within the various species, although there may be growth, expression and structural similarities. A number of fungal lipases seem to be quite similar in terms of specificity, stability and structural properties. Table 1 includes some of the most studied microbial lipases. A more extensive list of microbial lipases and possible industrial sources is given in Table 2.

Table 2. Sources of microbial lipases.

Microorganisms[a]	Industrial source[b]
Absidia corymbifera	
Absidia hyalospora	
Achinotobacter calcoaceticus	
Achinotobacter pseudoalcaligenes	
Achromobacter sp.	
Achromobacter lipolyticum	
Alcaligenes sp.	
Alcaligenes denitrificans	
Amylomyces rouxii	
Arthrobacter sp.	
Aspergillus awamori	
Aspergillus flavus	
Aspergillus fumigatus	
Aspergillus japonicus	
Aspergillus niger	Amano Novo Nordisk Biocatalysts Fluka
Aspergillus oryzae	
Bacillus cereus	
Bacillus megaterium	
Bacillus stearothermophilus	
Candida antarctica	Novo Nordisk
Candida auricularia	
Candida curvata	
Candida cylindracea (rugosa)	Meito Amano Biocatalysts Fluka
Candida deformans	
Candida foliorum	
Candida humicola	
Candida (Yarrowia) lipolytica	Fluka Amano Biocatalysts
Candida rugosa (cylindracea)	(See above)
Candida tsukubaensis	
Chaetomium thermofile	
Chromobacterium chocolatum	
Chromobacterium viscosum	Toyo Jozo Fluka Biocatalysts
Corynebacterium acnes	
Flavobacterium arborescens	
Fusarium oxysporum	
Fusarium solani	

Table 2. Continued.

Microorganisms[a]	Industrial source[b]
Geotrichum candidum	Biocatalysts
	Amano
Humicola grisea	
Humicola insolens	
Humicola (Thermomyces) lanuginosa	Novo Nordisk
	Amano
	Biocatalysts
Lactobacillus sp.	
Leishmania donovani	
Micrococcus freudenreichii	
Mucor sp.	
Mucor javanicus	Amano
	Biocatalysts
Mucor lipolyticus	Fluka
Mucor (Rhizomucor) miehei	Novo Nordisk
	Gist Brocades
	Biocatalysts
Mucor pusillus	
Myxococcus xantus	
Penicillium crustosum	
Penicillium cyclopium	Amano
	Biocatalysts
Penicillium roquefortii	Amano
	Fluka
	Biocatalysts
Phycomyces nitens	
Pichia miso	
Propionibacterium acnes	
Propionibacterium granulosum	
Proteus sp.	
Pseudomonas sp.	Fluka
	(lipoprotein lipase)
Pseudomonas aeruginosa	
Pseudomonas cepacia	
Pseudomonas fluorescens	Amano
	Biocatalysts
	Fluka
Pseudomonas fragi	
Pseudomonas pseudoalcaligenes	
Pseudomonas stutzeri	
Rhizopus sp.	
Rhizpus arrhizus	Biocatalysts
	Fluka
Rhizopus chinensis	
Rhizopus delemar	Amano
	Biocatalysts
	Fluka

Table 2. Continued.

Microorganisms[a]	Industrial source[b]
Rhizopus japonicus	Biocatalysts
Rhizopus microsporus	
Rhizopus niveus	Amano Biocatalysts Fluka
Rhizopus nodosus	
Rhizopus oligosporus	
Rhizopus oryzae	
Rhodotorula minuta	
Rhodotorula pilimonae	
Saccharomyces fragilis	
Saccharomyces lipolytica	
Schizosaccharomyces pombe	
Sporotrichum thermophile	
Staphylococcus aureus	
Staphylococcus carnosus	
Streptococcus lactis	
Streptomyces coelicolor	
Streptomyces fradiae	
Streptomyces panayensis	
Talaromyces thermophilus	
Thielavia minor	
Torula thermophila	

[a] Adapted from Godtfredsen [3] where references to the individual lipases are given, except *Pseudomonas cepacia* [55].
[b] Many lipases are supplied by distributors other than the producer, and product range, activity and availability vary. A relatively broad lipase range is available from Biocatalysts in the U.K., Amano in Japan and Fluka in Switzerland. Large scale lipase production is done by Novo Nordisk and Meito.

Lipase assays. Screening assays are well developed to detect lipase activity released from growing microorganisms. Agar plates containing lipid overlayers are usually used and give rise to clearing zones in lipid emulsions or fatty acid induced colour formation [52]. The standard activity assays are based on pH-stat or end-point titration of liberated fatty acids from emulsions of tributyrine or purified vegetable oils, usually olive oil [56].

The above methods can be quite lipase specific, so that esterase activity from other enzymes or other acid products from the microorganism are not assayed. Choice of lipid substrate, temperature, pH, etc., can be used to direct the screening towards desired lipase properties.

Lipase assay methodology has become a science in itself because of the complexity of two-phase enzyme kinetics [57]. Assays with immobilized lipases will be discussed in a later section.

Table 3. Homology between substrate binding regions in some lipases.

CCL	K	V	T	I	F	**G**	E	**S**	A	**G**	S	M	S	V
GCL	K	V	M	I	F	**G**	E	**S**	A	**G**	A	M	S	V
PFL	R	V	N	L	I	**G**	H	**S**	Q	**G**	A	L	T	A
SAL	K	V	H	L	V	**G**	H	**S**	M	**G**	G	Q	T	I
SHL	P	V	H	F	I	**G**	H	**S**	M	**G**	G	Q	T	I
RML	K	V	A	V	T	**G**	H	**S**	L	**G**	G	A	T	A
RLL	K	I	H	Y	V	**G**	H	**S**	Q	**G**	T	T	I	G
HGL	Q	L	H	Y	V	**G**	H	**S**	Q	**G**	T	T	I	G
HHL	K	V	H	L	I	**G**	Y	**S**	L	**G**	A	H	V	S
PPL	N	V	H	V	I	**G**	H	**S**	L	**G**	S	H	A	A
HLPL	N	V	H	L	L	**G**	Y	**S**	L	**G**	A	H	A	A
HLCAT	P	V	F	L	I	**G**	H	**S**	L	**G**	C	L	H	L

GCL: *C. cylindracea* lipase [62]
GCL: *G. candidum* lipase [63]
PFL: *P. fragi* lipase [64]
SAL: *S. aureus* lipase [65]
SHL: *S. hyicus* lipase [66]
RML: *R. miehei* lipase [67]
RLL: Rat lingual lipase [68]
HGL: Human gastric lipase [69]
HHL: Human hepatic lipase [70]
PPL: Porcine pancreatic lipase [61]
HLPL: Human lipoprotein lipase [71]
HLCAT: Human lecithin-cholesterol acyltransferase [72]

Amino acid sequence around the essential serine. One letter code: Ala/A, Arg/R, Asn/N, Asp/D, Cys/C, Gln/Q, Glu/E, Gly/G, His/H,. Ile/I, Leu/L, Lys/K, Met/M, Phe/F, Pro/P, Ser/S, Thr/T, Trp/W, Tyr/Y, Val/V.

2.1.3 Cloning, Expression and Structure

Molecular biology of lipases is an area of rapid development and recent break-throughs such as the establishment of high-yielding expression systems [58] and high-resolution three-dimensional structures [59, 60] are important to both our understanding and the optimal production and application of lipases. However, much information on lipase structure and function has been gathered using classical protein/enzyme chemical techniques. The first lipase protein sequence was reported for porcine pancreatic lipase by Desnuelle's group in 1981 [61]. Cloning and cDNA sequencing of microbial and mammalian lipases have now been done in a number of cases, showing active site sequence homologies within about 10 amino acids around one essential serine and two highly conserved glycines:

–Gly–X–Ser–X–Gly–

Table 3 shows this homology for 12 sequenced lipases. A recent short review gives property, structure and cloning data for some mammalian and microbial lipases [73].

The yield of lipase protein from most microorganisms is rather low (milligram per litre range), even from strains subjected to optimization by mutation and fermentation techniques. This has been a limiting factor in the large-scale applications of lipases and cloning of desired source enzymes into high-yielding host organisms has now been achieved. A general useful system has been described for *A. oryzae* [58] in which several fungal enzymes, e.g. *M. miehei* lipase [67], have been expressed. The lipases from *C. cylindracea* [62] and *G. candidum* [63] were recently cloned in *Escherichia coli*, and some bacterial lipases have also been cloned and sequenced [73]. The fungal lipases are usually moderately glycosylated and this may favour a good expression, whereas the carbohydrate might have less importance for stability and activity [67]. Several of mammalian lipases have been cloned [73] and human gastric lipase has been expressed in yeast [69].

High-resolution X-ray crystal structural data have only recently become available. It has been assumed for some time that the catalytic mechanism of lipases involved acyl-enzyme intermediates similar to the class of serine proteases. However, the binding of lipid substrate and interfacial activation needed additional structural explanation. From the *M. miehei* lipase structure [59] there seemed to be a catalytic active triad of Ser, His and Asp which are in a configuration similar to that known from trypsin. Furthermore the triad is covered by a structural element—a 'lid'—that might be movable during substrate binding/interfacial activation. These observations are consistent with the three-dimensional structures of another fungal lipase, *G. candidum* lipase [74], and human pancreatic lipase [60]. The fact that many lipases do not react easily with classical active site serine inhibitors like PMSF unless hydrophobic or interfacial conditions are provided might also be explained. A more detailed analysis of the binding of lipid substrates can be expected to give answers regarding the catalytic and specificity properties of lipases. Other structural features of importance to the use of lipases in lipid modifications will be revealed and changes by protein engineering are expected, because some desired properties have not yet been found in the known lipases. Work has been done to change proteases to lipases because good expression systems and crystal structures were available for the subtilisins, and the detergent potential for improved proteases created mutants anyway. The esterase/protease ratio for Subtilisin Carlsberg protease can also be increased by chemical modification [75].

2.1.4 *Specificity*

Lipase substrate specificity can be divided into various types:

glyceride- (tri-, di- or mono-glyceride)
positional- (on glycerides: 1,3-(regio), 2- or non-specific)
fatty acid- (chain length, double bonds)
alcohol- (primary or secondary –OH, polyols other than glycerol)
stereo- (*sn*-1 or -3 on glycerides, R/S on other substrates)

Methods for determining lipase specificity have been reviewed [76]. In many cases, specificity is not absolute and selectivity is a more adequate description.

1,3-Specific lipases. Highly 1,3-specific lipases are quite common and pancreatic lipase is used in the analysis of the positional distribution of fatty acids in triglycerides [77]. Many other mammalian lipases belong to this specificity class, e.g. lipoprotein lipase, gastric lipase, hepatic lipase and hormone sensitive lipase. Some of the many known microbial 1,3-specific lipases are listed in Table 1.

The positional specific microbial lipases have been exploited due to their ability to catalyse rearrangement reactions of triglycerides in a selective manner (as will be discussed in more detail later).

Assays for positional specificity are usually based on (1) the release of free fatty acids from synthetic triglycerides of the type XYX or YXY or (2) the formation of 1,2(2,3)- respectively 1,3-diglycerides from triglyceride [76]. Both methods are dependent on a short reaction time to minimize acyl migration reactions and the former demands very pure triglycerides. A non-hydrolytic assay based on the incorporation of lauric acid in cocoa butter (XOX) has been suggested [20].

It is likely that pure lipases show a variation in their degree of 1,3-specificity, although lipase components or a long reaction time may reduce the apparent specificity. The lipase structural elements responsible for this (and other types of) specificity are not as yet known. However, branching on the –OH bearing carbon atom in the alcohol substrate is known to reduce lipase activity, not only in the case of the 2-position in glycerol [78]. Lower activity towards di- and monoglycerides is also usually observed. Most 1,3-specific lipases do not readily form 2-monoglycerides as might be expected, but a mixture with diglyceride and free fatty acid as main products. This result may be due both to low activity towards diglycerides and inhibition by the free fatty acids accumulated at the interface, where they can block triglycerides or the lipase active site. However, 1,3-specific lipases can result in almost complete hydrolysis to glycerol and free fatty acids if sufficient enzyme and a long reaction time are provided.

Non-specific lipases. Lipases with pronounced activity towards all three positions in glycerides are less abundant. Among the mammalian enzymes, bile salt activated lipase and pancreatic carboxyl ester lipase (homologous to BSAL) are glyceride non-specific. The partial glyceride specific lipases mentioned below are positional non-specific. The lipase from *C. cylindracea* (sometimes classified as *C. rugosa*) is the most well-known microbial, non-specific lipase. *C. cylindracea* lipase hydrolyses triglycerides rapidly to glycerol and free fatty acids. This lipase is used industrially for fat splitting and many efforts have been made to apply this or find other, more thermostable, lipases and efficient processes for this application. A lipase component from *C. antarctica* (not related to *C. cylindracea*) has been found to be extremely thermostable and non-specific in interesterification, but with moderate hydrolytic activity [20]. The *C. cylindracea* lipase has recently been cloned [62], but no specificity studies on the expressed enzyme or structural data have been published that might explain its unusual hydrolytic properties.

Lipases with phospholipase activity. Lipases from guinea pig pancreas [14] and *S. hyiscus* [29] belong to this group. Some other purified microbial lipases have been shown to possess phospholipase activity [79].

2-Position specific lipases. The existence of such lipases is still a matter of controversy, although phospholipase A_2 has this type of specificity. A 2-specific lipase would be of considerable interest as a complementary catalyst to the 1,3-specific lipases for the preparation of structured triglycerides. Recent reports from Nippon Oils and Fats Co. claim that components with both 2- and 1,3-position activity can be derived from *C. cylindracea* lipase and total hydrolysis is achieved by synergy [80].

Partial glyceride lipases. These are enzymes with substantially higher activity towards mono- and/or diglycerides than triglycerides. Assays can be based on the rate of hydrolysis of purified mono-, di- and triglycerides. The known enzymes are positional non-specific. Applications for such lipases could be monoglyceride synthesis, removal of partial glycerides from oils or combination with 1,3-specific lipase for total hydrolysis. One lipase from *P. cyclopium* is known to have partial glyceride specificity [25].

A monoglyceride lipase is found in adipose tissue [81] and has been reported from *B. stearothermophilus* [82].

Fatty acid specific lipases. Most lipases are not equally active towards all fatty acids, but few have a very narrow fatty acid specificity. The preference for or reduced activity towards certain fatty acids can still be important for the function or application of the lipase. Assays for fatty acid specificity based on hydrolysis of monoacid triglycerides (XXX) are complicated due to the change in physical properties (melting point, emulsification) of triglycerides with different fatty acid structures [76]. A simple method based on alcoholysis of fatty acid esters has been reported [83]. Generally, lipases show reduced or no activity towards (1) acids shorter than about C4 and (2) acids with substituents or double bonds near the carboxyl group, both in hydrolysis and synthesis. However, good acid substrates are found outside the group of natural fatty acids [78]. Long-chain polyunsaturated fatty acids from fish oils are hydrolysed slowly by pancreatic and many microbial lipases, whereas some of the mammalian and microbial lipases have a preference for short-chain saturated fatty acids. Both disadvantages and use of these properties will be discussed later. The most pronounced fatty acid specific lipase reported is that of *G. candidum* [22]. It requires a *cis*-Δ-9 fatty acid structure without double bonds closer to the ester bond.

Alcohol specificity. Alcohols other than glycerol are often good lipase substrates. Like the acids, branching or functional groups near the hydroxyl group will usually reduce the activity [78]. Secondary alcohols are accepted by some lipases, but the specificity may be different than expected from glyceride reactions. Tertiary alcohols do not react. The possibility of making selective hydrolysis or esterification of polyols can be very useful, but the specificity of a given lipase is substrate dependent and a classification of lipases according to their alcohol specificity is difficult.

Stereospecificity. Strict stereospecificity towards glycerides (*sn*-1 or -3) has not been reported, but partial stereospecificity of both mammalian and microbial lipases is known [4]. As with the carboxyl and alcohol substrates few general guidelines can be given, but with the appropriate enzyme, substrate and process conditions a high degree of stereospecificity is sometimes found. Many examples have now been described concerning resolutions, where lipases are used to make stereoselective esterifications or hydrolyses [84]. A well studied enzyme is pig lever esterase [85]; however, microbial lipases from, for example, *C. cylindracea* [86] and *M. miehei* [87] have also been applied.

2.1.5 Other Lipase Properties

Except for the structural features and substrate specificities that are particularly associated with lipases, they are as diverse as enzymes in general with regard to other properties such as molecular weight, isoelectric point, glycosylation, specific activity, pH and temperature activity and stability, protease resistance, activation and deactivation mechanisms, etc. Due to the partial similarity to the serine proteases, serine-, histidine- and tryptophane-modifying chemicals can be expected to affect substrate binding and catalytic properties. Lipases are not dependent on co-factors for catalytic activity, but proteins, surfactants and metals may affect activity, e.g. pancreatic co-lipase (eliminate bile salt inhibition), BSAL (the opposite effect) and removal of free fatty acids by Ca^{2+} (apparent increase in hydrolytic activity). Moderate glycosylation is common among fungal lipases and may affect properties without being essential for structure of activity. Differently glycosylated isoenzymes of *A. niger* [18] and *M. miehei* [24] are such examples. Specific activities in the range up to 6000–12 000 U mg^{-1} protein (on emulsified triglycerides) have been reported for some highly purified lipases [88, 89]. Gastric lipase [15] is stable and active at low pH. Most lipases are optimally active in the neutral range, but some are active below pH 5. Alkaline bacterial [90] and fungal lipases [88] have been described, active at pH 11. The most thermostable lipase reported is the A-component from *C. antarctica*, stable at 80°C [20].

2.2 Phospholipases

2.2.1 *Phospholipase Types, Function and Occurrence*

Phospholipases share their substrates, the phosphoglycerides, commonly phospholipids, and their esterase activity. Otherwise they comprise different enzymes with structures and catalytic mechanisms that are not related to the lipases.

Four major groups exist: phospholipases A_1, A_2, C and D, dependent on their site of action on the phospholipid molecule. Phospholipids differ in structure from triglycerides in that the 3-position fatty acid is replaced by a negatively charged phosphate diester, linked to a variable polar head group. A_1 and A_2 are acyl hydrolases acting on the 1- and 2-position fatty acids, respectively, and C and D are phosphodiesterases acting on the left and right side of the phosphorus atom, respectively. Some other phospholipases are classified according to a more narrow substrate specificity or as lysophospholipases, acting on partially hydrolysed phospholipids. It was previously assumed that a single phospholipase B existed, with activity towards both acyl groups.

A common assay for phospholipases is based on the hydrolysis of egg yolk phospholipid (mainly PC) with pH-stat titration of released fatty acids [91]. As a proton is released by action of each type of phospholipase, titration may (in theory) always be used; however, it is sometimes not sufficiently specific. Some methods are based on the use of monocomponent phospholipids or chromogenic analogues with titration or colorimetric assays. Phosphate determination of, for example, phospholipase C released phosphocholine, assayed spectrophotometrically as a molybdate complex may be a useful method. Phospholipase D can be assayed with purified PC, using a choline oxidase/peroxidase colorimetric method [92].

Substrate specificity may be analysed with pure phospholipids or mixtures where chromatographic separation and quantification of phospholipids, including lyso-products, are made [93].

Phospholipases are as widespread and as biologically important as lipases

Table 4. Selected phospholipases and characteristics.

Source	Specificity	pH_{opt}	Other characteristics	References
Bacillus megaterium	A_1	6–7	thermolabile, no co-factor	[95]
Porcine pancreas	A_2	8–9	thermostable, Ca^{2+}-dependent	[91]
Penicillium notatum	Lyso, (B)	~4	no co-factor	[96]
Bacillus cereus	C	6–8	thermostable, Zn^{2+}-dependent	[97]
Cabbage	D	5–7	Ca^{2+}-dependent	[98]
Streptomyces chromofuscus	D	7–9	Ca^{2+}-dependent, thermostable	[99]

and many have been characterized in detail. A recent monograph by Waite is available [94].

Table 4 lists some of the different phospholipases including a few basic characteristics.

Some phospholipases have important physiological and regulatory functions, others have been studied due to their activity towards biological membranes and contribution to the toxicity of snake and insect venoms. In the following section emphasis will be on the phospholipases that are now used or most likely to be applied in the modification of phospholipids, i.e. A_2, C and D.

2.2.2 *Phospholipases A_2 (EC 3.1.1.4)*

General. These enzymes are found extracellularly from the pancreas and as components of snake and insect venoms. They are the most thoroughly characterized lipolytic enzymes [91, 94]. The pancreatic enzymes function to break down dietary phospholipids of micellar form. In contrast, venom enzymes can act on membrane phospholipids. this activity may be toxic in itself and can increase the effect of toxins. Recent studies have shown that a more hydrophobic surface, in the case of venom or chemically modified pancreatic enzymes, increases the activity towards bilayer substrates (membranes), possibly due to increased penetration [100].

The pancreatic enzymes are proenzymes. Porcine phospholipase A_2 is activated by tryptic cleavage of a heptapeptide from the N-terminal. The proenzyme is active only towards monomolecular substrates.

Structure. Phospholipases A_2 are highly homologous enzymes. They are single-chain, very thermostable proteins containing about 125 amino acids or about 14 000 Da, six or seven disulphide bridges, no free –SH and are not glycosylated. The p*I* is 6.3. Ca^{2+} is an essential, catalytic co-factor. Chelation or exchange with some other divalent cations (Zn, Cd, Cu, Pb, Ba, Sr) inhibits the enzymes.

The homologies extend to sequence and tertiary structure. High-resolution crystal structures of bovine pancreas and *Crotalus atrox* venom enzymes have been published [101]. The structural basis for binding and catalysis is partly understood, with involvement of Asp-99 and His-48 in catalysis [102]. Other amino acids take part in the binding of Ca^{2+}, and phosphate and acyl groups.

Activity and substrate specificity. The activity is low towards monomeric substrates, but is 10^3–10^4-fold increased with substrates present as organized lipid/water interfaces such as micelles. Long-chain phospholipids present in bilayer structures like membranes are very poor substrates. Specific activities on micellar substrates are of the order of 1000 U mg^{-1} protein. Optimal activity is around pH 8–9 and 50°C. Below pH 6 the enzyme is inactive but stable. At pH 8–9 the enzyme is stable at 60–80°C, mostly in the presence of phospholipid. Inactivation is possible with a combination of protease, pH and heat treatments [103].

The enzymes are stereospecific, acting on naturally occurring *sn*-3 phosphoglycerides, 2-position specific and fatty acid non-specific. Minimum structural requirements have been established for the porcine pancreatic enzyme [91]: R–CO–O–CH–CH–O–POO$^-$O–. Phosphate can be replaced by sulphonate and the C–O–P bond by C–P [104]. Anionic phospholipids are preferred but no pronounced polar head group specificity has been observed [91]. Kinetic studies with interfacial binding and activation have been reported [105].

Molecular biology, enzyme availability and properties. Porcine pancreatic phospholipase A_2 has been cloned and expressed in *E. coli* [106] and mammalian cells, the latter including mutant enzymes [107]. Microbial expression was only possible by fusion with other protein and subsequent cleavage to obtain small amounts of active enzyme.

At present the industrial applications of the enzyme is based on crude pancreatin or purified products, like Lecitase™ [103], and has to do with the production of lysolecithin emulsifiers. A microbial source might be attractive in the future, especially if a more thermolabile enzyme for food processing could be made.

Phospholipases A_1 could probably be used in some cases where A_2 enzymes are applied. Both microbial and mammalian enzymes are described that are not Ca^{2+}-dependent and less thermostable. However, these enzymes are not yet readily available or have not been characterized in detail.

2.2.3 Phospholipases C (EC 3.1.4.3)

General. Phospholipase C can hydrolyse C–O–P ester bonds in phosphoglycerides to 1,2-diglycerides and phosphate esters. Sphingomyelin is also a substrate for some enzymes. Substrate specificity is dependent on the polar head group: PC and PI specific enzymes are common. Occurrence is widespread with many extracellular microbial enzymes being reported. The mammalian enzymes have a physiological importance. Enzymes with different specificities are often produced by the same organism. The phospholipases C from *Clostridium perfringens*, one of which is toxic, and from *B. cereus* are the most studied.

B. cereus *phospholipase C.* Intensive characterization studies including cloning [108] and high-resolution crystal structure analysis have been published [109]. Being a *Bacillus* enzyme it is likely to become available on a large scale if desired, although neither high yields nor specific applications have yet been reported.

The enzyme is a 245 amino acid, 28 388 Da (excluding metals), all-helix protein, with three Zn^{2+} in the active site. There are no Cys. Zn is involved both in catalysis and stability. Asp-122 forms a bridge between Zn-1 and Zn-3. The substrate binding has been simulated with smaller phosphate analogues, showing that phosphate is located close to all three zinc atoms. A number of amino acids have been assigned in the metal coordination, but those essential for substrate binding and catalysis have not been finally established.

Activity increase by substrate aggregation is found also with this enzyme, with maximum specific acvitity around 300 U mg^{-1} [110]. The enzyme preferentially hydrolyses PC but is active towards PE and PS. PI is not hydrolysed. Other structural requirements are the acyl ester bonds, a certain fatty acid length without branching close to the ester bond and glycerol-C2 R-configuration [111].

2.2.4 *Phospholipases D (EC 3.1.4.4)*

Phospholipase D can hydrolyse P–O–C ester bonds in phosphoglycerides to diacyl glycerophosphate and the free polar head group, e.g. choline from PC. Sphingomyelin is also a substrate. The major interest in the D enzymes concerns their transphosphatidylation activity, whereby polar head group exchange can be performed in a one-step reaction with interesting application possibilities (as will be described in more detail later). These enzymes occur widely in plants, with that from cabbage being the most studied. Several extracellular phospholipases D are known from *Streptomyces* spp. Mammalian enzymes have also been described. Knowledge of their structure is less advanced than for the previously described A_2 and C enzymes. The cabbage and *Streptomyces* enzymes are both Ca^{2+}-dependent enzymes of about 50 000–60 000 Da (as monomer; the cabbage enzyme is dimeric) of relatively broad substrate specificity. However, PI is slowly or not hydrolysed. The pH optima for activity and stability are in the range from 6 to 9. Specific activities are in the range 100–1000 U mg^{-1} for the highly purified enzymes. A review on phospholipase D has been published by Heller [112]. The *S. chromofuscus* enzyme is commercially available [99]. It is stable at 50–60°C, and in this and other respects different to the cabbage enzyme. The enzymatic properties related to transphosphatidylation will be described in a later section.

2.3 Fatty Acid Modifying Enzymes and Reactions

The biosynthesis, metabolism and modification of fatty acids in different organisms have been well characterized; however, many of the reactions are dependent on complex, cell-associated enzyme systems that are not easily applied in biotechnological processes.

Some of the basic fatty acid biotransformations are shown in Figure 1(a). Mid-chain modifications may be desaturation, saturation, oxidation and hydroxylation. Terminal oxidation and reduction may occur at the CH_3– and COOH– part of the fatty acid, respectively. Elongation, along with desaturation, is an important part of the metabolism of essential fatty acids as shown in Figure 1(b).

A recent review of enzymic modification at the mid-chain of fatty acids with considerations regarding possible future industrial applications has been published by Hammond [113]. One of the interesting reactions discussed is biohydrogenation. Rumen bacteria can transform dietary C18-fatty acids into stearic acid by a combination of desaturases and isomerases as shown in Figure 1(c). There is no reason to believe, however, that a biotechnical process can be developed in the near future at a scale which could be considered for industrial hardening of vegetable oils. Today chemical catalysis is used extensively. A major problem would be the need for free fatty acids in the biological processes.

This and other examples of fatty acid modifying enzyme systems are summarized in Table 5. One of the few enzymes which are not co-factor dependent in this context is soybean lipoxygenase, which may be used as isolated enzyme. In most cases, processes based on the use of whole microbial cells must be considered.

2.4 Enzyme Immobilization

2.4.1 Lipases

Lipase catalysed modification of lipids is often dependent on the availability of suitable immobilized enzyme preparations. It seems therefore appropriate to discuss the technology briefly, especially because the interfacial properties of lipases, the water insoluble lipid substrates and the non-hydrolytic reactions have demanded new enzyme immobilization technology.

General enzyme immobilization methods, including a large number of examples, have been reviewed by Mosbach [117]. Cross-linking of cells or isolated enzymes, attachment to solid supports by precipitation, dispersion, adsorption or covalent coupling, and entrapment in gels are commonly used. Several of these methods have been applied to lipases as indicated in Table 6.

Although dried cells have been used, neither cross-linking of cells nor

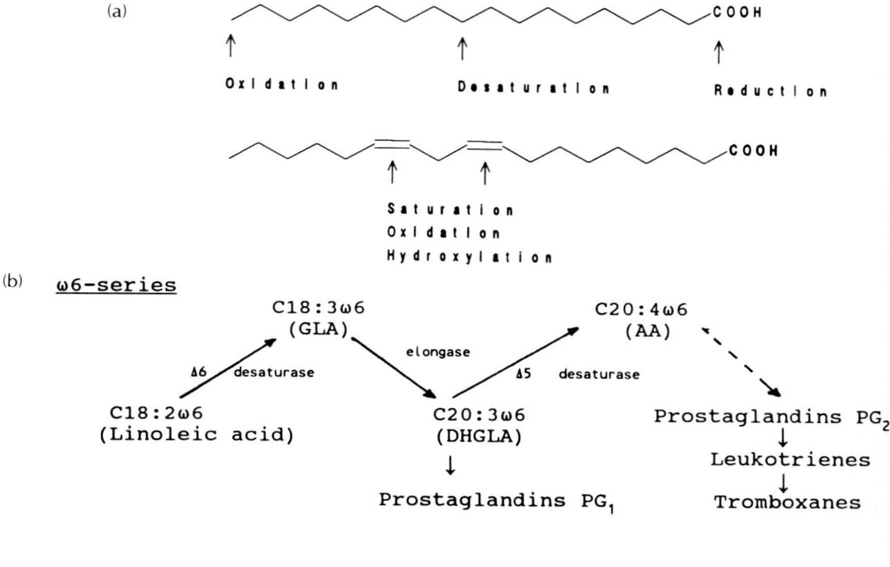

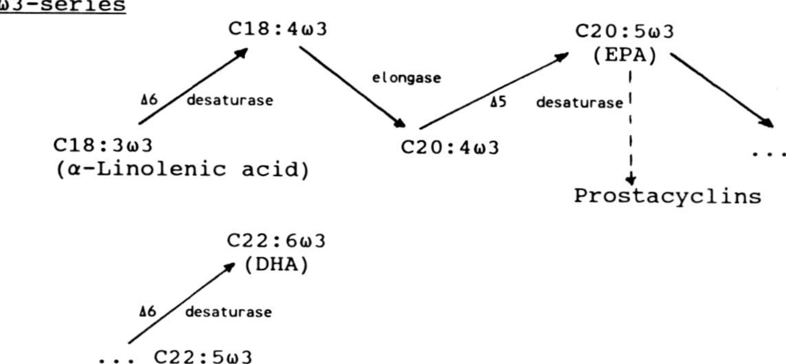

Figure 1. Some fatty acid modification reactions (also see examples in Table 5). (a) Basic fatty acid biotransformations (adapted from Werdelmann and Schmid [114]). (b) Essential fatty acids: desaturation and elongation (main pathways only; ω9 and ω7 series not included; taken from Moesch [115] and Brenner [116]). (c) Biohydrogenation reactions by rumen bacteria (adapted from Hammond [113]).

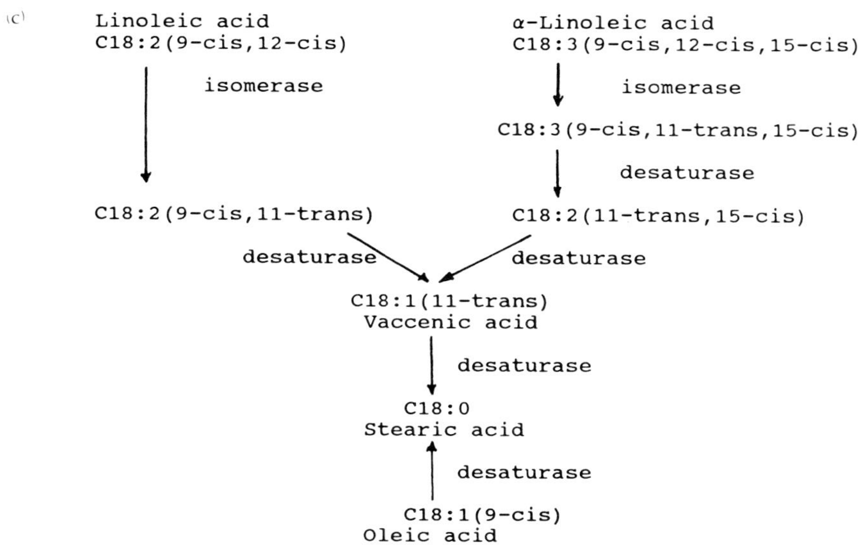

Figure 1. Continued

Table 5. Selected fatty acid modifying enzyme systems.

Enzyme (source)	Reaction example (simplified)	Characteristics
Lipoxygenase (Soybean)	Linoleic acid + O_2 → 9− and 13-hydroperoxy-derivatives	No co-factor Isoenzyme specificity variation
P-450 monooxygenase (*Bacillus megaterium* cells)	Palmitoleic + O_2 → 9,10-epoxy- and ω-1–3 hydroxy-derivatives	P-450 redox system Use of whole (microbial) cells Isoenzyme specificity variation
Oleate 12 hydroxylase (Castor bean)	Oleic acid + O_2 → Ricinoleic acid	NADH co-factor 2-oleyl-phospholipid substrate
Acyl-CoA (Δ^9) desaturase (Wide distribution)	Stearic acid + O_2 → Oleic acid	Complex enzyme/co-factor system Other desaturases (Δ^6, Δ^{12} etc.)
Saturases (Rumen bacteria)	Oleic acid + H_2 → Stearic acid	Several enzyme steps Mixed cultures, anaerobic NADPH co-factor C18 : 2 and α-C18 : 3 also substrates
Oleate hydratase (*Pseudomonas* sp.)	Oleic acid + H_2O → (R)-10-hydroxy stearic acid	Intracellular No co-factor Stereospecific

Adapted from Hammond [113].

Table 6. Lipase Immobilization Methods

Method	Examples	Reference
Dried cells	*Rhizopus arrhizus* dried mycelia (wax ester synthesis, hydrolysis)	[118, 119]
Precipitation or dispersion on inert supports	*Rhizopus arrhizus* lipase acetone precipitation on Celite (interesterification)	[120]
	Rhizopus niveus lipase dried onto Celite (interesterification)	[121]
Adsorption	*Mucor miehei* lipase on macroporous anion exchange resins (interesterification, synthesis)	[122]
	C. cylindracea lipase on microporous polypropylene membrane (hydrolysis)	[123]
	C. cylindracea lipase on powdered pig bone (hydrolysis)	[124]
	C. cylindracea lipase on porous glass beads (ester synthesis)	[125]
Covalent coupling insoluble	*C. cylindracea* lipase bound to aldehyde-activated Sepharose (stability studied)	[126]
Soluble or insoluble (magnetic) PEG-modification	*Pseudomonas fragi* lipase modified with polyethylene glycol (PEG) ± magnetit (ester synthesis)	[127, 128]
Chem. modification, soluble	*C. cylindracea* lipase modified with fluorine chemicals (asymmetric hydrolysis)	[129]
Entrapment	*Rhizopus delemar* lipase in photo-cross linkable urethane prepolymer (interesterification)	[130]
	R. chinensis cells grown in urethane foam and dried (interesterification)	[131]

purified lipases have been reported to give active and stable preparations as in the case of, for example, glutaraldehyde and glucose isomerase.

The nature of lipase catalysed reactions in non-aqueous systems has one advantage in terms of immobilization: preparations with water soluble lipase may be used, e.g. those based on Celite [120, 121]. The interfacial properties of lipases, however, are a complication when it comes to efficient immobilization methods. The activation and large surface area for reaction which characterize lipases in oil/water emulsions has been attempted in the immobilized state by choosing adsorption methods and supports of suitable pore structure along with the necessary lipase/water binding properties. One can say that not only the lipase, but also the interface has to be immobilized. Useful supports are particulate polymers (resins) or inorganics with typical pore diameters in the range of 300–3000 Å and surface areas of $10–100\ m^2\ g^{-1}$ [122]. They are usually referred to as macroporous, sometimes microporous, but with sufficiently large pores to accommodate enzymes and transport lipid substrates without major diffusion problems.

Table 7. Lipase reaction systems.

System	Examples	References
Stirred tank		
Emulsions and soluble lipase	Continuous hydrolysis with *C. cylindracea* lipase	[140]
Immobilized lipase in vacuum	Repeated batch esterifications in vacuum with immobilized *M. miehei* lipase (Lipozyme™) in solvent-free system	[78]
Packed bed	Continuous interesterification with immobilized *M. miehei* lipase on Celite in solvent system and on Duolite-resin (Lipozyme™) in solvent-free system	[141, 142]
Membrane reactor	Continuous hydrolysis with *C. cylindracea* lipase on polypropylene	[143]
	Continuous glyceride synthesis with *C. viscosum* lipase on polypropylene	[144]
Microemulsions (reverse micelles)	Glyceride synthesis with *C. viscosum* lipase in heptane/water/AOT[a] system	[145]
	Interesterification with *R. delemar* lipase in nonane/water/triethyleneglycol dodecyl ether	[146]
PEG-lipase (Table 6)	May function as lipase in microemulsion	
Supercritical carbon dioxide	Interesterification with immobilized *M. miehei* lipase (Lipozyme™)	[147, 148]

[a] AOT: Aerosol OT: sodium bis(2-ethyl-hexyl)sulphosuccinate.

The preparations described on Celite and certain other porous or powdered inert materials may have good initial activity but will often be difficult to handle or have insufficient enzymatic and physical stability in industrial processes. Dust formation, displacement of the enzyme from the support, high pressure drops in packed bed columns and other problems have been experienced.

Particulate or membrane supports based on phenolic, acrylic, propylene and silica materials of suitable porosity have been described which offer strong adsorption, high activity and stability of the lipase, and which make processes with industrial conditions possible.

Lipase immobilization methods have often been associated with specific reaction systems or processes, e.g. as in the case of hydrolysis and synthesis of glycerides in membrane reactors (Table 7).

2.4.2 *Assays with Immobilized Lipases*

Use of emulsified triglyceride substrates in the assay of immobilized lipase preparations is generally not feasible or even relevant. With porous, particulate preparations such systems would consist of three phases (oil/water/lipase-solid phase) and the reaction rates would be very low. Oil droplets will usually be larger than the particle pores.

However, in the assay of synthesis and interesterification activity, water concentration is sufficiently low to enable one-phase substrates with or without solvents. A useful acidolysis assay based on the initial rate of palmitic acid incorporation in triolein using GLC-FAME analysis has been described [132]. This and other immobilized lipase assays have been compared [133].

2.4.3 *Phospholipases*

Compared to the lipases, few reports are available regarding the immobilization of phospholipases, but approaches similar to those used for lipases in non-hydrolytic reactions may be found.

3. MODIFICATION REACTIONS

3.1 Lipase Catalysed Glyceride Reactions

Figure 2 is a schematic description of the primary glyceride reactions catalysed by lipases. It is subdivided according to hydrolytic, synthesis and interesterification reactions as well as to the specificity of the lipase. Interesterification is further divided into (1) acidolysis, i.e. acyl exchange between free fatty acid and triglyceride, (2) transesterification, i.e. acyl exchange between triglycerides, and (3) alcoholysis, i.e. alcohol exchange between free alcohol and triglyceride.

In order to keep the reaction schemes as short as possible, examples have been given and simplifications made: water and some of the glyceride intermediates and species are omitted, stoichiometry is not balanced, etc.

It has long been assumed that the catalytic action of lipases is displayed through fatty acid acyl-enzyme intermediates in a way similar to that of the serine proteases. This has recently been substantiated by the identification of a protease-like catalytic amino acid triad in two different lipases [59,60]. The first step—following substrate binding and interfacial activation—in the lipase catalysed reaction would then be fatty acid acyl-O-Ser-lipase formation. Dependent on the reaction mixture composition, water, partial glycerides or alcohols will release the free HO-Ser-lipase by formation of free fatty acid or ester and consumption or release of water. All are equilibrium reactions and when a given one is investigaged, the exact composition including water is of course very important for the outcome of the reaction.

Investigations on mass transfer and kinetics of immobilized *M. miehei* lipase have been the subject of two works [134, 135].

Another general point to be mentioned in connection with Figure 2 is that partial glycerides may be subject to non-enzymatic intramolecular fatty acid acyl migration and this can influence the composition of products from the lipase catalysed reaction. With long reaction times and 1,3-specific lipase, almost complete hydrolysis of triglycerides (reaction 2a) as well as significant 2-position incorporation in acidolysis (reaction 2h) can be the result of acyl

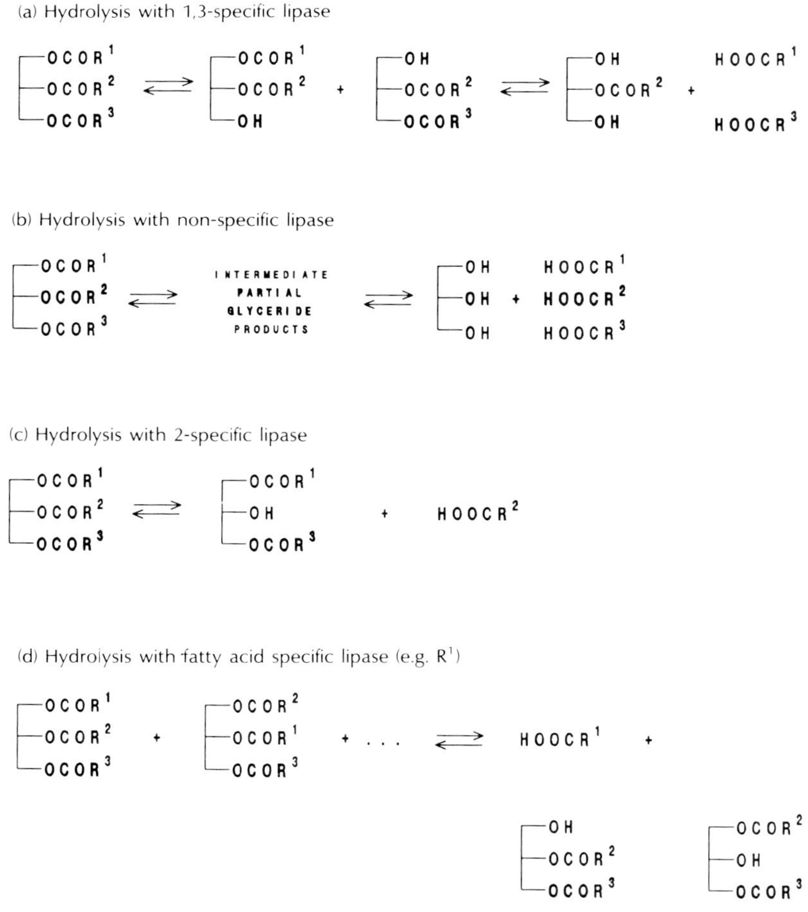

Figure 2. Continued over

(e) Hydrolysis with *sn*-1 or *sn*-3 stereospecifity

[OH, OCOR^2, OCOR^3] $\underset{}{\overset{sn\text{-}1}{\rightleftharpoons}}$ [OCOR^1, OCOR^2, OCOR^3] $\overset{sn\text{-}3}{\rightleftharpoons}$ [OCOR^1, OCOR^2, OH]

(f) Hydrolysis with partial glyceride specific lipase

[OCOR^1, OCOR^2, OCOR^3] + [OCOR^1, OCOR^2, OH] $\rightleftharpoons$

[OCOR^1, OCOR^2, OCOR^3] + [OH, OH, OH] + HOOCR^1 HOOCR^2

(g) Synthesis with non-specific lipase

[OH, OH, OH] + 3 HOOCR^1 $\rightleftharpoons$ INTERMEDIATE PARTIAL GLYCERIDE PRODUCTS $\rightleftharpoons$ [OCOR^1, OCOR^2, OCOR^3]

(h) Acidolysis with 1,3-specific lipase

[OCOR^1, OCOR^2, OCOR^3] + HOOCR^4 $\rightleftharpoons$ [OCOR^4, OCOR^2, OCOR^3] + [OCOR^1, OCOR^2, OCOR^4] $\rightleftharpoons$ [OCOR^4, OCOR^2, OCOR^4] + HOOCR^1 HOOCR^3

(i) *Trans*-esterification with non-specific lipase (randomization)

[OCOR^1, OCOR^2, OCOR^3] + [OCOR^4, OCOR^5, OCOR^6] + . . . $\rightleftharpoons$

[OCOR^4, OCOR^3, OCOR^2] + [OCOR^5, OCOR^6, OCOR^1] + . . . Triglyceride mixture with random distribution of fatty acids

(j) Alcoholysis with 1,3-specific lipase

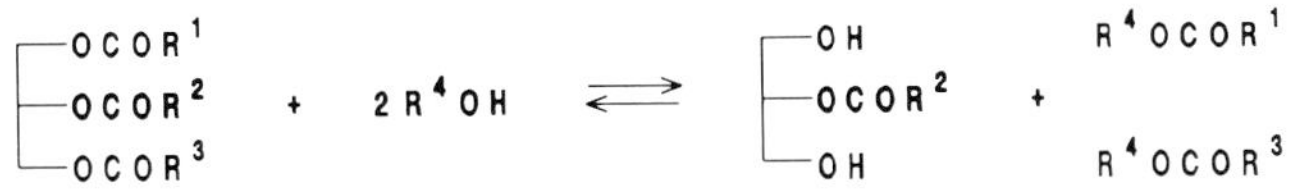

$$\begin{bmatrix} OCOR^1 \\ OCOR^2 \\ OCOR^3 \end{bmatrix} + 2R^4OH \rightleftharpoons \begin{bmatrix} OH \\ OCOR^2 \\ OH \end{bmatrix} + \begin{matrix} R^4OCOR^1 \\ \\ R^4OCOR^3 \end{matrix}$$

Figure 2. Lipase catalysed glyceride reactions.

migration. Some of the factors which may determine the rate of acyl migration are temperature and pH.

Characteristics of the lipases and applications of the reactions in Figure 2 are described in previous and following sections.

3.2 Fatty Acid Ester Synthesis and Hydrolysis Reactions

In many ways these reactions are very similar to those just described for the glycerides, just with another alcohol component than glycerol, and therefore only briefly described in Figure 3.

Lipase alcohol specificity is of particular interest in reactions with primary/secondary hydroxyls, branched alcohols and selective esterifications of polyols and carbohydrates. The application of lipases, especially for fatty acid ester synthesis, will be further discussed later.

Lipase stereo-specificity may be applied in resolution or stereoselective synthesis of (fatty) acids, alcohols and esters [40].

Fatty acid amides can be made by lipase catalysed reactions, e.g. by ε-amidation of lysine [136].

$$R\text{-}OH + HOOCR^1 \rightleftharpoons R\text{-}OCO\text{-}R^1 + H_2O$$

R : Primary or secondary alcohol, fatty alcohol, polyol, carbohydrate

R^1: Fatty acid

Amide synthesis: $R\text{-}NH_2 + HOOCR^1 \rightleftharpoons R\text{-}NH\text{-}CO\text{-}R^1 + H_2O$

Figure 3. Lipase catalysed fatty acid ester synthesis and hydrolysis.

3.3 Phospholipase Catalysed Reactions

Figure 4 describes the most common phospholipid substrates and the hydrolytic (a) and interesterification (b and c) reactions catalysed by individual phospholipases. Enzyme properties and applied reactions are largely discussed in the sections on Phospholipases and Phospholipids.

Use of phospholipases in 'reverse' for direct esterification has not been described, but polar head group exchange by transphosphatidylation can be done efficiently with soluble phospholipase D. Acidolysis reactions with immobilized phospholipase A_2 [137] and a number of microbial lipases [138, 139] have been described. Some of the 1,3-specific microbial lipases seem to have a relatively low but 1-specific activity towards phospholipids.

3.4 Reaction Systems and Process Technology

In the field of lipase technology, many interesting solutions to how the enzymes can function in various reactions have been found. Different examples of lipase reaction systems are given in Table 7. Some of these have been developed to industrial or pilot plant scale while others are more complicated. More details will be given for some of the examples in later sections. The very variable conditions under which lipases can perform efficiently make them versatile catalysts both for the more simple and specialized applications.

In making the choice of how to run a given reaction, the following points should be considered:

- Enzyme: specificity, stability, availability, cost, etc.
- Soluble or immobilized, hydrolysis or esterification
- Scale, batch or continuous mode, value of product
- Investments in process equipment
- Safety regarding both process and product

Simple hydrolytic reactions can be made with soluble lipase and emulsions in stirred tanks. If very large volumes are to be processed and economy based on enzyme re-use, batch systems are not very attractive. Particulate, immobilized lipases are also not attractive because of their poor reaction rates in such 3-phase systems. Solutions may be either the emulsion system described by Bühler and Wandrey [140] or a suitable form of the membrane systems, as reported by Brady *et al.* [143] or Hoq *et al.* [123]. Membrane supported lipases enable a large surface area for reaction and a stabilized interface with the two phases on either side and with less diffusion problems.

Esterification reactions can be made with particulate, immobilized lipases in vacuum with re-use of enzyme. Conditions can be found where fatty acid

(a) Hydrolytic action of different phospholipases

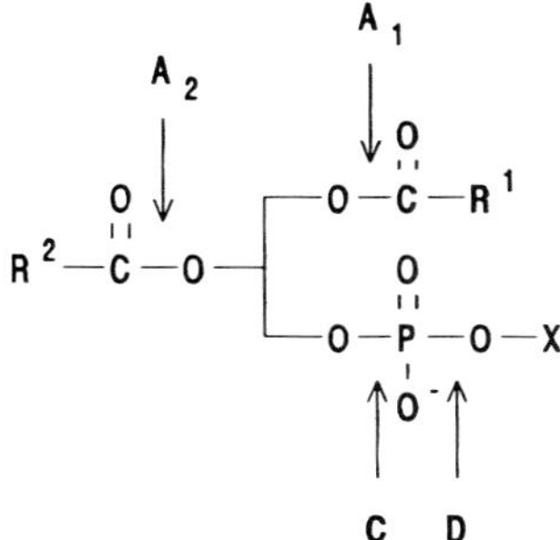

Enzyme	Product
A_1	2-acyl phosphoglyceride
A_2	1-acyl phosphoglyceride
Lyso	Glycerol phosphate ester
C	Diglyceride
D	Phosphatidic acid

X	Phospholipid
Choline	Phosphatidyl choline (PC)
Ethanolamine	Phosphatidyl ethanolamine (PE)
Inositol	Phosphatidyl inositol (PI)
Serine	Phosphatidyl serine (PS)
Glycerol	Phosphatidyl glycerol (PG)
H	Phosphatidic acid (PA)

(b) Transphosphatidylation with phospolipase D

$$\left[\begin{array}{l}OCOR^1\\OCOR^2\\OPO_3^- \cdot X\end{array}\right. + \; Y\cdot OH \rightleftharpoons \left[\begin{array}{l}OCOR^1\\OCOR^2\\OPO_3^- \cdot Y\end{array}\right. + \; X\cdot OH$$

(c) Acidolysis with phosphilipases and lipases with activity towards phospholipids

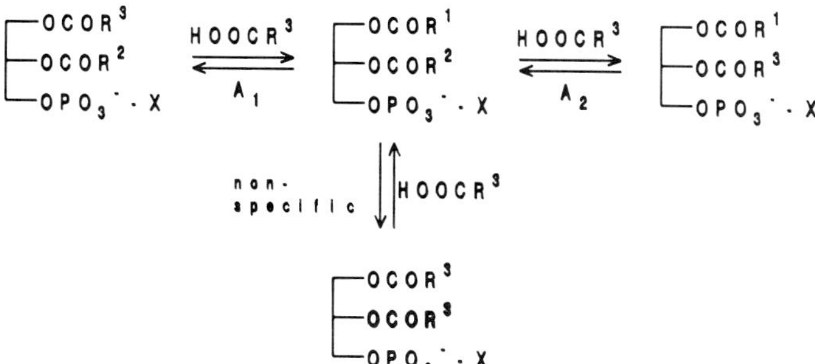

Figure 4. Phospholipase catalysed reactions.

esters are produced in quantitative yield, because water (or, for example, methanol in the case where fatty acid methyl esters are used) is removed from the reaction mixture without dehydration and deactivation of the lipase. The process is particularly simple and efficient when the alcohol component is less volatile than water [78], but even if that is not the case, useful vacuum systems can be designed.

Microemulsions or reverse micelles are systems with a continuous organic phase with water microdroplets stabilized by a surfactant monolayer. Such systems are transparent due to the small droplet size, thermodynamically stable and defined by the composition of the phases, with droplet size dependent on the molar ratio of water to surfactant. Proteins, including lipases, can be incorporated in the water phase and these reactions systems have been thoroughly studied [149].

Droplet size can be in the range where single enzyme molecules are accommodated and high reaction rates can be obtained with very large interface areas and no diffusion limitations. The technique could well be described as a type of immobilization. Various reaction types involving lipases have been reported [145, 150, 151]. When it comes to commercial potential, some problems appear: the microemulsions used so far have mainly been based on lipid in solvents, surfactants which are not food grade and there has been little concern about lipase re-use.

PEG-lipases, are lipases covalently modified wtih amphiphilic Poly Ethylene Glycol-monomethyl ether side chains, usually with a molecular weight of about 5000. Coupling is affected with trichlorotriazine, which reacts with a terminal –OH on the PEG and lysine-NH_2 on the lipase protein. Such lipases have been investigated by Inada [128] and claimed to be soluble in apolar solvents, and therefore useful in esterification and interesterification reactions.

It is questionable whether a derivatization which involves a toxic chemical, solvents and possibly significant activity losses is worthwhile given the number of alternatives. A more fundamental question is, whether it is desirable to avoid the interface—which may be important for high lipase activity—and whether PEG-lipases are in fact not soluble in organic solvents but in a transparent, microaqeuous environment much like the microemulsions, where the function of PEG is water binding and stabilization of droplets? No reports have documented, e.g. by light scattering analysis, that PEG-lipases are in solution.

The last system to be commented on is based on reaction media with supercritical carbon dioxide. In connection with lipases, this has been considered an alternative to solvent based reactions and with the advantage of low viscosity, no toxicity and without mass transfer problems. Another attraction is the possibility of varying the solubility of triglycerides and more polar lipid components with pressure and/or temperature, whereby reaction and fractionation can be combined.

Immobilized *M. miehei* lipase (Lipozyme™) has been studied in esterification [152] and interesterification [147, 148] reactions with supercritical carbon dioxide and found to be stable more than 180 h at 60°C, 100 atm [153]. Protein stability under supercritical carbon dioxide conditions has been studied [154].

Disadvantages are big investments in high-pressure equipment if the scale is large, moderate solubility of triglycerides (large volumes) and strong demands to eliminate oxygen and avoid lipid oxidation, due to the high pressure.

4. APPLICATIONS AND PRODUCTS

4.1 Flavour in food

Flavour development of cheeses is to a great extent due to the ability of lipases to modify milk fat by a partial hydrolysis. Specific free fatty acid profiles are generated by naturally occurring microbial lipases or enzymes added for flavour enhancement. In the first case, moulds involved in the ripening of certain variety of cheeses, e.g. *Penicillium* spp., produce lipases [39]. *P. roqueforti* lipases [155] have a short-chain fatty acid specificity and the liberation of a suitable balance of mainly short-chain, volatile fatty acids seems to be the general effect behind a characteristic flavour [156, 157]. Probably some of the fatty acids may be released as ethyl or other alcohol esters with flavour potential, by lipase-catalysed alcoholysis or esterification. Addition of mammalian and microbial lipases, like proteases, starter cultures etc., during the manufacturing process is also widely used. Pre-gastric esterases are used in the manufacture of Italian cheeses to produce the characteristic 'picante' flavours. The flavours are due to short-chain fatty acids, especially butyric, which are preferentially released from milk fat by these enzymes [39].

In some cases cheese flavour concentrates are made to be used as ingredients in foods like soups, biscuits, dressings, snacks, etc.

Flavour development is usually the result of a very small degree of lipid hydrolysis (few percent) whereas a higher degree may lead also to a change in the physical properties of the fat. Such partial hydrolysis of oils and fats can improve the palatability and digestibility of the lipid component in feeds and pet foods [158]. Hard fats may be softened due to mono- and diglycerides and more energy supplied to the animals if the lipid is 'pre-digested'.

The above mentioned beneficial effects of lipolysis and the off-flavours and spoilage of foods based on milk fat by microbial lipases have been reviewed [51].

4.2 Fatty Acid Production

4.2.1 Fat Splitting

The use of lipases for total hydrolysis of oils and fats to glycerol and free fatty acids has been a subject of extensive research and development. Fatty acids and glycerol are produced today in very large scale by a high temperature (250°C), high pressure (50 atm) steam splitting process (Colgate–Emery). About 1.6 million tons of fatty acids are produced annually worldwide [159]. It is a very efficient process, yielding more than 97% free fatty acid with 2–3 hours residence time [143], but with some disadvantages that has created the interest in enzymatic methods. At present only one large-scale enzyme based operation is known, that of Myoshi Oil & Fat Co. in Japan, where *C. cylindracea* lipase is used [160]. The problems generally associated with the Colgate–Emery process are: high investment and energy costs, colour formation and sensitivity of unsaturated or hydroxy fatty acids to the process conditions. More extensive use of a lipase based fat-splitting process would demand development of a cheap or reusable thermostable, non-specific lipase and a continuous reaction system to handle large volumes and give high conversion to free fatty acids.

The commercially available *C. cylindracea* lipase has a suitable specificity and is not inhibited by high free fatty acid concentrations, but thermostability is limited to 40–50°C, which is inadequate, at least for the more saturated fats. Enzyme price, which is currently quite high, could probably be reduced significantly if the demand was high. Non-specific, more thermostable lipases with the same hydrolytic efficiency have not been reported. However, significant advances as regards immobilization and reactor technology have been made recently, mainly based on the *C. cylindracea* lipase, although other lipases have been investigated.

Hydrolysis can be made in simple stirred tank reactors with 50–70% oil-in-water emulsions containing soluble lipase in the aqueous phase. Most oils will be emulsified sufficiently by the stirring and the partial glycerides and fatty acids formed will stabilize the emulsion. pH is important, because it will influence both lipase activity and ionization of the fatty acids. Addition of base and Ca^{2+} may improve hydrolysis above pH 6.5. One of the advantages of the *C. cylindracea* lipase is that it works optimally at slightly acidic conditions. Re-use of lipase in the above system is difficult due to accumulation in the interfacial layer.

The most advanced system reported on re-use of soluble lipase is a continuous, two-stage mixer–settler process operated in 15 l scale [140]. Here, 70% (w/w) soybean oil emulsion was hydrolysed with *C. cylindracea* lipase, 0.5 g kg^{-1} oil, at pH 5.6 (50 mM acetate). The degree of hydrolysis was up to 98% and yields 11 kg fatty acid l^{-1} day^{-1}. The conditions were selected so

Table 8. Comparison of reactor configurations for immobilized *C. cylindracea* lipase. (From [143].)

Reactor	Half-life (h)	Productivity (kg FFA/kg immobilized enzyme)
Fixed bed	196	100
Membrane	223	1700
Stirred tank	5680	1100

that about 90% of the pure aqueous phase containing glycerol (first stage) and about 90% of the pure fat phase containing fatty acids (second stage) were separated. By this series of incomplete separations, it was possible to recycle about 90% of the enzyme together with the interfacial layer.

Another approach is a continuous, multi-stage column reactor where oil droplets flow upwards through a lipase solution with a dispersing polymer membrane to increase the interfacial area [161]. This system, developed by Nippon Oils & Fats Co., is probably the first attempt to perform continuous fat splitting. Microemulsions [151, 162] and other reaction systems with solvents included [163] have been considered for lipase catalysed hydrolysis, but such systems are not likely to be used in large-scale fat splitting.

Processes based on stabilization and re-use of lipase by immobilization onto solid supports have shown promising results. Certain hydrophobic, porous particles and membrances made of polyethylene (PE) and polypropylene (PP) seem very useful. These materials are often referred to as 'microporous' although the pore size range is within that for materials usually named 'macroporous', i.e. 50–5000 Å pore diameter; sufficient for accommodation of enzyme and transport of substrate and products.

Various reactor configurations have been developed in order to obtain good performance in such immobilized lipase/water/oil three-phase systems.

Brady and co-workers [143] have compared *C. cylindracea* lipase immobilized on PE and PP (Accurel) powder and fibres in fixed bed, membrane and stirred tank reactors with olive oil, and claim that enzymatic fat splitting can be competitive with present energy costs for steam splitting (Table 8).

Continuous hydrolysis of olive oil in a polypropylene membrane reactor with the same immobilized lipase was first reported by Yamane and co-workers [123] and similar systems have been described since. In one modification, a hydrophilic, cellulose based, hollow fibre membrane (Cuprophan^{TM}) was used [159]. The lipase half-life was 43 days at 30°C.

The more thermostable *Thermomyces (Humicola) lanuginosus* lipase was immobilized on an acrylic membrane (a Gelman pleated capsule filter) and used for hydrolysis of tallow at 50°C with a half-life of 2 months [164]. This enzyme has also been immobilized on particulate acrylic resins (Amberlite

XAD-7) and used in batch and column reactions [165]. In both reports the degree of hydrolysis was small, 35–50%, which is not surprising as the enzyme is 1,3-specific.

Combined lipase systems with 1,3-specific lipase and 2-position- or partial glyceride-specific lipase [166] to give complete hydrolysis have been suggested. As discussed earlier, some reports claim that the commercial lipase preparation from *C. cylindracea* actually functions in this way. A castor oil hydrolysis process with 1,3- and monoglyceride-specific lipases has been reported to operate on a commercial scale [167].

4.2.2 Selective Hydrolysis

Cases have been reported where the fatty acid specificity of lipases enabled certain fatty acids to be enriched from others by selective hydrolysis, either as free fatty acids or partial glycerides. Unfortunately, few lipases, except that from *G. candidum*, have a very pronounced fatty acid specificity and the more interesting polyunsaturated fatty acids are not easily concentrated by use of the existing lipases. However, many lipases have quite low activity towards PUFA and the principle of glyceride enrichment is most often used, e.g. as described for the *C. cylindracea* lipase [168]. A lipase from *Arthrobacter aurefaciens* has been reported to increase PUFA by 40–50% in this way [169]. The reverse specificity was observed with a lipase preparation from cod used on marine oils, where PUFA were preferentially hydrolysed [50].

Fatty acids are usually not randomly distributed in triglycerides and phospholipids, and use of enzymes with positional specificity may lead to fatty acid enrichment. In marine phospholipids, EPA and DHA are mainly in the 2-position and phospholipase A_2 has been used to obtain a 24% EPA + 40% DHA concentrate in 60% overall yield from cod roe [170].

4.3 Detergent Lipases

4.3.1 General

Fatty and oily soilings are difficult to remove from fabrics at low washing temperatures. Spillage of lipid containing foods and accumulation of sebum from the skin are common problems. Therefore, intensive research efforts have been directed towards development of detergent compounds, surfactants and enzymes which could facilitate removal of this type of soiling under low temperature wash conditions. Such are used by tradition in large parts of the world, and in hot water wash areas, in particular Europe, the washing habits are fast moving towards application of temperatures of 30–40°C.

Fats which are solid at this temperature level are particularly difficult to remove. Washing above 60°C will remove most fats without the need for detergent improvements.

Hydrolysis of triglycerides and other fatty esters does not increase the water solubility of the lipids as readily as in the case of protein and starch hydrolysis. Fatty acids, mono/diglycerides and glycerol are removed more easily than triglycerides, but alkaline conditions are necessary to ionize the fatty acids and other surfactants will promote lipid removal. It has been shown that a mixture with partial glycerides and free fatty acids is more effectively removed from the fabric than free fatty acids alone [171], so that a non-specific lipase is not necessary or even preferable in washing. Surfactants that are common in detergents, e.g. synthetic anionic and non-ionic compounds, may have a pronounced, often inhibitory effect, on lipase activity.

4.3.2 Requirements of a Detergent Lipase

Based on the general observations above and knowledge of existing detergent products and washing procedures, the main requirements of a detergent lipase can be listed:

- It must be alkaline, with high activity and stability at pH 9–10 to make it compatible with common powder detergents.
- It must be thermostable, with high activity at 30–60°C during washing.
- It must be compatible with anionic surfactants to make it compatible with common powder detergents.
- It must be stable in protease containing detergents.
- It must be stable in the presence of activated bleach systems.
- It must be available in large amounts at low cost.

4.3.3 Detergent Lipase Developments

In 1913 attempts were made to incorporate pancreatic powder into detergents by Dr Otto Röhm [172]. However, none of the pancreatic enzymes, including lipase, had the necessary stability to make this idea viable. More recently, intensive screening programmes have been carried out to find microbial lipases which would fulfil some of the above requirements. A number of lipases from *Pseudomonas* spp. have been considered [173, 174]. Some of these are sufficiently thermostable, alkaline and detergent compatible, but protease sensitivity and production difficulties have so far limited their development. *C. cylindracea* lipase has been investigated at pH 7 without protease [175].

Lipases from *Humicola* spp. have been found to have a high general stability under conventional washing conditions [176].

Lipolase™. The first detergent lipase on the market was Lipolase™, introduced early in 1988. It was also the first industrial enzyme produced by recombinant DNA technology. It is a 1,3-specific *H. lanuginosa* lipase found by screening using the above mentioned criteria and expressed in *A. oryzae* [177].

Specific activity of the lipase is 4000–5000 LU mg^{-1}. The lipase has maximum activity above pH 10 (at 30°C) with tributyrine emulsion as well as olive oil coated PVC powder as substrate. High thermostability is seen in the range pH 5–10. The lipase also shows excellent stability in detergent solutions containing proteases. It is a glycoprotein of about 35 000 Da with a p*I* of 4.4. Specificity studies on triolein show that the lipase is 1,3-specific [178].

Lipolase™ washing performance. Lipolase™ contributes to the overall performance of a detergent and facilitates removal of very difficult stains such as fatty food stains and sebum residues on cuffs and collars. The performance of Lipolase™ is in particular pronounced after more than one wash or when it is used as a pre-spotter.

Effects are measured both by residual fatty matter on the fabric and reflectance (coloured matter will often be attached to the fat).

High lipase activity is found in formulations with alcohol ethoxylates (AE) alone or in mixtures with linear alkyl benzene sulphonates (LAS). Low activity is observed in binary alcohol ether sulphase (AES)/AE systems.

Results with liquid detergent or special pre-spotter formulations show high activity at surfactant concentrations between 25 and 45% [179].

The future directions in detergent lipase research will be towards enzymes of high efficiency during washing. The interactions between lipase and surfactants must be better understood in order to improve lipase properties, detergent formulations and washing procedures.

4.4 Pulp and Paper

Enzyme treatments, including hydrolitic application of lipases, have recently been taken up by the paper and pulp industry. Although lipids (resins) only constitute a very small part of wood, they may have important effects. Further, large volumes are processed.

In spruce, about 2% resin is found. The resin is a mixture of triglycerides, free fatty acids and other components [180]. The resin content is reduced to about 0.5% or less in existing pulping processes. However, more complete resin removal is possible by addition of lipase in the process with improved water absorption characteristics of the cellulose fibres as a result. This property is very important in diapers and other sanitary tissue products. A lipase preparation, Resinase™ A can be used under common alkaline pulping conditions [181].

The resin content may also influence other properties, e.g. paper strength or colour, and result in pitch deposits in paper making processes [182].

Lipases and other enzymes applied to pulping and bleaching show great promise in reducing both energy and the use of chemicals (e.g. chlorine) in this industry.

Printing inks may contain a vegetable oil base [183]. It is therefore possible that printed paper can be treated by lipase in order to facilitate removal of colour from the printing ink. If so, this principle might be used for improvement of return paper.

4.5 Fatty Acid Esters

4.5.1 General

Fatty acid esters comprise a wide range of industrial important compounds. In this section, the application of lipases for synthesis will be the main topic. Triacyl-glycerides and phosphoglycerides will be described in the following section. Iwai and co-workers were among the first to explore the use of microbial lipases in 'reverse' for ester synthesis [184]. A broad screening of microbial lipases for ester synthesis has been published by Lazar *et al.* [185]. The characteristics of *M. miehei* lipase in ester synthesis have been investigated by Miller *et al.* [78], showing a broad specificity of the immobilized lipase.

4.5.2 Flavour Esters

It was mentioned earlier that short-chain fatty acids released from milk fat by lipase influenced the flavour properties of, for example, cheese, either as free acids or ester derivatives. Furthermore, the majority of flavours and fragrances used in foods and perfumery are esters, some of which are fatty acid esters. Many of the naturally occurring esters are now being synthesized, and lipase biocatalysis could be advantageous due to specific and mild reactions. Although many flavour esters are composed of acids and alcohols very different from fatty acids and glycerol, i.e. branched, cyclic, aromatic, very short chain, lipases may still accept these substrates.

Synthesis of terpene alcohol esters by various lipases has been reported by Lazar *et al.* [185], Iwai *et al.* [186], Gatfield [187] and Langrand *et al.* [188].

Gillies *et al.* [189] have reported the continuous synthesis of ethyl butyrate (pineapple/banana flavour) by immobilized *C. cylindracea* lipase in hexane. A number of other flavour esters have also been produced.

Biochemical production of butter flavour can be made by alcoholysis of butter fat and ethanol, e.g. with *C. cylindracea* lipase [190]. This type of reaction will provide a mixture of fatty acid ethyl esters, which may be incorporated as natural flavour components in foods, e.g. margarine and cookies.

Fatty acid	Fatty alcohol	% synthesis		
		1h	2h	3h
myristic acid	myristyl alcohol	98.4	99.0	99.3
palmitic acid	stearyl alcohol	96.1	97.2	98.0
stearic acid	stearyl alcohol	96.8	>96.8	>96.8
oleic acid	oleyl alcohol	97.5	98.5	99.1

Substrate:
0.3 mol of fatty acid and fatty alcohol
T = 70°C Pressure: 0.04 bar
Enzymedosage: 12 g of Lipozyme

Figure 5. Synthesis of fatty acid–fatty alcohol esters in batch vaccum system with immobilized *M. miehei* lipase (Lipozyme™).

4.5.3 *Wax Esters*

Wax esters are mixtures of long-chain, fatty acid–fatty alcohol esters. Naturally occurring esters include wax from the sperm whale and oil from the jojoba plant. Such waxes have become established high-value products used in special lubricants or cosmetics. Further, wax esters have been considered as low-caloric fats [191]. Pure wax esters are difficult to make by chemical catalysis due to the long chain length and unsaturation. Synthesis of oleyl oleate and other wax esters with immobilized lipases and water removal by vacuum has been described by Hansen [192]. Almost quantitative yields can be obtained at mild and simple conditions. Figure 5 describes the synthesis of a range of wax esters.

4.5.4 *Surfactants*

This group consists of fatty acid monoesters of short-chain aliphatic alcohols, polyols including glycerol and glyceryl phosphate, carbohydrates and sterols. The more important industrial products like, for example, distilled monoglycerides and isopropyl myristate, are made by synthesis, but enzymatic processes are on their way.

All of these compounds have a hydrophilic and hydrophobic functionality, which can be varied according to the chemical structure. They may function as emulsifiers, detergents, viscosity builders, etc.

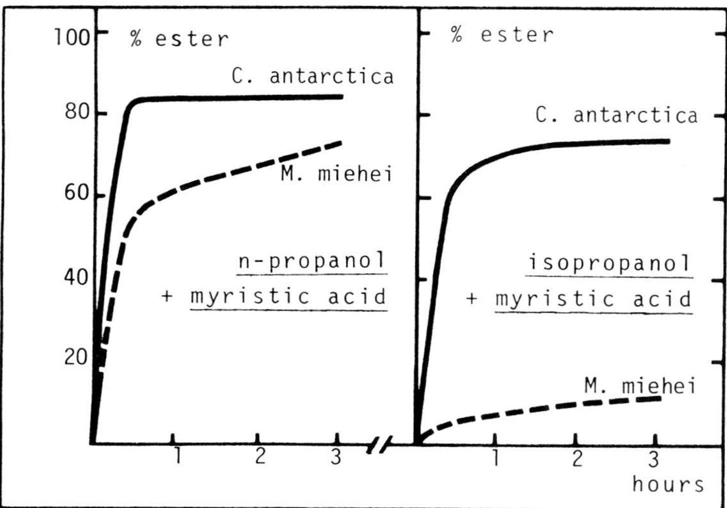

Figure 6. Synthesis of *n*- and iso-propyl myristate esters in batch system with immobilized *C. antarctica* and *M. miehei* lipases. Reaction conditions: 60°C, 0.05 mmol acid and alcohol, 1 g of immobilized lipase. From Heldt-Hansen *et al.* [20].

Use of immobilized *M. miehei* and *C. antarctica* lipases in the synthesis of *n*- and iso-propyl myristate is shown in Figure 6 [20], where the difference in alcohol specificity is seen.

Monoglycerides. These are manufactured by glycerolysis of fats of oils at 180–230°C under alkaline catalysis. After molecular distillation, 90% or more 1-monoglyceride is obtained. A number of monoglyceride derivatives are made with functional groups in the 3-position of glycerol [193].

Different enzymatic routes to monoglycerides have been reported. Glycerolysis with lipase catalysis has been made, but with low yield of monoglyceride [194]. The solubility of glycerol and triglyceride in each other is very small at lower temperatures and reasonable reaction rates therefore difficult to obtain. One would believe that hydrolysis of triglyceride with 1,3-specific lipase was a useful method. However, in this case, equilibrium mixtures are also formed which usually contain much more di- and tri- than monoglyceride (Figure 7). Direct esterification from glycerol and free fatty acid in a recycled system have been reported [195]. Lipases from *R. delemar*, *P. cyclopium* and *Pseudomonas* spp. were used in a two-phase solvent based system. Apparently, good yields and purity could be obtained, but again with long reaction times (about 7 h). The mono/diglyceride specific *Penicillium* lipase

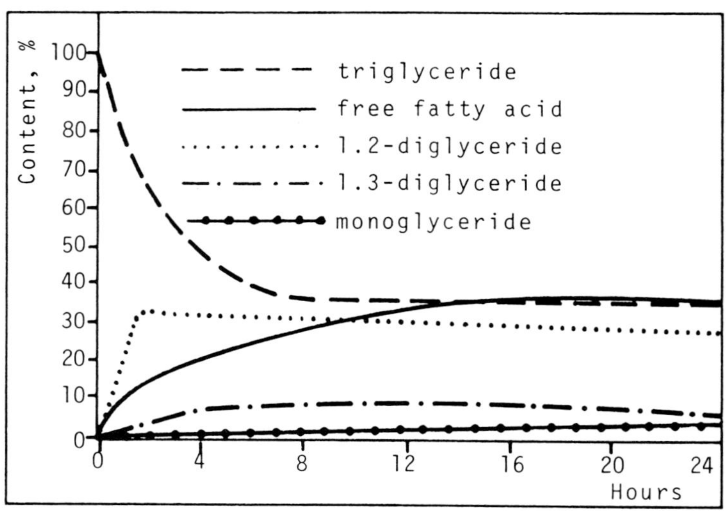

Figure 7. Composition of free fatty acids and glycerides during triglyceride hydrolysis with soluble 1,3-specific *M. miehei* lipase. Reaction conditions: 50% (w/w) olive oil emulsion at 45°C.

seems particularly useful and there are other reports on this enzyme for monoglyceride production [196].

Holmberg and co-workers have used lipases in microemulsions to increase reaction rates and thereby make monoglycerides more efficiently both in hydrolysis [197] and glycerolysis [198]. In the former reaction, an 80% yield of monoglyceride was obtained in 3 h at 35°C with *Rhizopus delemar* lipase.

A route to 100% monoacid monoglyceride has been developed by Godtfredsen *et al.* [199] (Figure 8). Isopropylidene glycerol is formed easily from glycerol and acetone by mild acid catalysis. This derivative has three advantages: (1) it has only one —OH available for esterification, (2) it is sufficiently non-polar to be mixed with a free fatty acid and (3) it is a good (*M. miehei*) lipase substrate. Esterification under reduced pressure enables full conversion to ester and the free 1-monoglyceride can be liberated by mild acid hydrolysis. The reaction has later been studied by Omar *et al.* [200].

Diglycerides. Immobilized *M. miehei* lipase has been used for direct diglyceride synthesis from glycerol and oleic acid [201]. 80% di-, 14% mono- and 6% triglyceride was obtained after 10 h at 40°C, 150 mm Hg. The diglyceride contained 95% 1,3- and 5% 1,2-diolein.

Other polyol esters. Use of lipase selectivity to make a range of diol monoesters has been reported by Cesti *et al.* [202]. Porcine pancreatic lipase and fatty acid ethyl esters were used at 30°C yielding 85–97% monoester in

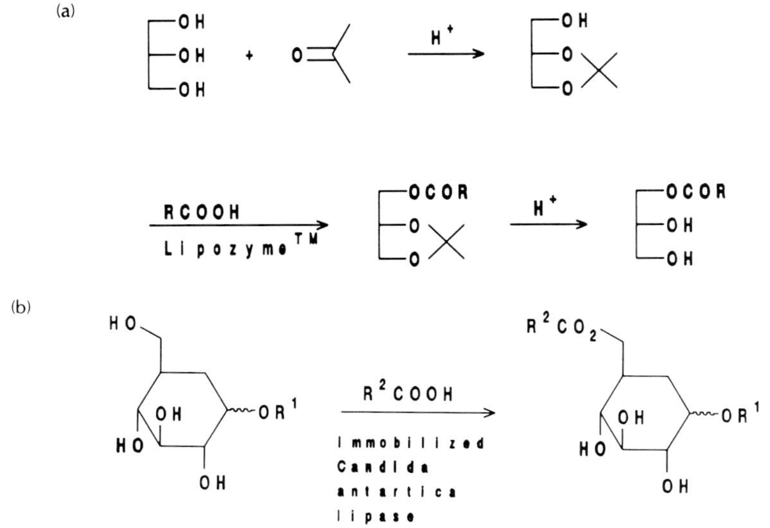

(a)

(b)

Figure 8. (a) Synthesis of pure monoglycerides. Esterification conditions: 50–60°C, water removal (e.g. by vacuum). From Godtfredsen et al. [199]. (b) Synthesis of glucoside fatty acid esters. R_1 = Me, Et, Pr^n, Pr^i, or Bu^n; R2 = C_7H_{15}, C_9H_{19}, $C_{11}H_{23}$, $C_{13}H_{27}$, $C_{15}H_{31}$ or $C_{17}H_{35}$. Reaction conditions: 70°C, 0.01 bar. From Adelhorst et al. [206].

16–44 h. In another reaction with *M. miehei* lipase and 1,2-propane diol, pure monoester and high conversion could not be obtained at the same time [203].

Glycolipids. A broad definition of glycolipids would be fatty esters or fatty glycosides of carbohydrates, usually mono- or disaccharides. Many of these compounds occur in nature and some of the microbial glycolipids have been considered as biosurfactants. In this section only the fatty acid monoesters will be discussed.

Due to the many hydroxyl groups, selective chemical esterification is difficult. Another problem is the mixture of polar carbohydrates and non-polar fatty acids. Chemical production of sucrose esters has been performed with rather expensive solvent and purification processes. Enzymatic methods based on the use of porcine pancreatic lipase in pyridine has been reported [204].

Seino *et al.* [205] has reported the synthesis of sucrose fatty acid esters with *C. cylindracea* lipase. The reaction is aqueous at 40°C for 3 days with subsequent freeze-drying and chloroform extraction (under which a significant part of the reaction may take place?).

The principle in the above mentioned monoglyceride process can also be adapted to glycolipids, e.g. in the esterification of di-isopropylidene galactose [203].

Adelhorst *et al.* [206] have devised a general and attractive process to make fatty acid monoesters of simple glucosides in high yields (Figure 8b). Ethyl glucosides, for example, in contrast to free glucose, are easily mixed and esterified without solvents. Esterification occur selectively in the 6-O-position with immobilized *C. antarctica* lipase. Variation of the fatty acid chain length enable production of pure, biodegradable surfactants of specific properties.

Lyso phospholipids. Phospholipids, like diglycerides, contain two fatty acid acyl groups and a polar moiety which make them surfactants. However, removal of one of the fatty acids to monoglyceride or lyso phospholipid improves the emulsifying properties.

Phospholipase A_2 is used to make 1-acyl phospholipids from natural phospholipids. Direct food processing examples are the improvement of egg yolk for mayonnaise [207] and dough for baking [208]. The enzyme is also used for production of lyso lecithin emulsifier from vegetable oil lecithin sources.

4.5.5 Other Fatty Acid Esters

Sterol esters. Sterols occur naturally also in the form of fatty acid esters. Lipase catalysed synthesis of cholesterol and other sterol fatty acid esters has been reported [209, 210]. Cholesterol has a secondary hydroxyl group and is more easily esterified with non-specific lipases. Cholesterol esterases do not generally have a narrow specificity according to their name, but are rather non-specific, such as pancreatic carboxyl ester lipase. This enzyme is important in the digestion of cholesterol esters and metabolism of cholesterol.

Polyesters. Synthesis of polyesters from dicarboxylic acids and diols [211] or polymerization of hydroxy acids [212] with lipase has been reported.

Hydrolysis of both synthetic [28] and natural [213] polyesters has been investigated. The former as part of a biodegradability study, where a large number of polyesters were hydrolysed with *Rhizopus delemar* lipase. The latter especially in connection with cutin wax hydrolysis. Cutinase, for example, e.g. from *Fusarium* spp. [214], seems to be important in fungal penetration of cutin wax on leaves and fruits. Digestion of cutin in the mammalian diet by pancreatic lipase has been studied [215].

4.6 Speciality Lipids

4.6.1 Triglycerides

Cocoa butter equivalents. The pioneering work by Coleman and Macrae about 15 years ago [120] on the use of 1,3-specific lipase catalysed triglyceride

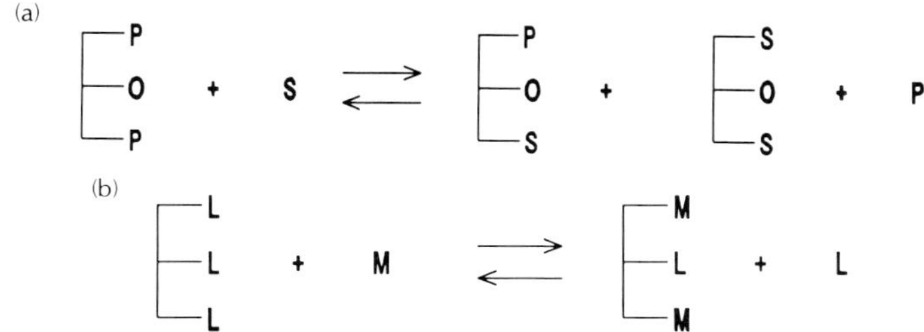

Figure 9. (a) Acidolysis reaction for CBE production with 1,3-specific lipase. Incorporation of stearic acid (S) in 2-position oleic acid rich vegetable oils, e.g. palm mid fraction or high oleic sunflower, to produce a mixture of triglycerides (POP, POS, SOS) similar to cocoa butter. CBE: cocoa butter equivalent. P and O: palmitic and oleic acid. Original principle by Coleman and Macrae [120]. (b) Acidolysis reaction for production of structured dietary lipids with 1,3-specific lipase. Incorporation of medium chain fatty acids (M), e.g. C8 and C10, in oils with long-chain unsaturated or polyunsaturated fatty acids (L) to produce lipids with specific nutritional properties. From Hansen [222].

rearrangement to make a fat composition similar to cocoa butter (CB) has stimulated both lipase technology and other specialty lipid developments.

Enzymatic routes to cocoa butter equivalents (CBE) have since been the subject of several patents and publications. The unique properties of CB are due to a balanced mixture of POP, POS and SOS triglycerides. This accounts for the desired melting and crystallization properties in confectionery fats. Supply of natural CB is from selected tropical areas with laborious manual harvesting and the products are subject to variations in quality and amounts. CB is therefore much more expensive than the 'large' vegetable oils.

Some of these do contain substantial amounts of XOX triglycerides which can be transformed into the CB triglyceride composition by acidolysis with palmitic/stearic acid (Figures 9a and 10).

The commonly described sources are palm oil mid fraction (PMF) or high oleic sunflower oil (HOSO). The fatty acids may also be of palm oil origin.

Continuous processes with columns of lipase dispersed on diatomaceous earth (Celite) and oil/fatty acid in petroleum ether have been developed by Unilever [141]. Similar lipase preparations have been used in systems with oil/fatty acid methyl or ethyl esters without solvents by Fuji Oil Co. [121]. Powdered, immobilized lipase preparations packed in columns demand a very low substrate viscosity in order to make a continuous system without excessive pressure drops. A solvent and/or use of fatty acid esters is a solution at moderate temperatures, especially if the down-stream refining can benefit from these components.

The substrate composition and reaction time/flow rate can be varied to reach the desired composition of the CBE (Figure 10). Fatty acids or their

Tri-glyce-ride	Reaction time					
	0h	1h	2h	4h	6h	24h
POP	39.5	22.6	16.3	10.6	8.4	5.6
POS	8.7	23.4	25.8	25.8	23.7	19.9
SOS	0.5	9.9	13.2	17.7	19.6	19.9

<u>Substrate</u>:
5.0 g of palm midfraction and
5.0 g of stearic acid

<u>T = 70°C</u>

<u>Enzymedosage</u>: 1.0 g of Lipozyme

Figure 10. Composition of main triglycerides during batch acidolysis of palm oil mid fraction with stearic acid using immobilized 1,3-specific *M. miehei* lipase (Lipozyme™).

esters will then have to be removed, either by distillation, solvent extraction or fractionation. Alkali refining is not sufficient because of the high fatty acid concentration.

Particulate, macroporous resin immobilized lipase preparations enable solvent-free, continuous reactions at higher temperatures [122]. A strong binding of lipase to this type of support will further improve stability in handling and use. In contrast to lipase dispersed or precipitated on powdered, inert supports, the lipase will remain immobilized and stable also in systems with significant amounts of water or other hydrophilic components.

In principle, any immobilized 1,3-specific lipase could be used to make CBE, but for economic reasons the more stable and readily available preparations will be preferred. Lipases from *Rhizopus* spp. and *M. miehei* have satisfactory specificity and stability properties, and the latter is commerically available in immobilized form (Lipozyme™) [216, 217]. Some characteristics of this enzyme in batch interesterification are described in Figure 11.

Set-up and performance in a continuous system is described in Posorske *et al.* [142] and in Figure 12.

Margarine. Lipase rearrangement to change the physical properties of triglycerides may be applied in other areas of edible oils and fats. Chemical catalysis with sodium metal or methoxide is used extensively in random interesterification of triglyceride mixtures for the production of margarine and similar products. Margarine is usually based on a mixture of natural, hydrogenated and interesterified vegetable and animal/marine oils. The

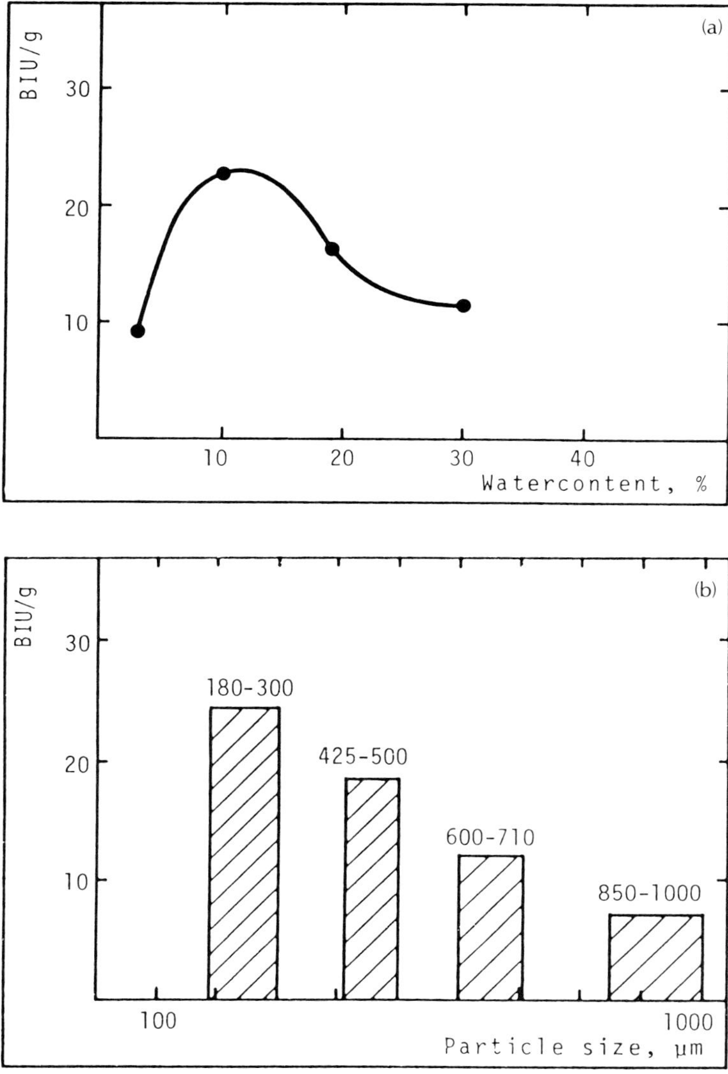

Figure 11. Some characteristics of Lipozyme™: Immobilized *M. miehei* lipase. From Eigtved and co-workers [216, 217]. (a) Batch interesterification activity (BIU g^{-1}) [132] as a function of immobilized lipase hydration. (b) Batch interesterification activity as a function of particle size. (c) Relative batch interesterification activity as a function of temperature.

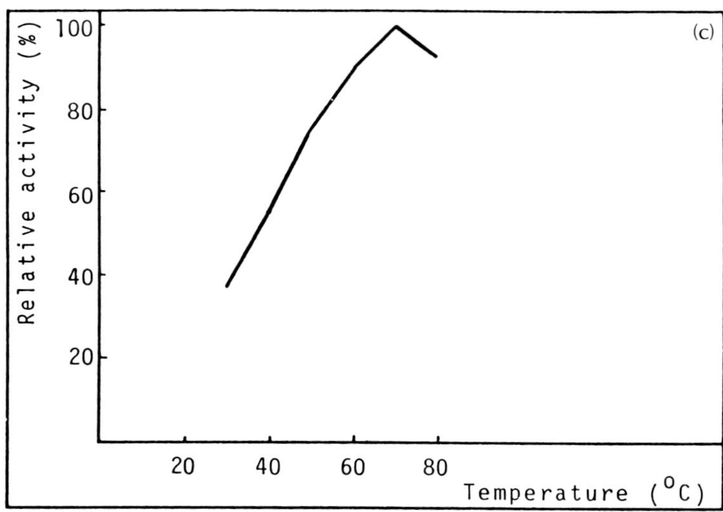

Figure 11. Continued

interesterification can improve the temperature–consistency profile (solid fat index, melting point, crystallization).

A thermostable, non-specific lipase from *C. antarctica* has been developed which can perform random interesterification similar to sodium methoxide [20]. The activity and stability of the immobilized enzyme in batch and column interesterification is shown in Figure 13. Although the chemical process is quite simple and therefore less likely to be substituted with enzyme in the near future, it does depend on corrosive chemicals, absence of polar components and removal of by-products, all of which might be avoided with the proper immobilized lipase. Even though moderate temperatures can be used in chemical interesterification, the use of polyunsaturated oils in larger amounts will be difficult due to their reactivity. The enzymatic reaction should decrease the need for hydrogenation, and triglyceride mixtures based mainly on their nutritional value with physical properties determined by incorporation of natural, saturated fats and enzyme rearrangement.

The 1,3-specific lipases and corresponding rearrangements may also be adequate or even preferred in the production of functional fat ingredients for margarine or similar products [218]. An example of such a reaction with a mixture of palm stearin and coconut oil is shown in Figure 14 with the change in triglyceride composition correlated to melting profile.

It is known that some diglyceride formation is associated with lipase transesterification and that this may affect the melting profile, but it has been shown that even quite high diglyceride levels have a small influence compared

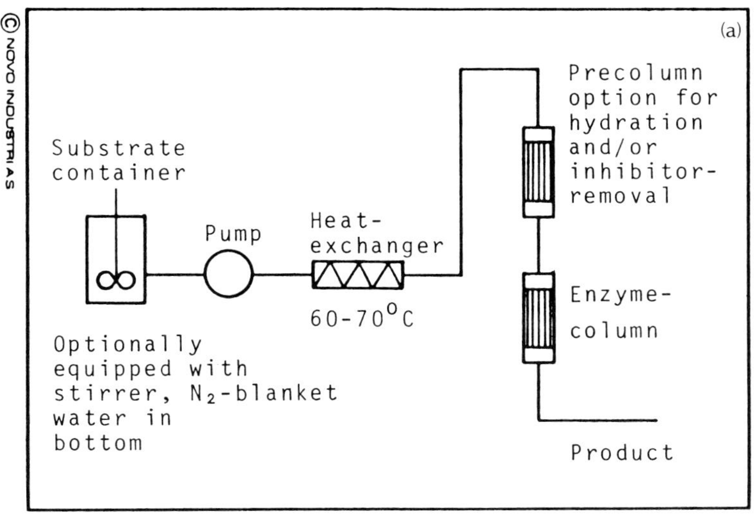

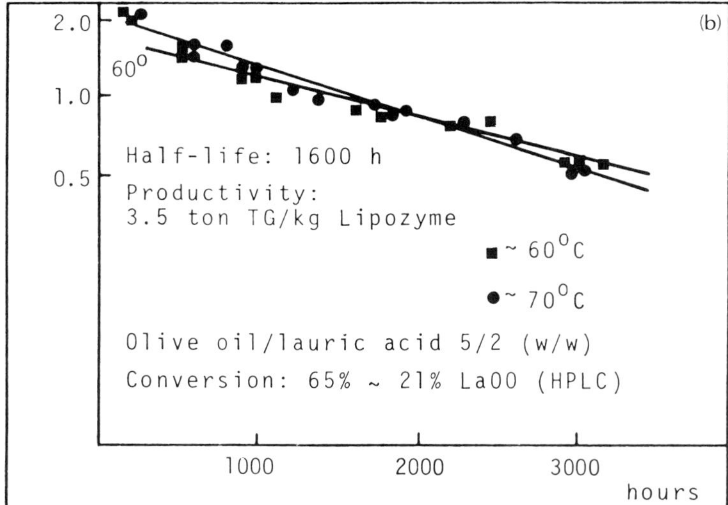

Figure 12. Examples of set-up and performance of immobilized *M. miehei* lipase (Lipozyme™) used in continuous acidolysis. In part from Hansen and Eigtved [217]. (a) Flow sheet. (b) Performance of a packed bed column of Lipozyme using olive oil/lauric acid system. Activity in ln(g triglyceride (TG) h^{-1} g $Lipozyme^{-1}$) (ordinate units) as a function of time at constant conversion.

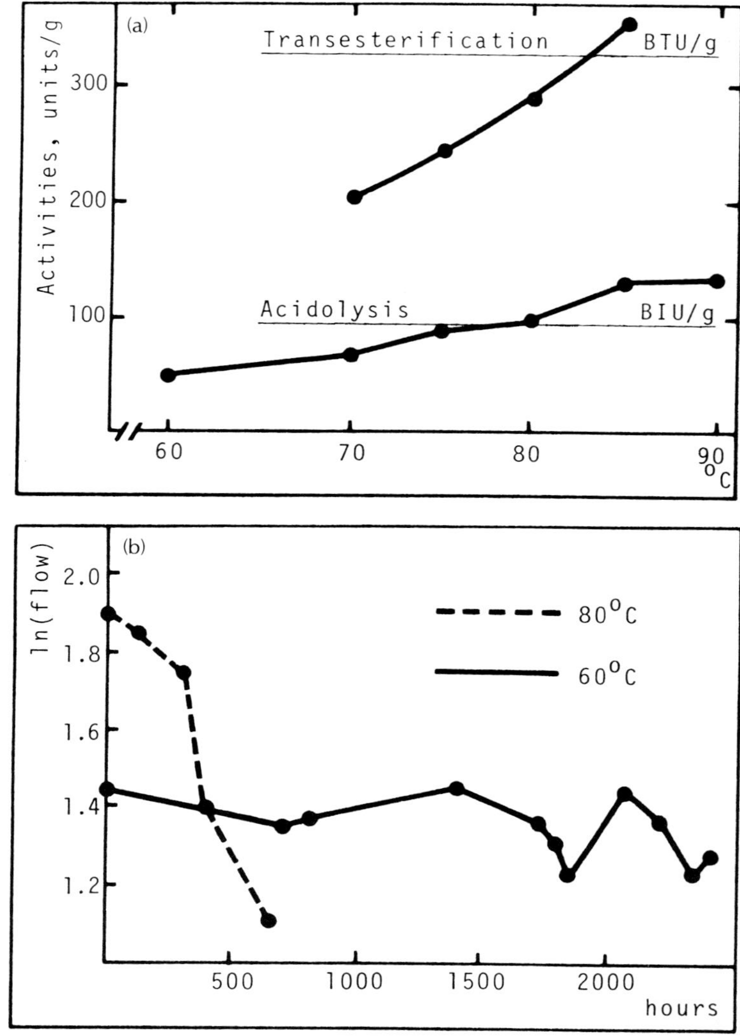

Figure 13. Interesterification characteristics of the immobilized, non-specific, thermostable *C. antarctica* lipase. (a) Batch interesterification activities in acidolysis (BIU/g) and transesterification (BTU/g) as a function of temperature. BIU-assay: Incorporation of palmitic acid (P) in triolein (OOO) measured by FAME-GLC (μmol min^{-1}) [132]. BTU-assay: Formation of POO and PPO from equimolar OOO and PPP triglycerides measured by HPLC (μmol min^{-1} POO and PPO). (b) Continuous acidolysis experiments at 60 and 80°C in packed bed reactors. Conditions: Olive oil/lauric acid 71/29 (w/w), conversion corresponding to 14% (w/w) lauric acid incorporation at variable flow rate. Activity, ln(flow), as ln(g triglyceride^{-1}g immobilized lipase^{-1}) [20].

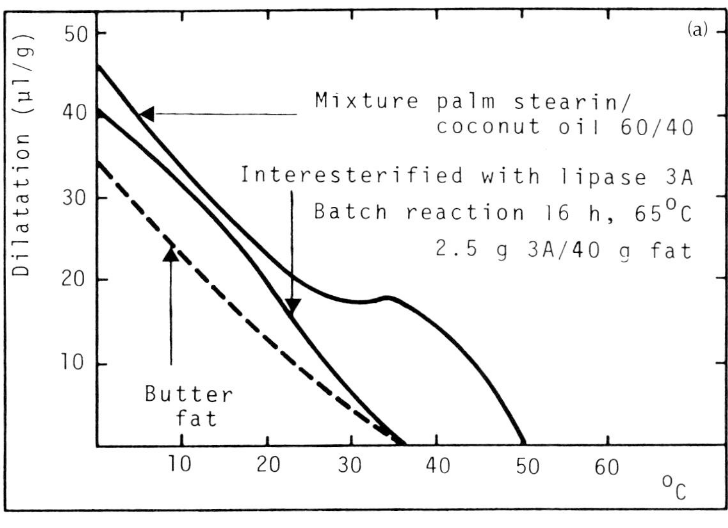

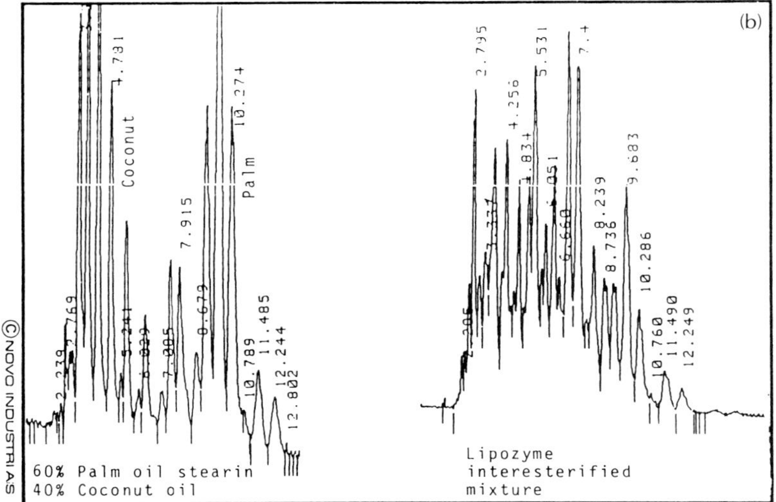

Figure 14. Batch interesterification of palm oil stearin/coconut oil mixture with immobilized *M. miehei* lipase (Lipozyme^{TM}). (a) Melting profiles of mixture, interesterified mixture and butter fat (comparison). Lipase '3A' = Lipozyme^{TM}. Dilatation (Solid Fat Index) measured according to AOCS Official Method Cd 10–57. (b) HPLC triglyceride spectrum of mixture and interesterified mixture. From Eigtved *et al.* [216].

to the change in triglyceride composition [142]. Further, mono/diglycerides are common components added to margarine and may not present a problem. Free fatty acids formed as hydrolytic by-products along with the diglycerides will have to be removed, but a refining step will be needed also in existing processes.

Dietary and pharmaceutical lipids. Although all lipids present in the diet may be considered as 'dietary', the meaning of dietary lipids is usually triglycerides of a specific fatty acid and triglyceride composition and intended for selected human nutrition purposes. However, the field is very broad, ranging from lipids with a pharmaceutical effect to consumer products.

The concern about fat type and amount in the human diet and consequences in terms of health and disease have stimulated the development of $n-3$ fatty acid enriched products. Although many issues remain unsettled, it seems well-documented from the studies of eskimo diets by Bang and Dyerberg [219] and later, independent studies, that a diet rich in $n-3$ fatty acids can reduce the risk of coronary heart disease significantly. While a diet with regular intake of fat fish may be adequate, a desire for more readily available and concentrated products have emerged. These are typically based on marine oils which have been subjected to a number of chemical and physical process steps. Hereby the content of desired $n-3$ fatty acids, mainly EPA (C20:5) and DHA (C22:6), can be increased compared to the natural marine oil content, either in the form of triglycerides or as fatty acid esters. Products have been introduced in the form of gelatin capsules, liquid oils stabilized with antioxidants or microincapsulated oil-in-gelatin powders [220].

Due to the difficulty in processing highly unsaturated oils, especially to avoid oxidation, enzymatic enrichment procedures have been developed. EPA and DHA can be incorporated into cod liver oil with immobilized *M. miehei* lipase (Lipozyme™) to give 40% EPA + 25% DHA in the oil [221]. The process conditions were 60–65°C with 10% Lipozyme™ and stirring under nitrogen for 48–72 h. The cod liver oil and fatty acid or fatty acid ester ratio was 1:3 (w/w). Due to the long reaction time, acyl migration occurred and incorporation was not restricted to the 1- and 3-positions.

The relative incorporation rates of the two fatty acids have been studied [222] (Figure 15). It is clear that this lipase has a relatively low activity towards DHA.

The $n-3$ enriched triglycerides are usually valued only on basis of EPA/DHA concentration and purity. Especially DHA is mainly located in the 2-position of marine oils [221], and the physical, chemical and biological effects of different fatty acid distributions within the triglyceride may be important—not only in connection with marine oil based products. Pancreatic lipase is 1,3-specific and the primary products formed during lipid digestion are fatty acid monoglycerides derived from the 2-position and free fatty acids.

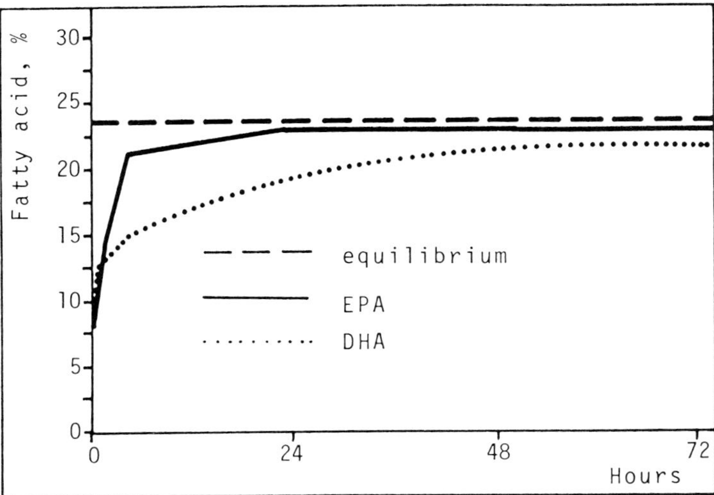

Figure 15. Incorporation of EPA and DHA into cod liver oil by batch acidolysis with immobilized *M. miehei* lipase (Lipozyme™). Reaction conditions: Cod liver oil/free fatty acid 1/3 (w/w), 60°C, 10% Lipozyme™ (w/w of substrate).

(Other lipolysis products are formed due to the combined action of gastric and pancreatic lipase followed by acyl migrations.) Free fatty acids are metabolized more easily if they are medium or short chain, i.e. C10 or below, in which case they provide a rapid energy uptake. Long-chain fatty acid monoglyceride can be absorbed directly and essential or desired fatty acids are therefore most efficiently utilized from the 2-position in glycerides. A structured fatty acid composition may thus be advantageous in certain cases [223].

Experiments with chemically synthesized 2-linoleyl-1,3-dioctanoyl glycerol and 2-oleyl-1,3-dioctanoyl glycerol have demonstrated fast absorption in rats [224]. It is, however, very difficult to make such XYX structures chemically and the use of 1,3-specific lipase in analogy to the CBE process is much more attractive.

Recently some publications have appeared on the enzymatic production of structured, dietary lipids:

- A composition based on high oleic sunflower oil with 82% C8/C10 fatty acid incorporation in the 1- and 3-positions (MLM) has been described [222] (Figure 9b).
- An example which challenges the rationale of improved bioavailability of MLM structures is based on the reverse type of structure, LML, with claimed superior digestibility and adsorptivity [225].

- A substitute milk fat of XPX-type, related to human milk triglycerides and intended for infant formula, has been described [226].

In all cases, immobilized *M. miehei* lipase was used.

The present pharmaceutical dietary lipids are divided in *enteral* and *parenteral* products. Both are intended for patients which cannot take or digest a normal diet and they may be administered with non-lipid nutrients. Enteral lipid products are based on medium-chain triglycerides (MCT), usually re-synthesized from coconut or palm kernel fatty acid fractions and supplemented with essential fatty acid containing vegetable oil [227]. Parenteral lipid products are administered intravenously in the form of lipid emulsions, largely soybean oil or soybean oil mixed with MCT [228].

Fatty acids and triglycerides with specific biological effects (when taken in small amounts) have been proposed among the polyunsaturated fatty acids. Apart from the $n-3$ fatty acids described above, $n-6$ fatty acids other than the main dietary essential fatty acid, i.e. linoleic acid, might be useful as dietary supplementation in certain cases.

Gamma-linolenic acid (GLA) is formed by desaturation of linoleic acid (see Figure 1b), but the delta-6 desaturase activity may be reduced in a number of disease conditions. Promising clinical effects following intake of GLA-containing evening primrose oil have been obtained in patients with atopic eczema [229].

Further steps in the $n-3$ series are conversion to dihomo-gammalinolenic acid (DHGLA) and arachidonic acid (ARA). The latter is important as precursor for a number of the eicosanoids and might have pharmacological potential.

Apart from the advantages of $n-3$ fatty acids mentioned above, DHA is known to be important for the early development of a.o. brain and retina tissue. This fatty acid is mainly administered to the newborn through the human milk phospholipids and is of interest in connection with infant formula, especially for premature babies.

4.6.2 Phospholipids

Enzymatically modified phospholipids are of potential interest in connection with technical, food and pharmaceutical products. The fatty acid composition of phospholipids will generally be quite similar to the corresponding triglyceride fraction from the same source, but exceptions like human milk (as mentioned above) occur. Furthermore, phospholipids have interesting functional properties, e.g. in biological membranes, and may provide interesting fatty acids from abundant sources.

Apart from the lyso-phospholipid emulsifiers mentioned previously, fatty acid or polar head group exchanges can be made enzymatically (see Figure 4.2) to give functional phospholipids.

Liposomes are usually spherical, phospholipid bilayer structures with a hydrophilic core for inclusion of pharmaceuticals or other substances. Synthetic, pure phospholipids, e.g. di–C18:0—PC, have been preferred, but they are expensive to make and enzyme interesterification of natural PC with pure fatty acids would be attractive. Such reactions have been reported with lipases and phospholipases [137–139], but are still less efficient than the corresponding triglyceride rearrangements. This may be due in part to the amphiphilic nature of phospholipids, which can make reactions in low water systems with immobilized enzyme difficult.

The 1,3-specific lipases and phospholipase A_2 seem to exchange only one equivalent of fatty acid in acidolysis, as would be expected when reaction is limited to the 1- or 2-position, respectively. Transphosphatidylation reactions (see Figure 4.2) may provide specialty phospholipids. Phosphatidyl glycerol (PG) has been suggested both as a suitable emulsifier in acidic foods [230] and as the preferred phospholipid type in liposomes [231]. It can be made from natural PC and glycerol with phospholipase D (PLD). This and similar reactions to make PS or PE with cabbage and microbial PLD have been studied in detail by Kudo [230] and Juneja *et al.* [232–234]. An example with 100% conversion of PC to PG was obtained with cabbage PLD-glycerol-buffer/PC-ether emulsion at 30°C in 4.5 h. Initial concentrations of PC and glycerol were 1.5 and 30% [232].

Vegetable and marine phospholipids may be attractive sources of speciality fatty acid or phosphoglyceride fractions, and specific enzymatic hydrolysis can be applied. One example is the use of phospholipase A_2 to produce 60% EPA/DHA fatty acid concentrate from cod roe [170]. The narrow substrate specificity of some PLC enzymes might be used in phospholipid fractionations.

4.6.3 Microbial Lipids, Biotransformations and Plant Cloning

As indicated in the Introduction, these fields are too large to cover in detail here. However, it is fair to consider not only isolated enzyme transformations but related and alternative biochemical routes to speciality fatty acids and lipids.

Plant cloning. Plant breeding for selected lipids is the topic of a future chapter [235]. In the past much success has been achieved in obtaining high vegetable oil yields and specific compositions, e.g. low eurucic rape seed and high oleic sunflower oils. Further developments in plant genetic enginering towards specific fatty acid or triglyceride containing oils will be based on the isolation of fatty acid biosynthesis genes [236].

Lipid biotransformations. Steroid biotransformation has become a specialized field of considerable industrial significance and extensive reviews on the subject are available, e.g. Vezina [237].

A number of fatty acid modifying reactions were described in Section 2.3,

Figure 1 and Table 5, but due to their general complex nature, none are yet used on a larger scale.

Microbial lipids. The most promising developments for the near future seem to be within the area of microbial lipids. This subject has been reviewed recently by Ratledge [6]. Microbial processes, especially fermentation of *Mortierella* spp., for the production of pharmacologically and dietary important PUFAs have been studied in detail by Shimizu and co-workers [238, 239]. Microalgae as a source of EPA and DHA has been investigated by Kyle *et al.* [240]. Although EPA/DHA and GLA can be obtained from marine and vegetable oils, concentration and purity may be a problem. Furthermore, DHGLA and ARA are not readily available and fermentations with *M. alpina* have yielded 2–4 $g\,l^{-1}$ of these fatty acids in triglyceride form. The PUFA content of the fatty acids was 23–35%.

Microbial, ca. 20% GLA-oils are produced commercially by Sturge in the UK (*Múcor javanicus*) and Idemitzu in Japan (*Mortierella isebellina*) [241].

Algae demand long fermentation times and usually large surfaces and light. However, some microalgae do not require light and could be used in submerged fermentations with good yields. It has been shown that either EPA or DHA can be produced [242].

4.7 Oil Refining

4.7.1 Removal of Free Fatty Acids and Partial Glycerides

In the processing of oils and fats, a number of unit operations are carried out in order to remove components other than the desired pure triglycerides of the refined oil. Enzyme technology has been investigated as a means of oil refining in a number of cases.

Free fatty acids, partial glycerides and glycerol may be the result of lipase action during harvest, storage and transportation of the crude oils. They account for a loss in oil yield—and quality—if they are not removed. Conventional processes based on alkali treatment or distillation can remove small amounts of free fatty acids efficiently, whereas diglycerides, for example, are more difficult to separate from triglycerides. Two different solutions based on lipases have been reported: (1) re-esterification of free fatty acids and diglycerides in a vacuum system, and (2) selective hydrolysis of partial glycerides.

Investigations of the first type are described by Kurashige [243], who used Celite-immobilized *P. fluorescens* lipase. The triglyceride content of palm olein was increased within 3 days from 87 to 95% at 100–20 ppm moisture, 60°C, 1–3 mm Hg. Although the reaction time was long, it is remarkable that a reasonable lipase activity remains at this low water level. Optimal activity was obtained with a combination of purified lipase and lecithin.

The other principle is based on the partial glyceride specific *P. cyclopium* lipase [244]. In a mixture with triglycerides, mono- and diglycerides are hydrolysed much more rapidly and subsequent separation of free fatty acids and glycerol can be made.

4.7.2 Removal of Phospholipids

Among the minor polar components of crude oils, phospholipids constitute a significant part. Removal of vegetable oil lecithins—degumming—to very low levels is necessary a.o. to avoid poisoning of the hydrogenation catalysts. Some of the less polar phospholipids are not easily removed by conventional water/phosphoric acid extraction and phospholipase treatments have been suggested to increase the hydrophilicity and thereby improve the degumming process.

Phospholipases A_2 and C have been investigated for this purpose but effects have so far been too small for any further development [245].

An interesting *carbohydrate* refining process [246] concerns the use of phospholipase/lysophospholipase to improve filterability and clarity of phospholipid containing wheat starch hydrolysate.

4.7.3 Other Processes

It is likely that other refining steps might benefit from the use of enzymes, e.g. colour removal with peroxidases. At present, colour is removed by adsorption (bleaching) and steam distillation in vacuum (desodourization), but both steps have other useful functions and enzymes may rather be introduced for specific than general refining processes.

4.8 Digestive and Analytical Applications

4.8.1 Digestive Lipases

As mentioned earlier, porcine pancreatic enzymes (pancreatin), including lipase, are used extensively as a pharmaceutical. Microbial enzyme preparations are also marketed by pharmaceutical companies, but mainly used as digestive aids by healthy people. The trend has been to make better pancreatin preparations suited for oral administration by entero-coating, but lipase activity and stability is not adequate [41, 42]. Developments towards recombinant mammalian gastric [15] and bile salt activated [16] lipases or microbial lipases with increased stability and high purity might improve lipid digestion in patients dependent on enzyme supplementation.

4.8.2 Analytical Uses

Clinical analysis of serum triglycerides, phospholipids and cholesterol are dependent on enzyme based assays [247]. In the investigation of positional distribution of fatty acids in triglycerides, porcine pancreatic lipase is used due to its high degree of 1,3-specificity [77].

REFERENCES

[1] Gunstone, F. D., Harwood, J. L. and Padley, F. (eds), *The Lipid Handbook*. Chapman and Hall, London, 1986.

[2] Borgström, B. and Brockman, H. L. (eds), *Lipases*. Elsevier, Amsterdam, 1984.

[3] Godtfredsen, S. E. Microbial lipases. In Fogarty, W. M. and Kelly, C. T. (eds), *Microbial Enzymes and Biotechnology*, 2nd edn. Elsevier, London, 1990, pp. 255–274.

[4] Macrae, A. R. and Hammond, R. C. *Biotechnol. Gen. Eng. Rev.*, 3 (1985) 193–217.

[5] International News on Fats, Oils and Related Materials (INFORM), 1 (1990) 103.

[6] Ratledge, C. 'Biotechnology of oils and fats.' In Ratledge, C. and Wilkinson, S. G. (eds), *Microbial Lipids*. Academic Press, London, 1989, pp. 567–668.

[7] Applewhite, T. H. (ed.), *Proc. World Conference on Biotechnology for the Fats and Oils Industry*, Hamburg, FRG, September 27–October 2, 1987, Published by the American Oil Chemists' Society, 1988.

[8] *J. Am. Oil. Chem. Soc.*, 66 (1989) 651–653.

[9] *J. Am. Oil. Chem. Soc.*, 66 (1989) 1572–1576.

[10] Baumann, H., Bühler, M., Fochem, H., Hirsinger, F., Zoebelein, H. and Falbe, J. *Angew. Chem. Int. Ed. Engl.*, 27 (1988) 41–62.

[11] Critchley, P. *Spec. Publ. Royal Soc. Chem.* (Chemical Aspects of Food Enzymes), 63 (1987) 149–155.

[12] Falbe, J. and Schmid, R. D. *Fette Seifen Anstrichmittel*, 6 (1986) 203–212.

[13] Sarda, L. and Desnuelle, P. *Biochim. Biophys. Acta*, 30 (1958) 513–521.

[14] Verger, R., *Pancreatic lipases*. In Borgström, B. and Brockman, H. L. (eds), *Lipases*. Elsevier, Amsterdam, 1984, pp. 83–150.

[15] Gargouri, Y., Moreau, H. and Verger, R. *Biochim. Biophys. Acta*, 106 (1989) 255–271.

[16] O'Connor, C. J. and Cleverly, D. R. In Cambie, R. C. (ed.), *Fats for the Future*. Ellis Horwood, Chichester, 1989, pp. 109–128.

[17] Borensztajn, J. (ed.), *Lipoprotein Lipase*. Evener, Chicago, 1987.

[18] Hofelmann, M., Hartmann, J., Zink, A. and Schreier, P. *J. Food Sci.*, 50 (1985) 1721–1731.

[19] Tomizuka, N., Ota, Y. and Yamada, K. *Agric. Biol. Chem.*, 30 (1966) 576–584.

[20] Heldt-Hansen, H. P., Ishii, M., Patkar, S. A., Hansen, T. T. and Eigtved, P. In Whitaker, J. R. and Sonnet, P. E. (eds), *Biocatalysis in Agricultural Biotechnology* (*ACS Symposium Series 389*) American Chemical Society, Washington, DC, 1989, pp. 158–172.

[21] Sugiura, M. and Isobe, M. *Chem. Pharm. Bull.*, 23 (1975) 1226–1230.

[22] Jensen, R. G. *Lipids*, 9 (1974) 149–157.

[23] Omar, I. C., Hayashi, M. and Nagai, S. *Agric. Biol. Chem.*, 51 (1987) 37–45.

[24] Huge-Jensen, B., Galluzzo, D. R. and Jensen, R. G. *Lipids*, 22 (1987) 559–565.

[25] Okumura, S. Iwai, M. and Tsujisaka, Y. *J. Biochem.*, 87 (1980) 205–211.

[26] Sugiura, M. and Isobe, M. *Chem. Pharm. Bull.*, 23 (1975) 681–682.

[27] Verger, R. 'Pancreatic lipases.' In Borgström, B. and Brockman, H. L. (eds), *Lipases*, Elsevier, Amsterdam, 1984, p. 94.

[28] Tokiwa, Y. and Suzuki, T. *Nature*, 270 (1977) 76–78.

[29] van Oort, M. G., Deveer, A. M. T. J., Dijkman, R., Tjeenk, M. L., Verheij, H. M., de Haas, G. H., Wenzig, E. and Götz, F. *Biochemistry*, 28 (1989) 9278–85.

[30] Huang, A. H. C. In Borgström, B. and Brockman, H. L. (eds) *Lipases*. Elsevier, Amsterdam, 1984, pp. 419–442.

[31] Chabane-Ben Hamida, J. and Mazliak, P. *Année Biologique*, 24 (1985) 201–232.
[32] Abigor, D. R., Opute, F. I., Opoku, A. R. and Osagie, A. U. *J. Sci. Food Agric.*, 36 (1985) 599–606.
[33] Hassanien, F. R. and Mukherjee, K. D. *J. Am. Oil. Chem. Soc.*, 63 (1986) 893–897.
[34] Hills, M. J., Kiewitt, I. and Mukherjee, K. D. *Biochim. Biophys. Acta*, 1042 (1990) 237–240.
[35] Hills, M. J., Kiewitt, I. and Mukherjee, K. D. *Biotechnol. Lett.*, 11 (1989) 629–632.
[36] Lin, Y.-H., Wimer, L. T. and Huang, A. H. C. *Plant Physiol.*, 73 (1983) 460–463.
[37] Weslake, R. J., Thomson, L. W., Tenaschuk, D. and MacKenzie, S. L. *Plant Physiol.*, 91 (1989) 1303–1307.
[38] Borgström, B. and Brockman, H. L. (eds), *Lipases*, Elsevier, Amsterdam, 1984, pp. 47–416.
[39] Deeth, H. C. and Fitz-Gerald, C. H. In Fox, P. F. (ed.), *Developments in Dairy Chemistry —2*. Applied Science Publishers, Barking, 1983, pp. 206–207.
[40] Whitesides, G. M. and Wong, C.-H. *Angew. Chem. Int. Ed. Engl.*, 24 (1985) 617–638.
[41] Robinson, P. J., Olinsky, A., Smith; A. L. and Chitravanshi, S. B. *Arch. Dis. Childhood*, 64 (1989) 143–145.
[42] Griffin, S. M., Alderson, D. and Farndon, J. R. *Gut*, 30 (1989) 1012–1015.
[43] Moreau, H., Laugier, R., Gargouri, Y. Ferrato, F. and Verger, R. *Gastroenteroloty*, 95 (1988) 1221–1226.
[44] Moreau, H., Gargouri, Y., Lecat, D., Junien, J. L. and Verger, R. *Biochim. Biophys. Acta*, 959 (1988) 247–252.
[45] Wang, C.-S., Martindale, M. E., King, M. and Tang, J. *Am. J. Clin. Nutr.*, 49 (1989) 457–463.
[46] Tang, J. and Wang, C.-S. European Patent Application 317 355 (1989).
[47] Wang, C.-S. *Biochem. Biophys. Res. Commun.*, 155 (1988) 950–955.
[48] Olivecrona, T. and Bengtsson-Olivecrona, G. 'Lipoprotein lipase and emulsion clearance.' In Hartig, W., Dietze, G., Weiner, R., Füst, P. (eds), *Nutrition in Clinical Practice*. Karger, Basel, 1989, pp. 77–82.
[49] Taskinen, M.-R., In Borensztajn, J. (ed.), *Lipoprotein Lipase*, Evener, Chicago, 1987, pp. 204–205.
[50] Lie, Ö. and Lambertsen, G. *Comp. Biochem. Physiol.*, 80B (1985) 447–450.
[51] Stead, D. *J. Dairy Res.*, 53 (1986) 481–505.
[52] Rollof, J., Hedström, S. Å. and Nilsson-Ehle, P. *AMPIS*, 97 (1989) 9–13.
[53] Ingham, E., Holland, K. T., Gowland, G. and Cunliffe, W. J. *J. Gen. Microbiol.*, 124 (1981) 393–401.
[54] Deeth, H. C. and Fitz-Gerald, C. H. In Fox, P. F. (ed.), *Developments in Dairy Chemistry —2*, Applied Science Publishers, Barking, 1983, pp. 204–206.
[55] Lonon, M. K., Woods, D. E. and Strauss, D. C. *J. Clin. Microbiol.*, 26 (1988) 979–84.
[56] Jensen, R. G. *Lipids*, 18 (1983) 650–657.
[57] Verger, R. *Methods Enzymol.*, 64 (1980) 340–392.
[58] Christensen, T., Wøldike, H., Boel, E., Mortensen, S. B., Hjortshøj, K., Thim, L. and Hansen, M. T. *Bio/Technology*, 6 (1988) 1419–1422.
[59] Brady, L., Brzozowski, A. M., Derewenda, Z. S., Dodson, E., Dodson, G., Tolley, S., Turkenburg, J. P., Christiansen, L., Huge-Jensen, B., Nørskov, L., Thim, L. and Menge, U. *Nature*, 343 (1990) 767–770.
[60] Winkler, F. K., D'Arcy, A. and Hunziker, W. *Nature*, 343 (1990) 771–774.
[61] De Caro, J., Boudouard, M., Bonicel, J., Guidoni, A., Desnuelle, P. and Rovery, M. *Biochim. Biophys. Acta*, 671 (1981) 129–138.
[62] Kawaguchi, Y., Honda, H., Taniguchi-Morimura, J. and Iwasaki, S. *Nature*, 341 (1989) 164–166.

[63] Shimada, Y., Sugihara, A., Tominaga, Y., Iizumi, T. and Tsunasawa, S. *J. Biochem.*, 106 (1989) 383–388.

[64] Kugimiya, W., Otani, Y., Hashimoto, Y. and Takagi, Y. *Biochem. Biophys. Res. Commun.*, 141 (1986) 185–190.

[65] Tyski, S., Hryniewicz, W. and Jeljaszewicz, J. *Biochim. Biophys. Acta*, 749 (1983) 312–317.

[66] Götz, F., Popp, F., Korn, E. and Schleifer, K. H. *Nucleic Acids Res.*, 13 (1985) 5895–5906.

[67] Boel, E., Huge-Jensen, B., Christensen, M., Thim, L. and Fiil, N. P. *Lipids*, 23 (1988) 701–706.

[68] Docherty, A. J. P., Bodmer, M. W., Angal, S., Verger, R., Rivière, C., Lowe, P. A., Lyons, A., Emtage, J. S. and Harris, T. J. R. *Nucleic Acids Res.*, 13 (1985) 1891–1903.

[69] Bodmer, M. W., Angal, S., Yarranton, G. T., Harris, T. J. R., Lyons, A., King, D. J., Piéroni, G., Rivière, C., Verger, R. and Lowe, P. A. *Biochim. Biophys. Acta*, 909 (1987) 237–244.

[70] Datta, S., Luo, C., Li, W., Van Tuinen, P., Ledbetter, D. H., Brown, M. A., Chen, S., Liu, S. and Chan, L. *J. Biol. Chem.*, 263 (1988) 1107–1110.

[71] Wion, K. L., Kirchgessner, T. G., Lusis, A. J., Schotz, M. C. and Lawn, R. M. *Science*, 235 (1987) 1638–1641.

[72] McLean, J., Fielding, C., Drayna, D., Dieplinger, H., Baer, B., Kohr, W., Henzel, W. and Lawn, R. M. *Proc. Natl. Acad. Sci. USA*, 83 (1986) 2335–1106.

[74] Hata, Y., Sugihara, A., Shimada, Y., Tominaga, Y., Katsube, Y. and Kakudu, M. *Abstracts of the 2nd International Conference on Protein Engineering*, Kobe, Japan, August 20–25, 1989. The Protein Engineering Society of Japan, p. 138.

[75] Svendsen, I. *Carlsberg Res. Commun.*, 41 (1976) 237–291.

[76] Jensen, R. G., DeJong, F. A. and Clark, R. N. *Lipids*, 18 (1983) 239–252.

[77] Paquot, C. and Hautfenne, A. *IUPAC Standard Methods for the Analysis of Oils, Fats and Derivatives*, 7th edn. Blackwell Scientific Publications, Oxford, 1987, pp. 111–115.

[78] Miller, C., Austin, H., Posorske, L. and Gonzalez, J. *J. Am. Oil Chem. Soc.*, 65 (1988) 927–931.

[79] Pedersen, K. B. Unpublished results.

[80] Nakata, M., Hiranto, J. and Funada, T., *Abstracts of the ISF–JOCS World Congress*, Tokyo, Japan, September 26–30, 1988, Japan Oil Chemists Society, p. 162.

[81] Belfrage, P., Fredrikson, G., Strålfors, P. and Tornqvist, H., In Borgström, B. and Brockman, H. L. (eds), *Lipases*. Elsevier, Amsterdam, 1984, pp. 388–393.

[82] Imamura, S., Takahashi, M., Misaki, H. and Matsuura, K. European Patent Application 285 101 (1988).

[83] Rangheard, M.-S., Langrand, G., Triantaphylides, C. and Baratti, J. *Biochim. Biophys. Acta*, 1004 (1989) 20–28.

[84] Chen, C.-S. and Sih, C. J. *Angew. Chem. Int. Ed. Engl.*, 28 (1989) 695–707.

[85] Jones, J. B. *Biol. Chem. Hoppe-Seyler*, 370 (1989) 984.

[86] Cambou, B. and Klibanov, A. M., *J. Am. Chem. Soc.*, 106 (1984) 2687–2692.

[87] Sonnet, P. and Baillargeon, M. W. *J. Chem. Ecology*, 13 (1987) 1279–1292.

[88] Huge-Jensen, B., *Abstract of the ISF–JOCS World Congress*, Tokyo, Japan, September 26–30, 1988. Japan Oil Chemists Society, p. 294.

[89] Huge-Jensen, B., Boel, E., Thim, L., Christensen, M., Andreasen, F. and Christensen, T. In Shukla, V. K. S. and Hølmer, G. (eds), *Proceedings of the 15th Scandinavian Symposium on Lipids*, Rebild, Denmark, 11–15 June, 1989, Lipidforum, pp. 293–304.

[90] Wantanabe, N., Ota, Y., Minoda, Y. and Yamada, K. *Agric. Biol. Chem.*, 41 (1977) 1353–1358.

[91] de Haas, G. H., Postema, N. M., Nieuwenhuizen, W. and van Deenen, L. L. M. *Biochim. Biophys. Acta*, 159 (1968) 103–129.

[92] Imamura, S. and Horiuti, Y. *J. Biochem.*, 83 (1978) 677–680.
[93] Nasner, A. and Kraus, L. *Fette Seifen Anstrichmittel*, 2 (1981) 70–73.
[94] Waite, M., *The Phospholipases (Handbook of Lipid Research, Vol. 5)*, Plenum Press, New York, 1987.
[95] Raybin, D. M., Bertsch, L. L. and Kornberg, A. *Biochemistry*, 11 (1972) 1754–1760.
[96] Kawasaki, N. and Saito, K. *Biochim. Biophys. Acta*, 296 (1973) 426–430.
[97] Little, C. *Methods Enzymol.*, 17 (1981) 725–730.
[98] Waite, M., *The Phospholipases (Handbook of Lipid Research, Vol. 5)*, Plenum Press, New York, 1987, pp. 65–66.
[99] Phospholipase, D. Product Information T-07 from Toyo Jozo Co., Japan, December 1979.
[100] Van der Wiele, F. C., Atsma, W., Roelofsen, B., van Linde, M., Van Binsbergen, J., Radvanyi, F., Raykova, D., Slotboom, A. J. and de Haas, G. H. *Biochemistry*, 27 (1988) 1688–1694.
[101] Renetseder, R., Brunie, S., Dijkstra, B. W., Drenth, J. and Sigler, P. B. *J. Biol. Chem.*, 260 (1985) 11627–11634.
[102] Slotboom, A. J., Verheij, H. M. and de Haas, G. H. On the mechanism of phospholipase A_2. In Hawthorne, J. N. and Ansell, G. B. (eds), *Phospholipids*, Elsevier Biomedical, Amsterdam, 1982, Vol. 4, pp. 359–434.
[103] Lecitase™. Product Information B 226 (1989) from Novo Nordisk A/S.
[104] Brockerhoff, H. and Jensen, R. G. *Lipolytic Enzymes*, Academic Press, New York, 1974.
[105] Verger, R. and de Haas, G. H. *Annu. Rev. Biophys. Bioeng.*, 5 (1976) 77–117.
[106] de Geus, P., van den Bergh, C. J., Kuipers, O., Verheij, H. M., Hoekstra, W. P. M. and de Haas, G. H. *Nucleic Acids Res.*, 15 (1987) 3743–3759.
[107] de Geus, P., Kuipers, O. P., Heuvel, M. Van, Verheij, H. M. and de Haas, G. H. *Chimicaoggi*, October (1987) 73–75.
[108] Johansen, T., Holm, T., Guddal, P. H., Sletten, K., Haugli, F. B. and Little, C. *Gene*, 65 (1988) 293–304.
[109] Hough, E., Hansen, L. K., Birknes, B., Jynge, K., Hansen, S., Hordvik, A., Little, C., Dodson, E. and Derewenda, Z. *Nature*, 338 (1989) 357–360.
[110] El-Sayed, M. Y., DeBose, C. D., Coury, L. A. and Roberts, M. F. *Biochim. Biophys. Acta*, 837 (1985) 325–335.
[111] Hough, E., Hansen, L. K., Hansen, S., Hordvik, A. I., Jynge, K. and Little, C., In Shukla, V. K. S. and Hølmer, G. (eds), *Proceedings of the 15th Scandinavian Symposium on Lipids*, Rebild, Denmark, 11–15 June, 1989. Lipidforum, pp. 323–335.
[112] Heller, M. *Adv. Lipid Res.*, 16 (1978) 267–326.
[113] Hammond, R. C. *Fat Sci. Technol.*, (1988) 18–27.
[114] Werdelmann, B. W. and Schmid, R. D. *Fette Seifen Anstrichmittel*, 84 (1982) 436–443.
[115] Moesch, H. *Lipids in Foods*. NESTEC, La-Tour-de-Peilz, 1984.
[116] Brenner, R. R. Factors Influencing Fatty Acid Chain Elongation and Desaturation. In Vergroesen, A. J. and Crawford, M. (eds), *The Role of Fats in Human Nutrition*, 2nd edn. Academic Press, London, 1989, pp. 45–79.
[117] Mosbach, K. (ed) *Methods in Enzymology*, Vol. 135 (Part B). Academic Press, Orlando, 1987.
[118] Knox, T. and Cliffe, K. R. *Process Biochem.*, October (1984) 188–192.
[119] Gancet, C. and Guignard, C. In Laane, C., Tramper, J. and Lilly, M. D. (eds), *Biocatalysis in Organic Media (Studies in Organic Chemistry 29)*, Elsevier, Amsterdam, 1987, pp. 261–266.
[120] Coleman, M. H. and Macrae, A. R. British Patent Specification 1 577 933 (1980).
[121] Matsuo, T., Sawamura, N., Hashimoto, Y. and Hashida, W. US Patent 4 416 991 (1983).
[122] Eigtved, P. European Patent Specification 140 542 (1989).

[123] Hoq, M. M., Yamane, T., Shimizu, S., Funada, T. and Ishida, S. *J. Am. Oil Chem. Soc.*, 62 (1985) 1016–1021.

[124] Negishi, S., Sato, S., Mukataka, S. and Takahashi, J. *J. Ferment, Bioeng.*, 67 (1989) 350–355.

[125] Marlot, C., Langrand, G., Triantaphylides, C. and Baratti, *J. Biotechnol. Lett.*, 7 (1985) 647–650.

[126] Otero, C., Ballesteros, A. and Guisan, J. M. *Appl. Biochem. Biotechnol.*, 19 (188) 163–175.

[127] Nishio, T., Takahashi, K., Tsuzuki, T., Yoshimoto, T., Kodera, Y., Matsushima, A., Saito, Y. and Inada, Y. *J. Biotechnol.*, 8 (1988) 39–44.

[128] Inada, Y., Takahashi, K., Yoshimoto, T., Kodera, Y., Matsushima, A. and Saito, Y. *TIBTECH*, 6 (1988) 131–134.

[129] Kitazume, T., Murata, K. and Ikeya, T. *J. Fluorine Chem.*, 32 (1986) 233–238.

[130] Yokozeki, K., Yamanaka, S., Takinami, K., Hirose, Y., Tanaka, A., Sonomoto, K. and Fukui, S. *Eur. J. Appl. Microbiol. Biotechnol.*, 14 (1982) 1–5.

[131] Nakashima, T., Fukuda, H., Kyotani, S. and Morikawa, H. *J. Ferment. Technol.*, 66 (1988) 441–448.

[132] Novo Method for the Determination of Lipase Batch Interesterification Activity. Analytical Method AF 206 (1986) from Novo Industri A/S.

[133] Graille, J., Muderhwa, J. and Pina, M. *Fat. Sci. Technol.*, 6 (1987) 224–226.

[134] Ison, A. *Mass Transfer Effects in Fat Interesterification Reactions Catalysed by Immobilized Lipase*, Ph.D. thesis, Dept. Chem. Biochem. Eng., University College, London, 1987.

[135] Luck, T. *Einfluss von Stofftransport und Kinetik immobiliserter Lipasen in heterogenen Stoffsystemen auf Auslegung und Entwicklung von Enzymreaktoren* (Fortscr.-Ber. VDI Reihe 3 Nr. 188, VDI-Verlag, Düsseldorf, 1989), Lehrstuhl für Prozesstechnik der Lebensmittel der Tecnischen Universität München, 1989.

[136] Montet, D., Pina, M., Graille, J., Renerd, G. and Grimaud, J. *Fat Sci. Technol.*, 1 (1989) 14–18.

[137] Pedersen, K. B. Unpublished results.

[138] Svensson, I., Adlercreutz, P. and Mattiasson, B. *Appl. Microbiol. Biotechnol.*, 33 (1990) 255–258.

[139] Yagi, T., Nokanishi, T., Yoshizansa, Y. and Fukui, F. *J. Ferment. Bioeng.*, 69 (1990) 23–25.

[140] Bühler, M. and Wandrey, C. In Applewhite, T. H. (ed.), *Proc. World Conference on Biotechnology for the Fats and Oils Industry*, Hamburg, FRG, September 27–October 2, 1987, Published by the American Oil Chemists' Society, 1988, pp. 230–237.

[141] Macrae, A. R. In Tramper, J. and van der Plas, H. C. (eds), *Biocatalysts in Organic Syntheses (Studies in Organic Chemistry 22)*, Elsevier, Amsterdam, 1985, pp. 195–208.

[142] Posorske, L. H., LeFebvre, G. K., Miller, C. A., Hansen, T. T. and Glenvig, B. L. *J. Am. Oil. Chem. Soc..*, 65 (1988) 922–926.

[143] Brady, C., Metcalfe, L., Slaboszewski, D. and Frank, D. *J. Am. Oil. Chem. Soc.*, 65 (1988) 917–921.

[144] Hoq, M. M., Yamane, T., Shimizu, S., Funada, T. and Ishida, S. *J. Am. Oil Chem. Soc.*, 61 (1984) 776–781.

[145] Fletcher, P. D. I., Freedman, R. B., Robinson, B. H., Rees, G. D. and Schomäcker, R. *Biochim. Biophys. Acta*, 912 (1987) 278–282.

[146] Holmberg, K. and Österberg, E. *Progr. Colloid Polym. Sci.*, 74 (1987) 98–102.

[147] Nakamura, K., Chi, Y. M. and Yano, T. In Perrut, M. *Proceedings of the International Symposium on Supercitical Fluids*, Société Français de Chimie, 1988, pp. 925–931.

[148] van Eijs, A. M. M., de Jong, J. P. L., Doddema, H. J. and Lindeboom, D. R. In Perrut, M. *Proceedings of the International Symposium on Supercritical Fluids*, Société Français de Chimie, 1988, pp. 933–942.

[149] Luisi, P. L. and Laane, C. *TIBTECH*, June (1986) 153–161.
[150] Fletcher, P. D. I., Robinson, B. H., Freedmann, R. B. and Oldfield, C. *J. Chem. Soc.*, Faraday Trans. I, 81 (1985) 2667–2679.
[151] Holmberg, K. In Shukla, V. K. S. and Hølmer, G. (eds), *Proceedings of the 15th Scandinavian Symposium on Lipids*, Rebild, Denmark, 11–15 June, 1989. Lipidforum, pp. 249–260.
[152] Marty, A., Chulalaksananukul, W., Condoret, J. S., Willemot, R. M. and Durand, G. *Biotechnol. Lett.*, 12 (1990) 11–16.
[153] Fujimoto, K., Shishikura, A., Arai, K. and Saito, S. *Abstracts of the ISF–JOCS World Congress*, Tokyo, Japan, September 26–30, 1988. Japan Oil Chemists Society, p. 286.
[154] Kasche, V., Schlothauer, R. and Brunner, G. *Biotechnol. Lett.*, 10 (1988) 569–574.
[155] Lamberet, G. and Menassa, A. *J. Dairy Res.*, 50 (1983) 459–468.
[156] Woo, A. H. and Lindsay, R. C. *J. Dairy Sci.*, 67 (1984) 960–968.
[157] Law, B. A. and Wigmore, A. S. *J. Soc. Dairy Technol.*, 38 (1985) 86–88.
[158] Lipozyme™ 10.000 L. Product Information B 299 (1990) from Novo Nordisk A/S.
[159] Pronk, W., Kerkhof, P. J. A. M., van Helden, C. and van't Riet, K. *Biotech. Bioeng.*, 32 (1988) 512–518.
[160] Japan Chemical Week, 14 May (1981) 2.
[161] Ishida, S. In Baldwin, A. R. (ed.), *Proceedings of the World Conference on Emerging Technologies in The Fats and Oils Industry*, Cannes, France, November 3–8, 1985. American Oil Chemists Society, 1986, pp. 359–364.
[162] Kang, S. T. and Rhee, J. S. *Biotechnol. Lett.*, 11 (1989) 37–42.
[163] Kim, K. H., Kwon, D. Y. and Rhee, J. S. *Lipids*, 19 (1984) 975–977.
[164] Taylor, F., Panzer, C. C., Craig, J. C. and O'Brien, D. J. *Biotech. Bioeng.*, 28 (1986) 1318–1322.
[165] Omar, E. C., Sacki, H., Nishio, N. and Nagai, S. *Agric. Biol. Chem.*, 52 (1988) 99–105.
[166] Park, Y. K., Pastore, G. M. and Almedia, M. M. *INFORM*, 1 (1990) 304.
[167] Yamane, T. In Shukla, V. K. S. and Hølmer, G. (eds), *Proceedings of the 15th Scandinavian Symposium on Lipids*, Rebild, Denmark, 11–15 June, 1989. Lipidforum, pp. 216–229.
[168] Lie, Ø. and Lambertsen, G. *Fette Seifen Anstrichmittel*, 9 (1986) 365–367.
[169] Kao Corp., Japanese Patent Application 134 446 (1986).
[170] Tocher, D.R., Webster, A. and Sargent, J. R. *Biotechnol. Appl. Biochem.*, 8 (1986) 83–95.
[171] Gormsen, E. and Elvig, N. Paper presented at the *AOCS Annual Meeting*, Honolulu, USA, May 14–18, 1986.
[172] Dambmann, C., Holm, P., Jensen, V. and Nielsen, M. H. *Dev. Ind. Microbiol.*, 12 (1971) 11–23.
[173] Thom, D., Swarthoff, T. and Maat, J. US Patent 4 707 291 (1987).
[174] Enomoto, M. and Riisgaard, S. Danish Patent Application 3589/85 (1986).
[175] Fujii, T., Tatara, T. and Minagawa, M. *J. Am. Oil Chem. Soc.*, 63 (1986) 796–799.
[176] Huge-Jensen, B. and Gormsen, E. US Patent 4 810 414 (1989).
[177] Boel, E. and Huge-Jensen, B. European Patent Application 305 216 (1989).
[178] Huge-Jensen, B., Paper presented at the *ISF-JOCS World Congress*, Tokyo, Japan, September 26–30, 1988 Japan Oil Chemists Society.
[179] Gormsen, E. and Stentebjerg-Olesen, B., Paper presented at the 21st Detergent Symposium Tokyo, Japan, October 19–20, 1989 Japan Oil Chemists Society.
[180] Fengel, D. and Wegener, G. *Wood*, W. de Gruyter, Berlin, New York, 1989, pp. 192–205.
[181] Resinase™ A. Product Information B 354a (1990) from Novo Nordisk A/S.
[182] Irie, Y., Usui, M., Matsukara, M. and Hata, K. In *Proceedings of the 1990 Papermakers Conf.*, Atlanta, USA, April 23–25, 1990 Tappi Press, Atlanta, 1990, pp. 1–10.
[183] INFORM, 1 (1990) 434–435.
[184] Iwai, M., Tsujisaka, Y. and Fukumoto, J. *J. Gen. Appl. Microbiol.*, 10 (1964) 13–22.
[185] Lazar, G., Weiss, A. and Schmid, R. D. In Baldwin, A. R. (ed.), *Proceedings of the World*

Conference on Emerging Technologies in The Fats and Oils Industry, Cannes, France, November 3–8, 1985. American Oil Chemists Society, 1986, pp. 346–354.
[186] Iwai, M., Okumura, S. and Tsujisaka, Y. *Agric. Biol. Chem.*, 44 (1980) 2731–2732.
[187] Gatfield, I. L. Lebensm.-Wiss, u.-Technol., 19 (1986) 87–88.
[188] Langrand, G., Triantaphylides, C. and Baratti, J. *Biotechnol. Lett.*, 10 (1988) 549–554.
[189] Gillies, B., Yamazaki, H. and Armstrong, D. W. *Biotechnol. Lett.*, 9 (1987) 709–714.
[190] Kanisawa, T. *Chem. Abstr.*, 100 (1984) 33406y.
[191] Anantharaman, K. *Abstracts of the IUFoST International Symposium on New Aspects of Dietary Lipids*, Göteborg, Sweden, September 17–20, 1989. SIK, Göteborg, p. 14.
[192] Hansen, T. T., US Patent 4 826 767 (1989).
[193] Krog, N. Surfactants and Emulsifiers, In Gunstone, F. D., Harwood, J. L. and Padley, F. (eds), *The Lipid Handbook*. Chapman and Hall, London, 1986, pp. 226–235.
[194] Yamane, T., Hoq, M. M., Itoh, S. and Shimizu, S. *J. Jpn. Oil Chem. Soc.*, 35 (1986) 632–636.
[195] Schröder, R. and Oba, K. European Patent Application 293 001 (1988).
[196] AMANO Pharm. Co. Japanese Patent Application 021 546 (1986).
[197] Holmberg, K. and Österberg, E. *J. Am. Oil. Chem. Soc.*, 65 (1988) 1544–1548.
[198] Holmberg, K., Lassen, B. and Stark, M.-B. *J. Am. Oil Chem. Soc.*, 66 (1989) 1796–1800.
[199] Godtfredsen, S. E., Bundgård, P and Andresen, O. International Patent Application (PCT) WO 86/05186 (1986).
[200] Omar, I. C., Saeki, H., Nichio, N. and Nagai, S. *Biotechnol. Lett.*, 11 (1989) 161–166.
[201] Hirota, Y., Kohori, J. and Kawakara, Y. European Patent Application 307 154 (1989).
[202] Cesti, P., Zaks, A. and Klibanov, A. M. *Appl. Biochem. Biotechnol.*, 11 (1985) 401–407.
[203] Eigtved, P., Hansen, T. T. and Miller, C. A. In Applewhite, T. H. (ed.), *Proc. World Conference on Biotechnology for the Fats and Oils Industry*, Hamburg, FRG, September 27–October 2, 1987, Published by the American Oil Chemists' Society, 1988, pp. 134–137.
[204] Therisod, M. and Klibanov, A. M., *J. Am. Chem. Soc.*, 108 (1986) 5638–5640.
[205] Seino, H., Uchibori, T., Inamasu, S. and Nishitani, T., US Patent 4 616 718 (1986).
[206] Adelhorst, K., Björkling, F., Godtfredsen, S. E. and Kirk, O. Synthesis, 2 (1990) 112–115.
[207] van Dam, A. F., US Patent 4 034 124 (1977).
[208] Inoue, S. and Ota, S., European Patent Application 109 244 (1984).
[209] Myojo, K., Matsufine, Y. and Yoshikawa, S. *Chem. Abstr.*, 107 (1987) 196401b.
[210] Osnai, S. *J. Jpn. Oil Chem. Soc.*, 35 (1986) 955–957.
[211] Okamura, S., Iwai, M. and Tominaga, Y. *Agric. Biol. Chem.*, 48 (1984) 2805–2808.
[212] Matsumara, S. and Takahashi, J. *Macromol. Chem., Rapid Commun.*, 7 (1986) 369–373.
[213] Iwai, M. and Wagano, N. *J. Jpn. Oil Chem. Soc.*, 36 (1987) 862–866.
[214] Kolattukudy, P. E. Cutinase from fungi and pollen. In Borgström, B. and Brockman, H. L. (eds), *Lipases*. Elsevier, Amsterdam, 1984, pp. 471–504.
[215] Brown, A. J. and Kolattukudy, P. E. *Arch. Biochem. Biophys.*, 190 (1978) 17–26.
[216] Eigtved, P., Hansen, T. T. and Huge-Jensen, B. In Marcuse, R. (ed.), *Proceedings of the 13th Scandinavian Symposium on Lipids*, Reykjavik, Iceland, June 30–July 3, 1985. Lipidforum/SIK, Göteborg, 1986, pp. 135–148.
[217] Hansen, T. T. and Eigtved, P. In Baldwin, A. R. (ed.), *Proceedings of the world Conference on Emerging Technologies in The Fats and Oils Industry*, Cannes, France, November 3–8, 1985. American Oil Chemists Society, 1986, pp. 365–369.
[218] Crook, S., Macrae, A. R. and Moore, H. European Patent Application 170 431 (1986).
[219] Bang, H. O., Dyerberg, J. and Sinclair, H. M. *Am. J. Clin. Nutr.*, 33 (1980) 2657–2661.
[220] Winning, M., In Shuka, V. K. S. and Hølmer, G. (eds), *Proceedings of the 15th Scandinavian Symposium on Lipids*, Rebild, Denmark, 11–15 June, 1989. Lipidforum, pp. 443–449.

[221] Haraldsson, G. G., Höskuldsson, P. A., Sigurdsson, S. T., Thorsteinsson, F. and Gudbjarnason, S. *Tetrahedron Lett.* (1989) 1671–1674.
[222] Hansen, T. T. *Abstract of the IUFoST International Symposium on New Aspects of Dietary Lipids*, Göteborg, Sweden, September 17–20, 1989. SIK, Göteborg, p. 25.
[223] Babayan, V. K. *Lipids*, 22 (1987) 417–420.
[224] Jandacek, R. J., Whiteside, J. A., Holcombe, B. N., Volpenhein, R. A. and Taulbee, J. D. *Am. J. Clin. Nutr.*, 45 (1987) 940–945.
[225] Seto, A. and Yamada, O. European Patent Application 265 699 (1988).
[226] King, D. M. and Padley, F. European Patent Application 209 327 (1987).
[227] The complete nutrition diet 'top up'. Product information (1988) from CSL-NOVO Pty. Ltd.
[228] Dennison, A. R., Ball, M., Crowe, P. J., White, K., Hands, L., Watkins, R. M. and Kettlewell, M. *Ann. Royal Coll. Surgeons Eng.*, 68 (1986) 119–121.
[229] Horrobin, D. F. Abstracts of the IUFoST International Symposium on New Aspects of Dietary Lipids, Göteborg, Sweden, September 17–20, 1989. SIK, Göteborg, p. 16.
[230] Kudo, S. In Applewhite, T. H. (ed.), *Proc. World Conference on Biotechnology for the Fats and Oils Industry*, Hamburg, FRG, September 27–October 2, 1987, Published by the American Oil Chemists' Society, 1988,pp. 195–201.
[231] Martin, J. F., Paper presented at the *AOCS Short Course on Lecithins*, Phoenix, USA, May 4–7, 1988.
[232] Juneja, L. R., Hibi, N., Inagaki, N., Yamane, T. and Shimizu, S. *Enzyme Microb. Technol.*, 9 (1987) 350–354.
[233] Juneja, L. R., Katzuoka, T., Yamane, T. and Shimizu, S. *Biochim. Biophys. Acta*, 960 (1988) 334–341.
[234] Juneja, L. R., Katzuoka, T., Goto, N., Yamane, T. and Shimizu, S., *Biochim. Biophys. Acta*, 1003 (1989) 277–283.
[235] Röbbelen, G. In Padley, F. (ed.) *Advances in Applied Lipid Research.* JAI Press, London, Vol. 1, forthcoming.
[236] Bedbrook, J. R. *Abstracts of the IUFoST International Symposium on New Aspects of Dietary Lipids*, Göteborg, Sweden, September 17–20, 1989. SIK, Göteborg, p. 30.
[237] Vezina, C., In Bu'Lock, J. D. and Kristiansen, B. (eds), *Basic Biotechnology*, Academic Press, London and San Diego, 1987, pp. 463–482.
[238] Shimizu, S. and Yamada, H. In Shukla, V. K. S. and Hølmer, G. (eds), *Proceedings of the 15th Scandanavian Symposium on Lipids*, Rebild, Denmark, 11–15 June, 1989. Lipidforum, pp. 230–244.
[239] Yamada, H., Shimizu, S., Shinmen, Y., Kawashima, H. and Akimoto, K. In Applewhite, T. H. (ed.), *Proc. World Conference on Biotechnology for the Fats and Oils Industry*, Hamburg, FRG, September 27–October 2, 1987, Published by the American Oil Chemists' Society, 1988, pp. 173–177.
[240] Kyle, D., Behrens, P., Bingham, S., Arnett, K. and Lieberman, D. In Applewhite, T. H. (ed.), *Proc. World Conference on Biotechnology for the Fats and Oils Industry*, Hamburg, FRG, September 27–October 2, 1987, Published by the American Oil Chemists' Society, 1988, pp. 117–122.
[241] Ratledge, C. in Ratledge, C. and Wilkinson, S. G. (eds), *Microbial Lipids.* Academic Press, London, 1989, pp. 117–122.
[242] Gladue, R., Hoeksema, S., Chen, H., Chantler, V.,Lieberman, D., Pai, N., Prichard, B., Rutten, J., Kyle, D. J. In Hayano, S. (ed.), *Proceedings of the ISF–JOCS World Confress 1988*, Tokyo, Japan, 26–30 September, 1988. The Japan Oil Chemists' Society, 1989, pp. 1007–1013.
[243] Kurashige, J. In Applewhite, T. H. (ed.), *Proc. World Conference on Biotechnology for the Fats and Oils Industry*, Hamburg, FRG, September 2&—October 2, 1987, Published by the American Oil Chemists' Society, 1988, pp. 138–141.

[244] Yamaguchi, S., Mase, T. and Asada, B. Japanese Patent Application Sho 62-287/A (1987).

[245] Novo Nordisk A/S. Unpublished results.

[246] Derez, F. G., Reeve, A. L. and De Sadeleer, J. W. G. C. European Patent Application 219 269 (1987).

[247] Uwajima, T. and Terada, O. *Trends Anal. Chem.*, 1 (1981) 55–58.

THE INTERACTION OF OXIDIZED LIPIDS WITH PROTEINS

Peter N. Gillatt and J. Barry Rossell

OUTLINE

Advances in Applied Lipid Research, Volume 1, pages 65–118

ISBN: 1-55938-317-8

1. INTRODUCTION

With the exception of water, proteins and lipids are possibly the most important chemical constituents of foods. Within the structure of living systems, lipids are protected from oxygen. Therefore, lipid oxidation and subsequent damage to protein do not occur readily. However, this is not the case in many food systems and, although it may have been thought that oxidized lipids and proteins do not interact, clear evidence to the contrary is provided in the discussion that follows.

Initiation of lipid oxidation occurs more readily in food systems due, for example, to processing techniques that result in tissue damage, with subsequent release of lipid and the enhanced exposure of the fat to oxygen and light.

It has been suggested [1] that oxidized lipids cause disruption of structural proteins and enzyme systems, a view that has been supported by the observation that membrane-bound proteins and enzymes are more susceptible to damage by peroxidizing lipids than are free proteins. In particular, sulphydryl enzymes [2], structural proteins [3] and cytochromes are sensitive to damage [4].

As there are complex interactions between oxidized lipids and proteins it is worth reviewing the oxidation of lipids. The scheme shown in Figure 1 was presented by Kochhar and Meara [5] for the oxidation of linoleic acid. In this

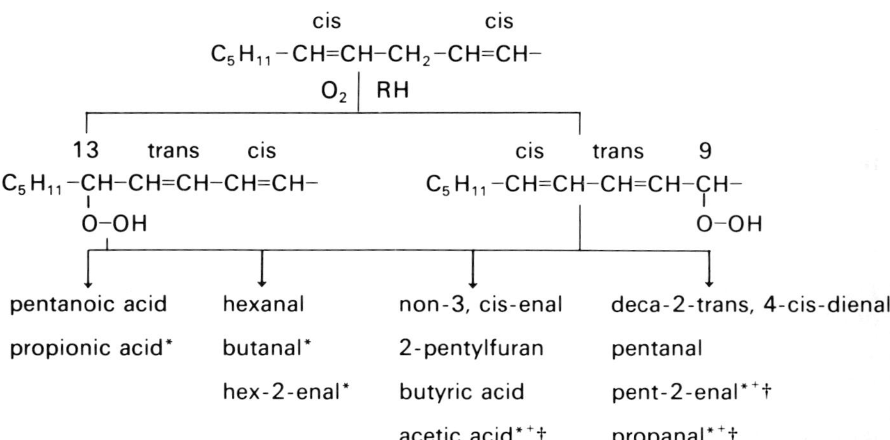

Figure 1. Pathway of expected volatile products from autoxidation of linoleate ester. (Courtesy of Leatherhead Food RA.)

scheme free radical initiation is followed by reaction with oxygen to form hydroperoxides. These decompose to form secondary oxidation products, in particular aldehydes and ketones. It is these compounds that give rise to the off-flavours that are associated with rancid foods.

Lipid oxidation is an autocatalytic process, in which the oxidation products themselves catalyse the reaction, resulting in the rate of reaction increasing with time. Lipids at an intermediate stage of oxidation may react differently from those already peroxidized and it is important to recognize the difference. The complex reactions and interactions that follow include those of hydroperoxides and their breakdown products, with proteins, with the production of chemical covalent bonds and more weakly bound non-covalent complexes.

Studies have, in the main, been undertaken using model systems to determine how oxidized lipids react with proteins. In food systems, however, the situation is more complex, but it has been observed that oxidized lipids cause denaturation and loss of protein solubility, deterioration of texture and rehydratability characteristics [6, 7], alteration of protein functional properties, loss of nutritional value [4, 8], and non-enzymic browning [9].

The information presented in this chapter deals with the reactions and interactions that take place in model or food systems. It should, however, be remembered that oxidized lipids are also found in living systems and that such reactive compounds might react with the surrounding proteins. Free radicals produced by lipid auto-oxidation induce cross-linking of polypeptide chains and reactions of lipoperoxides are thought to belong to important mechanisms of ageing phenomena [10]. The interaction of oxidized lipids with proteins proceeds very slowly *in vivo* producing brown insoluble deposits called ceroids. The production of these deposits is enhanced if the diet contains large quantities of polyunsaturated fatty acids but few antioxidants such as tocopherols [11]. Ceroid is deposited mainly in the brain and neuronal tissue but may also be found in the liver and uterus. It is considered that the formation of ceroids is caused by reactions of peroxidizing lipids intracellularly with cell proteins [12].

Free radicals produced by lipid oxidation and their secondary breakdown products such as aldehydes are thought to have the potential to be precursors of cancer. Reiss and Tappel [13] detected fluorescent compounds produced by reactions of peroxidizing arachidonic acid with DNA, which substantiates the view that hydroperoxides have carcinogenic activity.

The amount of literature regarding the effect of oxidized lipids *in vivo* is limited and, as a consequence, it is not dealt with in any detail in the following chapter.

The following text examines the reactions of proteins and oxidized lipids from the point of view of the products formed, and concludes with a section dealing with reactions of oxidized lipids and proteins in a number of edible materials. The mechanisms of the reactions are very dependent on the water activity of the system and, therefore, different reaction products may be found in moist and dry forms of the same food. In addition, oxidized lipids are sometimes regenerated to the original unoxidized form by interaction with protein, and thus act as catalysts of protein oxidation, i.e.

$$\text{lipid} \xrightarrow{O_2/\text{catalysts}} \text{oxidized lipid}$$

$$\text{oxidized lipid} + \text{protein} \rightarrow \text{lipid} + \text{oxidized protein}$$

As a consequence, an apparently insignificant amount of polyunsaturated lipid may then promote oxidation of a much larger quantity of protein.

Although some of the interactions described are beneficial, in that they impart attractive flavours or give improved volume of baked products, in the main they damage protein and/or introduce unpleasant off-flavours.

The work on model systems is reviewed in some detail so that food manufacturers will be able to interpret the knowledge in relation to their own products.

Four appendices are included to aid the reader. The first gives chemical structures of the most important and frequently referred to amino acids and the second illustrates a number of oxidized sulphur compounds and their nomenclature. The third and fourth give details of abbreviations used and a guide to the oxidation of oils and fats, respectively.

2. REACTIONS OF OXIDIZED LIPIDS WITH PROTEINS

2.1 Non-covalent Links

The interactions between oxidized lipids and proteins usually lead to a number of covalently bound species. A large proportion of the lipid, however, complexes to the protein through hydrophobic association and hydrogen bonding. It was suggested [14–16] that egg albumen physically binds oxidized lipids through hydrogen bonds.

It has been possible to establish quantitatively the type and strength of bonds holding lipid and protein together by a series of extraction steps [17]. The weakly or hydrophobically bound lipid (probably unoxidized lipid) was extracted by a weakly polar solvent such as diethyl ether. The hydrogen bound lipid, which contained the greatest quantity of oxygenated material, required a more polar solvent such as chloroform [18]. Most of the peroxidized lipid can be extracted with chloroform or, as Kanner and Karel [19] suggested, by breaking the complex using urea and sodium dodecyl sulphate (SDS). Lipid that can only be extracted by more vigorous means is more strongly bound than lipid extracted by solvents of low polarity, and is therefore probably covalently bound.

Harmuth-Hoene and Delincée [20] indicated that oxidized lipid–protein aggregates were formed when casein was stored in the presence of corn oil and cocoa fat. The non-covalently bound fat was easily extracted with acetone and diethyl ether, suggesting that weak non-covalent forces were holding the aggregate together. The biological activity of the protein had been significantly reduced, but this returned following extraction of the fat and remained even when new unoxidized fat was added. However, addition of the extracted, oxidized fat to fresh casein did not reduce the protein's nutritional value, suggesting that the formation of non-covalent complexes between oxidized lipids and proteins is a slow process.

It is important to take non-covalently bound complexes into consideration when interpreting results. For example, Matsushita and Kobayashi [21] found that linoleic acid was as effective at inhibiting a number of enzymes as was its hydroperoxide. This may be explained by the fatty acid binding to the hydrophobic region of the protein/enzyme, as opposed to its chemically active region. The fact that effective blank determinations should be undertaken in such experiments was illustrated by Matsushita et al. [22], who found that pepsin could be activated by linoleic acid hydroperoxide but not by the acid itself. This indicates that the hydroperoxide reacted with the enzyme chemically rather than physically.

Since aldehydes and ketones are secondary oxidation products having low flavour thresholds, they have been studied extensively. Using a leaf protein concentrate, Franzen and Kinsella [23] showed that addition of aldehydes and methyl ketones utilized the hydrophobic sites and thus reduced the binding capacity of the concentrate. The weak unspecified forces that bound aldehydes, alcohols and ketones to bovine serum albumin or soya protein have been attributed to electrostatic and hydrophobic attractions [24].

Although the above demonstrates that aldehydes and ketones can bind physically to proteins, it is more usual for such reactive compounds to form chemical bonds to proteins via the formation of Schiff's bases.

2.2 Protein–Lipid Cross-links

A general mechanism proposed by Schaich [25] for the production of a covalent bond between lipid and protein is as follows:

(1) Hydroperoxide formation:
   (i) $L^{\cdot} + O_2 \rightarrow LOO^{\cdot}$
   (ii) $LOOH \rightarrow LO^{\cdot} + {}^{\cdot}OH$
   where L is an unsaturated lipid and $L^{\cdot}$ is its free radical, etc.

(2) Hydroperoxide reaction with protein (PH):
   (i) $LOO^{\cdot} + PH \rightarrow LOOH + P^{\cdot}$
   (ii) $LO^{\cdot} + PH \rightarrow P^{\cdot} + LOH$

(3) Addition of lipid to protein:
 (i) $LOO^{\cdot} + PH \rightarrow {}^{\cdot}LOOPH$
 (ii) $LO^{\cdot} + PH \rightarrow {}^{\cdot}LOPH$

(4) Polymerization and cross-linking:
 (i) ${}^{\cdot}LOPH + O_2 \rightarrow {}^{\cdot}OOLOPH$
 (ii) ${}^{\cdot}LOOPH + O_2 \rightarrow {}^{\cdot}OOLOOPH$

(5) These species can then react further:
 (i) ${}^{\cdot}OOLOPH + P \rightarrow {}^{\cdot}POOLOPH$
 (ii) ${}^{\cdot}OOLOOPH + P \rightarrow {}^{\cdot}POOLOOPH$

The postulation that radicals are involved in the formation of lipid–protein cross-links was furthered by work undertaken by Gardner et al. [26, 27] and Gardner [28]. Gardner [26] found that the amino acid cysteine, which contains a sulphydryl group (–SH), was covalently bound to 13-hydroperoxylinoleic acid. The reaction was catalysed by 10^{-5} M ferric chloride in aqueous ethanol (1 : 4). It was found that in the absence of oxygen either cysteine or *N*-acetylcysteine became bound via a thiyl bond to the cis-9 olefinic carbon of 13-hydroperoxylinoleic acid. The 13-hydroperoxy group was reduced to a 13-hydroxy group, and the 10-olefinic carbon became substituted with an ethoxy or hydroxy group. Thus the products of the reaction are 9-*S*-cysteine-13-hydroxy-10-ethoxy-*trans*-11-octadecenoic acid (B) and 9-*S*-cysteine-10,13-dihydroxy-*trans*-11-octadecenoic acid (A). The product formation, A and B, was followed by thin-layer densitogram as illustrated in Figure 2.

Under nitrogen, the quantity of the dihydroxy substituted compound was approximately a quarter of that of the ethoxy compound. This is a direct consequence of the use of a 1 : 4 water/ethanol solvent system. The proposed reaction mechanism is as shown in Figure 3.

Substitution by water at C-10 would produce the dihydroxy derivative.

It was important that for production of the above species, the reaction was undertaken under nitrogen, as shown by the thin-layer densitogram (Figure 2). When cysteine and 13-hydroperoxylinoleic acid reacted in the presence of oxygen a large number of oxygenated fatty acids, of which Gardner et al. [29] identified nine, were produced instead of the compounds illustrated in Figure 3. The absence of the cysteine fatty acid complex and the production of oxygenated fatty acids can only be explained by the active participation of oxygen in the reaction. Oxygen is an effective scavenger for radicals and would react with them preferentially, thereby quenching the reaction of fatty acid radicals with other compounds, such as cysteine. This is illustrated in Figure 4.

Although no similar cysteine-fatty acid compounds were formed when a mixture of 9- and 13-hydroperoxides was treated with cysteine in air, a complex was formed with production of 9-oxo-nonanoic acid ($HO_2C(CH_2)_7CHO$).

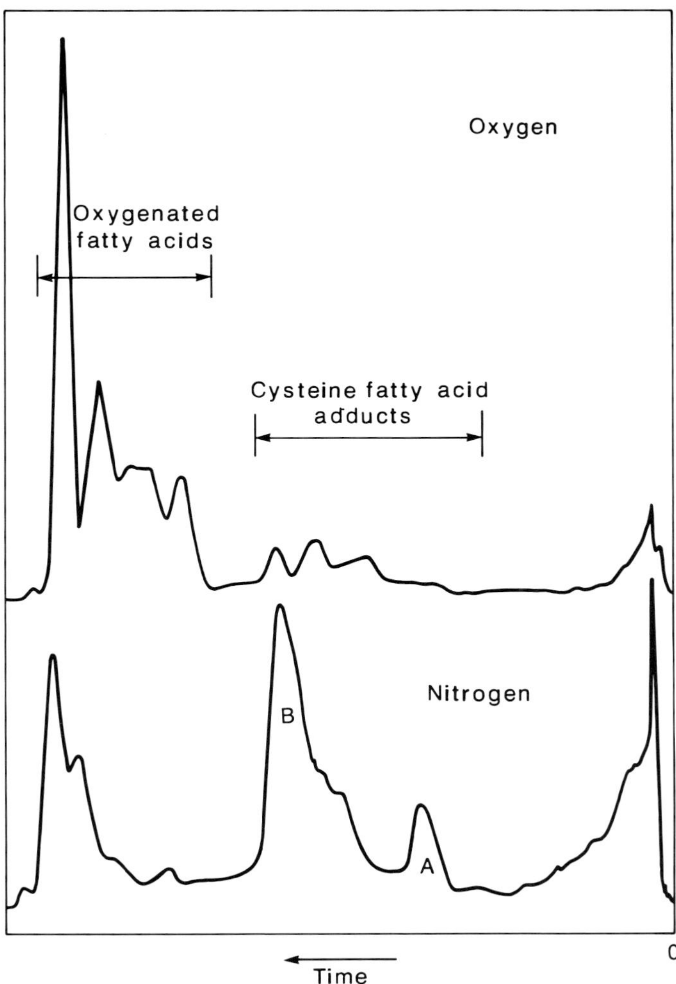

Figure 2. Thin-layer densitogram of products obtained from incubation of 13-hydroperoxylinoleic acid and cysteine. (From Gardner [28].)

Formation of the complex required participation of the free amino group and would not, therefore, proceed with the *N*-acetyl derivative of cysteine. The compound produced is shown in Figure 5.

Gardner et al. [27] demonstrated that thiol oxidation by peroxidized lipids can proceed via free radicals rather than with lipid hydroperoxide breakdown products such as malondialdehyde. This is in line with the findings of Buttkus [9], who specifically noted that reaction of sulphydryl groups with peroxides is more favourable than with aldehydes. Previous work by Kanawaza et al. [30, 31] led Kanawaza et al. [32] to investigate the binding effects of linoleic acid and its hydroperoxide and secondary oxidation products with casein. Incubation of casein with linoleic acid indicated that the protein had anti-

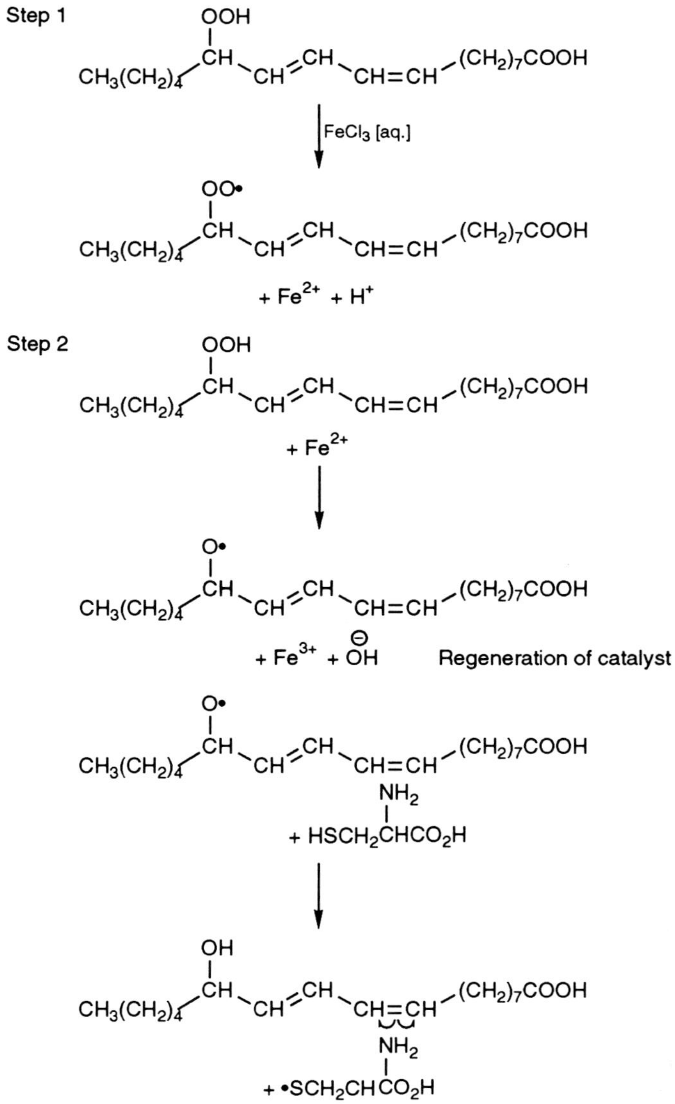

Step 1

$CH_3(CH_2)_4CH(OOH)CH{=}CHCH{=}CH(CH_2)_7COOH$

$\downarrow$ $FeCl_3$ [aq.]

$CH_3(CH_2)_4CH(OO\bullet)CH{=}CHCH{=}CH(CH_2)_7COOH$

$+ Fe^{2+} + H^+$

Step 2

$CH_3(CH_2)_4CH(OOH)CH{=}CHCH{=}CH(CH_2)_7COOH$

$+ Fe^{2+}$

$\downarrow$

$CH_3(CH_2)_4CH(O\bullet)CH{=}CHCH{=}CH(CH_2)_7COOH$

$+ Fe^{3+} + OH^{\ominus}$ Regeneration of catalyst

$CH_3(CH_2)_4CH(O\bullet)CH{=}CHCH{=}CH(CH_2)_7COOH$

$+ HSCH_2CH(NH_2)CO_2H$

$\downarrow$

$CH_3(CH_2)_4CH(OH)CH{=}CHCH{=}CH(CH_2)_7COOH$

$+ \bullet SCH_2CH(NH_2)CO_2H$

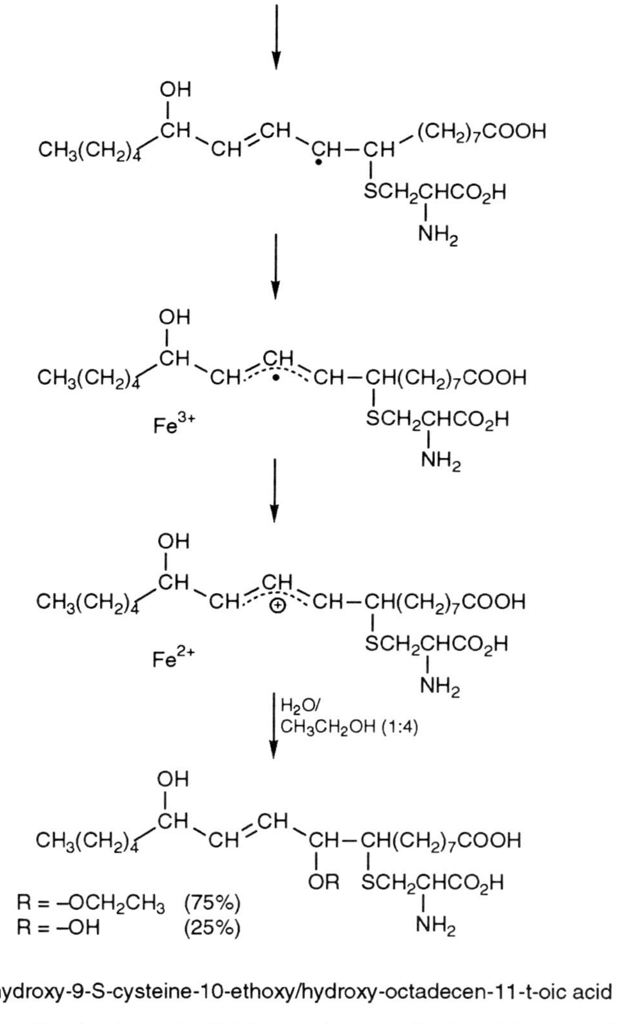

OH
|
$CH_3(CH_2)_4$–CH–CH=CH–$\dot{C}$H–CH–$(CH_2)_7COOH$
|
SCH_2CHCO_2H
|
NH_2

↓

OH
|
$CH_3(CH_2)_4$–CH–CH⋯CH(•)⋯CH–CH$(CH_2)_7COOH$
|
Fe^{3+} SCH_2CHCO_2H
|
NH_2

↓

OH
|
$CH_3(CH_2)_4$–CH–CH⋯CH(⊕)⋯CH–CH$(CH_2)_7COOH$
|
Fe^{2+} SCH_2CHCO_2H
|
NH_2

↓ H_2O/ CH_3CH_2OH (1:4)

OH
|
$CH_3(CH_2)_4$–CH–CH=CH–CH–CH$(CH_2)_7COOH$
| |
OR SCH_2CHCO_2H
|
NH_2

R = $-OCH_2CH_3$ (75%)
R = $-OH$ (25%)

13-hydroxy-9-S-cysteine-10-ethoxy/hydroxy-octadecen-11-t-oic acid

Figure 3. Production of a lipid–protein complex involving cysteine.

oxidant properties, because the peroxide (PV) and thiobarbituric acid value (TBA) maxima were reached after 7 days with casein but after 3 days without. In addition, the PV and TBA maxima were lower for casein-containing solutions. It was found that after 5 min, 15% of the linoleic acid, 30% of the hydroperoxide and 75% of the secondary breakdown products had been bound to the casein.

Such binding resulted in marked changes to the solubility as well as the digestibility (as determined by the digestive enzymes trypsin, chymotrypsin

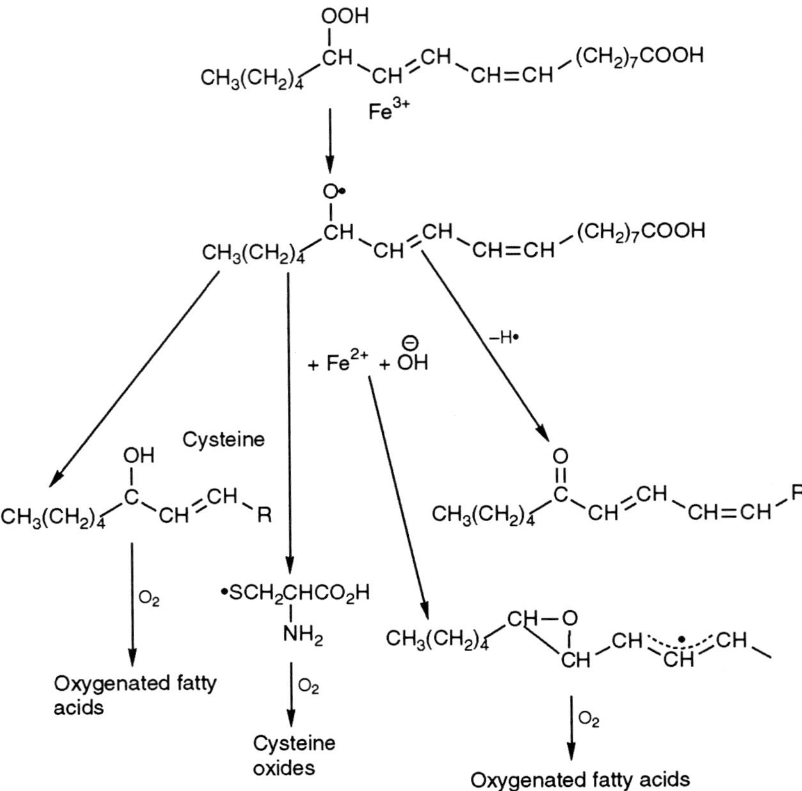

Figure 4. Inhibition of protein–lipid complex formation in the presence of oxygen [28].

and pepsin) of the protein. Incubation of casein with linoleic acid hydroperoxide and secondary breakdown products reduced the solubility of casein by 60% after 4 days and 90% after 2 days, respectively. Addition of linoleic acid did not change casein solubility.

The digestibility of casein increased after incubation with linoleic acid, but

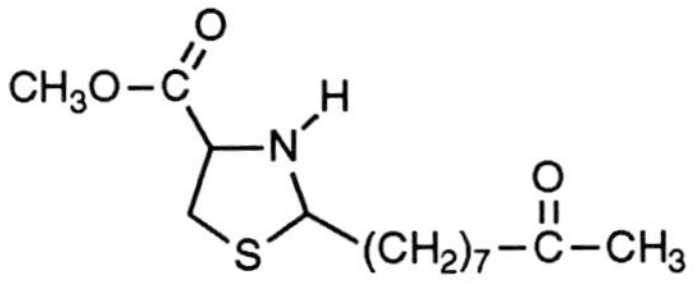

Figure 5. A reaction product of cysteine and hydroperoxides in air.

Table 1. Effect of interaction of albumin on oxidized ethyl linoleate.

Material analysed	Oxidized fatty acids	Unoxidized fatty acids
Original oxidized ethyl linoleate (A)	26.3	73.7
(A) following 1 h reaction with albumin at 60°C	58.0	42.0
(A) following 64 h reaction with albumin at 60°C	92.6	7.5

was reduced after incubation with its hydroperoxide and to a greater extent after incubation with secondary breakdown products.

It has been stated that the interactions of oxidized lipids with proteins, particularly when covalently bonded lipoproteins are formed, decrease the nutritional value of food since certain, often essential, amino acids in the protein are converted into indigestible products.

The mechanism of interactions of oxidized lipids with protein is unclear and complicated. It has, however, been shown that lipoproteins produced in this way contain lipids bound by hydrogen bonds [16, 17] as well as by covalent links [17, 33].

Pokorny et al. [34] undertook work in which they attempted to determine the stage at which the reaction of oxidized lipids caused the loss of available lysine in proteins.

The basis of their experiment involved the exposure of finely ground albumin to oxidized ethyl linoleate in a ratio 4 : 1. The mixture was stored at 60°C.

In dry mixtures of egg albumin and oxidized ethyl linoleate (OEL), it was found that the total oxidized fatty acids consisted of several subfractions; indeed the original OEL contained mainly unchanged ethyl linoleate. However, after 1 h storage of the mixture at 60°C, the quantity of unoxidized ethyl linoleate was reduced by half and following incubation for 64 h virtually no unreacted linoleate remained. The amount of oxidized fatty acids rose accordingly, as may be seen from Table 1.

It was further noted that following the peroxide value reaching its maximum, the oxygen absorption, which initially had been rapid, ceased almost totally.

Oxidized free lipid reacted with protein-forming, at first, hydrogen, bonded lipoproteins; the latter were slowly converted into covalently bonded complexes. However, at the same time part of the fatty moiety was cleaved with the production of volatile components.

Table 2. Changes to amino acids during the interactions of oxidized ethyl linoleate with albumin.

Material analysed	Lysine	Methionine	Arginine	Tyrosine
	(g/16 g N)			
Original oxidized ethyl linoleate (A)	8.90	2.56	1.89	2.32
(A) following 1 h storage with albumin at 60°C	8.42	2.38	1.78	2.06
(A) following 64 h storage with albumin at 60°C	7.00	2.25	1.52	1.39

The data presented in Table 2 illustrate the effect that incubation of EOL had on the concentration of reactive amino acids.

From the sequence of hydroperoxide and carbonyl maxima it would appear that both the former and latter are precursors of protein oligomerization. Pokorny et al. [34] put forward the theory that decomposition of lipid hydroperoxides produced free radicals and that these were passed from the lipid to protein. This was in line with the thinking of Roubal [35].

The content of covalently bound lipid (CBL) was in direct correlation with that of protein oligomers (PL), as they were found in greater amounts in the stage following the maximum of hydrogen-bonded lipoprotens, i.e. later than the maxima of decomposition rates of hydroperoxides and carbonylic derivatives when protein oligomers were produced at the fastest rate. Pokorny stated, therefore, that a linear regression could be obtained, whereby

$$\mathrm{CBL} = 1.70\ \mathrm{PL} - 1.28$$

The formation of covalently bound lipids (CBL) was correlated linearly with that of volatile decomposition products (VDP), whereby

$$\mathrm{CBL} = 0.88\ \mathrm{VDP} + 4.72$$

It was suggested that the VDP were produced during the process in which the lipids were bound into covalently bonded lipoproteins.

It is unclear exactly how this process takes place. Pokorny et al. [36] showed that a rapid reaction between saturated alkanals and albumin takes place; similarly between unsaturated 2-alkenals and proteins. Gardner [28] and Frankel [37] have both described other routes, including the fact that unsaturated aldehydes readily auto-oxidize to dicarbonylic products, such as malondialdehyde (MDA) and that these products react to become chemically bound to protein.

Pokorny et al. [34] showed that the formation of CBL was accompanied by the rapid reduction in available lysine. If unavailable lysine is represented by ULY then

$$CBL = 4.31\ ULY - 5.79$$

It was considered that if, as Roubal [38] had stated, lipoprotein formation involved the reaction of free radicals being transformed to MDA, which in turn reacted with free amino groups of lysine to form a Schiffs base, then the loss of unavailable lysine (ULY) would correspond with the formation of protein oligomers (PL). This was found to be the case with a linear relationship, such that

$$ULY = 0.36\ PL + 2.12$$

It must be stressed that unavailable lysine derivatives resulting from the reactions of protein with oxidized lipids cannot be broken down to available lysine by the action of digestive enzymes and, therefore, the reaction of oxidized lipids with proteins can reduce the nutritional quality of food.

2.3 Protein–Protein Cross-links

In aqueous solution or at high water activities (a_w), cross-links can be formed between two protin molecules in the presence of peroxidizing lipid. Roubal and Tappel [4] suggested the following polymerization mechanism, without the incorporation of lipid material.

$$ROO^{\cdot} + PH \rightarrow P^{\cdot} + ROOH$$

$$RO^{\cdot} + PH \rightarrow P^{\cdot} + ROH$$

$$R^{\cdot} + PH \rightarrow P^{\cdot} + RH$$

$$HO^{\cdot} + PH \rightarrow H_2O + P^{\cdot}$$

$$P^{\cdot} + P^{\cdot} \rightarrow P\text{–}P$$

$$P^{\cdot} + PH \rightarrow HP\text{–}P^{\cdot}$$

$$HP\text{–}P^{\cdot} + P^{\cdot} \rightarrow HP\text{–}P\text{–}P^{\cdot}$$

where PH = protein and RH = lipid.

This is a bimolecular homolytic substitution or radical displacement reaction, where the lipid hydroperoxide breaks down to produce a radical, which is transferred to the protein with subsequent protein polymerization and regeneration of the lipid. Therefore, the lipid is a catalyst.

Bimolecular homolytic substitution reactions have a high activation energy, and therefore are unlikely to occur at an sp^3-hybridized four-coordinated carbon atom. Such substitution reactions are liable to occur on aromatic amino acid residues.

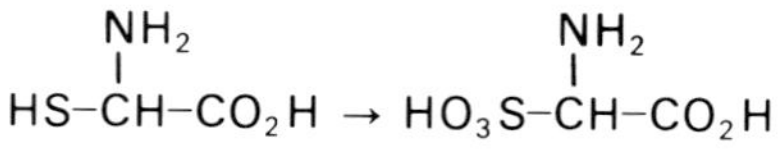

$$HS{-}\underset{|}{\overset{NH_2}{C}}H{-}CO_2H \rightarrow HO_3S{-}\underset{|}{\overset{NH_2}{C}}H{-}CO_2H$$

cysteine cysteic acid

Figure 6. Oxidation of cysteine.

Schaich and Karel [39] proposed that protein–protein cross-linking would occur through radical termination reactions, which therefore limits the reaction to the collision of two protein radicals, i.e.

$$P^{\cdot} + P^{\cdot} \rightarrow P{-}P$$

Finley, Walker and Wheeler [40] demonstrated that mildly oxidized proteins are more likely to form cross-links than reduced or highly oxidized proteins, and showed that when present most of the cysteine was oxidized to alanine sulphinic or cysteic acids (see Figure 6).

The intermediate oxidation products, i.e. cystine mono- and dioxides, were very susceptible to lysinoalanine formation, even under mild heating or alkaline conditions [40, 41] suggesting that such oxidations take place in food systems where there is a limited amount of oxidant present.

Finley and Lundin [41] demonstrated that the level of sulphur amino acids in α-lactalbumin decreased with increasing concentration of hydrogen peroxide, which correlated with a decrease in the protein efficiency ratio of the protein. This suggested that the more highly oxidized systems did not form detectable levels of lysinoalanine, consistent with Finley's previous work. It was concluded that the cystine mono- or dioxide present in proteins readily undergoes β-elimination, producing dehydroalanine, and that this cross-links with the ε-amino group of lysine to form lysinoalanine (see Figure 7).

The results of the reaction of glutathione (GSH) with hydrogen peroxide are given in Table 3 and show a rapid increase in the quantity of oxidized disulphides as the pH was increased.

The results in Table 3 were obtained following measurement by proton NMR and an analogous technique was employed to follow the effects of lipid hydroperoxides on the oxidation of glutathione. The results are shown in Table 4.

Table 4 illustrates that the oxidation of the –SH in glutathione by hydrogen peroxide proceeds more quickly than with linoleic acid hydroperoxide but in the latter case a greater proportion of sulphoxy compound may be formed as shown by reaction products after 60 min. It was suggested that this is due to the limited solubility of the lipid hydroperoxide. Nevertheless, similar product ratios were produced by both oxidants and there was an increase in oxidation of protein with time.

Leake and Karel [42] provided further evidence of the production of protein–protein cross-links via polymerization reactions, following the reaction

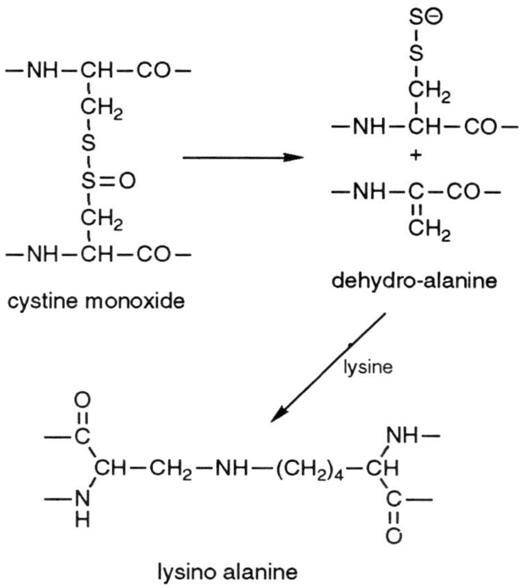

Figure 7. β-elimination of cysteine monoxide to form dehydroalanine followed by reaction with lysine to produce lysinoalanine [41].

of lysozyme with peroxidizing lipids. Equal weights of methyl linoleate and lysozyme were emulsified in water and freeze-dried. The resulting protein–lipid foams ($a_w = 0.75$) were exposed to air at 22°C for periods of between 10 and 20 days, after which lipid was removed. The lipid oxidation reached a maximum after 6 days. Protein radicals were detected using the electron spin resonance (ESR) method of Schaich and Karel [39]. Separation of the oxidized lysozyme by gel filtration on Biogel P-60 (see Figure 8) and con-

Table 3. Reaction of glutathione with equimolar hydrogen peroxide at various pH values.

pH	Per cent of each derivative after 2 min				
	$-SH$	$-SS-$	$-S(=O)S-$ + $-S(=O)_2-S-$	$-SO_2$	$-SO_3$
2.6	91	4	4	0	0
3.6	90	3	6	0	0
4.6	64	16	18	2	0
5.6	25	33	36	5	0
6.6	23	30	40	4	4
7.6	19	31	45	3	3

Table 4. Oxidation of glutathione with equimolar linoleic acid hydroperoxide and hydrogen peroxide at pH 6.0.

	Oxidation time (min)	Per cent of each derivative				
		$-SH$	$-SS-$	$-S(=O)S-$ + $-S(=O)_2-S-$	$-SO_2H$	$-SO_3H$
Hydrogen peroxide	2	25	32	39	3	1
	60	19	34	41	3	3
Linoleic acid hydroperoxide	2	85	1	11	3	0
	10	66	10	16	6	2
	50	39	14	41	4	4
	500	12	25	55	3	5

firmation by SDS electrophoresis demonstrated that the peaks A–E could be designated as follows:

A: lysozyme trimers and higher oligomers
B: denatured lysozyme dimer
C: native lysozyme dimer
D: denatured lysozyme monomer
E: native lysozyme.

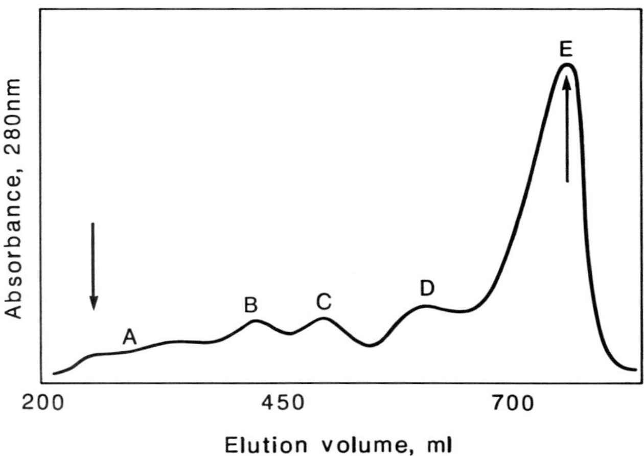

Figure 8. Separation of oxidized lysozyme on a 3.1 × 110 cm column of Biogel P-60 eluted with 1 M acetic acid. (↓ = void volume of the column; ↑ = elution volume of untreated lysozyme; letters A–E denote the fractions collected.) (From Leake and Kavel [42].)

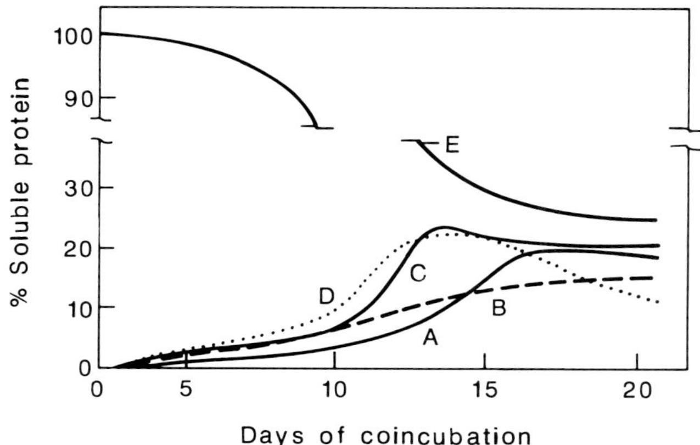

Figure 9. Formation kinetics of the lysozyme fractions (labelling as for Figure 7). (From Leake and Kavel [42].)

The kinetics of the formation of fractions A–E (Figure 9) show that the denatured monomer D and the native dimer C are formed simultaneously. Leake and Karel [42] concluded that denaturation and polymerization are independent processes, and that monomer denaturation is not a prerequisite for polymerization. Leake suggested the following mechanism for the polymerization of lysozyme, where N = native lysozyme, D = a partially denatured lysozyme molecule and L = a reactive lipid oxidation intermediate.

$$\mathrm{N} \rightarrow \mathrm{D} \text{ (fraction D)}$$

$$\mathrm{N} + \mathrm{L}^{\cdot} \rightarrow \mathrm{N}^{\cdot} + \mathrm{L}$$

$$\mathrm{D} + \mathrm{L}^{\cdot} \rightarrow \mathrm{D}^{\cdot} + \mathrm{L}$$

$$\mathrm{N}^{\cdot} + \mathrm{N}^{\cdot} \rightarrow \mathrm{N{-}N} \text{ (fraction C)}$$

$$\mathrm{N}^{\cdot} + \mathrm{D}^{\cdot} \rightarrow \mathrm{N{-}D} \text{ (fraction B)}$$

The polymerization reaction could continue, resulting in higher oligomers, as found in fraction A.

The enzymic activity of the lysozyme decreased after exposure to oxidizing lipids. The native monomer, fraction E, showed little change in activity. However, the dimer fraction C and denatured fractions B and D lost approximately 50% of their activity. Fraction A, the most polymerized, contained little residual enzymic activity (Figure 10).

There was a decrease in tryptophan residues (see Figure 11), which correlated with increased polymerization and reduced biological activity.

Schaich and Karel [43] suggested that tryptophan radicals would be formed in non-sulphydryl proteins, such as lysozyme. The evidence produced by Leake agrees with this.

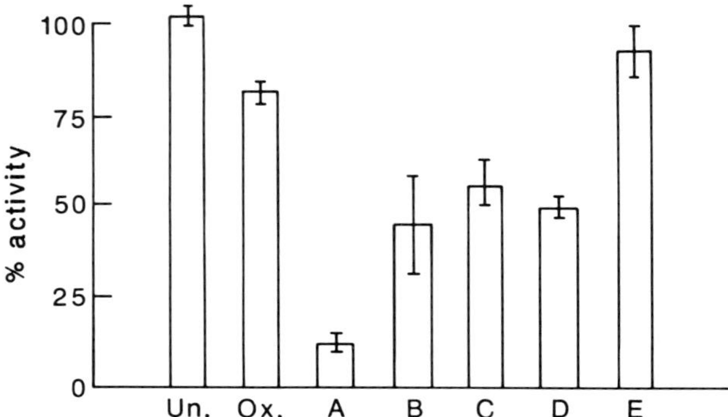

Figure 10. Reduction in enzyme activity of lysozyme with exposure to lipid hydroperoxide [41].

Aqueous emulsions containing hen-egg lysozyme (5 or 0.5%) and methyl esters (5 or 0.5%) of oleic, linoleic and linolenic acids, in a weight ratio 1 : 1 were prepared [44]. Oxidation was carried out in the dark at 37°C and determined by measurement of TBA and PV. The samples were analysed over a 10-day period. The TBA value increased with the degree of lipid unsaturation.

The effect of freeze drying the emulsions was to reduce the TBA value, indicating that this process must remove part of the malondialdehyde and/or other volatile TBA-reactive products. Freeze drying was also associated with PV reduction, indicating loss of peroxide oxygen.

Lysozyme decomposition was followed by measuring the disappearance of the monomer and formation of lysozyme dimers, by scanning gels after electrophoretic separation and by determining the total soluble protein after lipid extraction. The relative concentrations were expressed as a fraction. Figure 12 illustrates the course of lysozyme dimerization and monomer disappearance in aqueous solutions containing oleic (O), linoleic (L) or linolenic (Ln) acid methyl esters. The monomer fraction decreases continuously, as monomers are being polymerized to higher oligomers. The

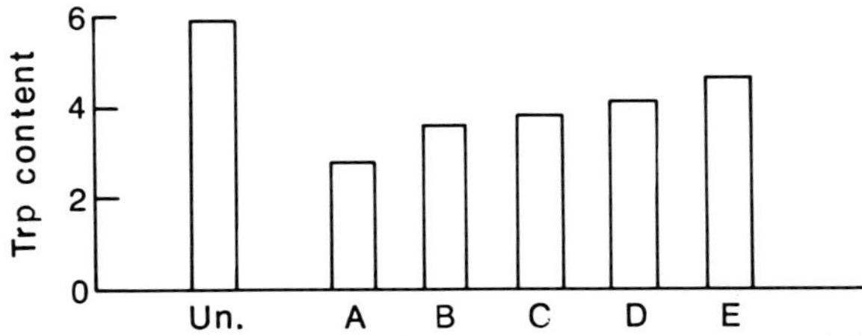

Figure 11. Estimated tryptophan content of lysozyme fractions [42].

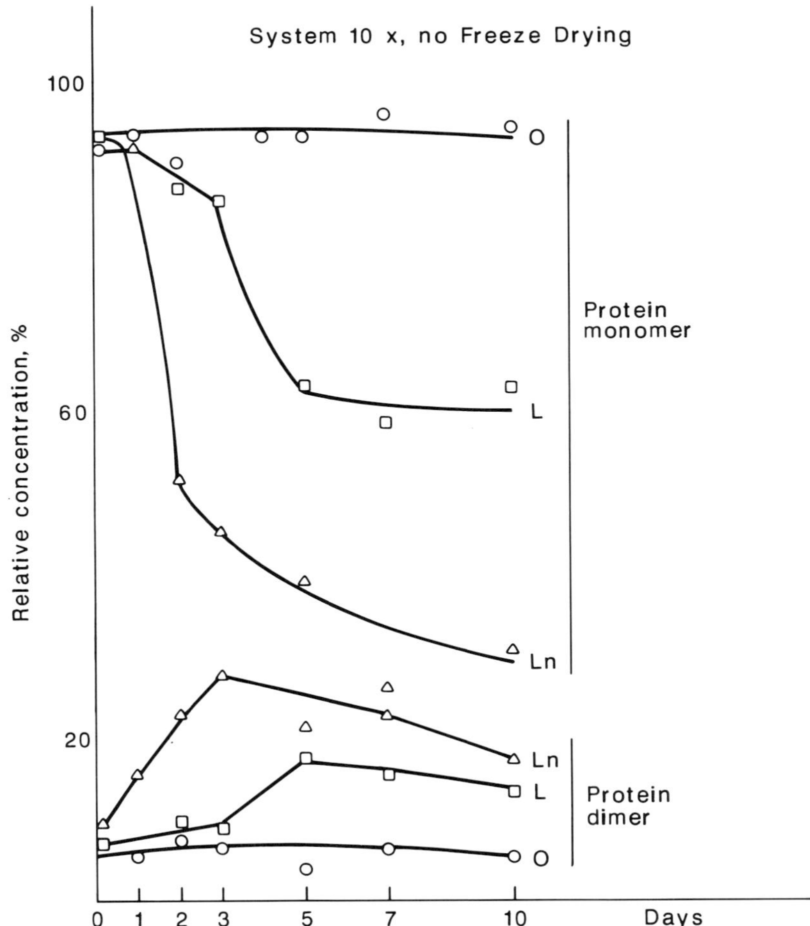

Figure 12. LYS polymerization as measured by dimer formation and monomer disappearance in LYS–lipid emulsion; systems 10X [5% lysozyme/5% methyl fatty acid] were not freeze-dried [44].

initial increase in the dimer fraction is followed by a decrease, due to further reactions to produce trimers and higher polymers. Figure 12 clearly demonstrates the increasing amount of polymerization with increasing degree of lipid unsaturation.

Freeze drying or increasing the concentration of both enzyme and peroxidizing lipid increased dimerization. In this case linolenic acid induced dimerization was maximized after 2 days and for linoleic acid (less reactive) after 3–5 days of oxidation.

$$\mathrm{LOOH + PH \longrightarrow [LOOH - - - - HP]}$$

$$\mathrm{LO^{\cdot} + P^{\cdot} + H_2O} \qquad \mathrm{LO^{\cdot} + {}^{\cdot}OH + PH}$$

$$\mathrm{PH^{\cdot} + {}^{\cdot}OH \rightarrow P^{\cdot} + LOH}$$

Figure 13. Protein polymerization via complex formation.

The maximum differences between the disappearance of monomer and formation of dimer appeared at the maximum concentration of lipid hydroperoxide (as measured by PV). Therefore, polymerization of protein may be promoted by an acceleration of lipid hydroperoxide decomposition produced by freeze drying. If peroxidizing lipid was extracted followed by freeze drying there was no increase in the extent of dimerization. Therefore, the induction of lysozyme polymerization appears to require the presence of peroxidizing lipid during freeze drying.

The biological activity of the enzyme decreased with increasing incubation time and also with the degree of unsaturation of lipid material.

Funes et al. [44] considered that polymerization would involve complex formation between radical-producing species and proteins, and that this would precede the transfer of free radicals from the lipid to the protein, as shown in Figure 13.

Funes et al. [44] have shown that freeze drying can affect the quality of food systems where the lipid material is already oxidized.

Delincée and Paul [45] examined the effect that irradiation and storage of proteins with oxidizing lipids had on the protein aggregates. Results indicated that if myoglobin, in the presence of polyunsaturated lipids, is irradiated and stored under aerobic conditions, aggregates arise due to the interactions with oxidized lipids. Differences between myoglobin aggregates induced by irradiation or oxidized lipids were noted. Protein polymerization proceeded in a stepwise manner through dimer, trimer, etc., following irradiation. Oxidizing lipids produced higher polymers of myoglobin at the onset of storage together with dimers and trimers, which then gradually increased during storage. The autocatalytic oxidation properties of the heme group in the myoglobin may be very important, because serum albumin did not form aggregates with oxidizing lipids. In a non-irradiated myoglobin–sunflowerseed oil mixture, and TBA value increased from 50 to 85 to 159 after 1, 7 and 14 days of storage, respectively. However, with serum albumin or without protein, the TBA value remained constant at 1.5.

Delincée suggested that aggregates formed by the effects of irradiation or oxidizing lipids could be attributed to the protein being cross-linked through covalent bonds. Both mechanisms are consistent with the observation that in purified polymerized products no lipid was covalently bound.

Exposure to oxidizing lipid (20% m/m) at a relative humidity of 80%

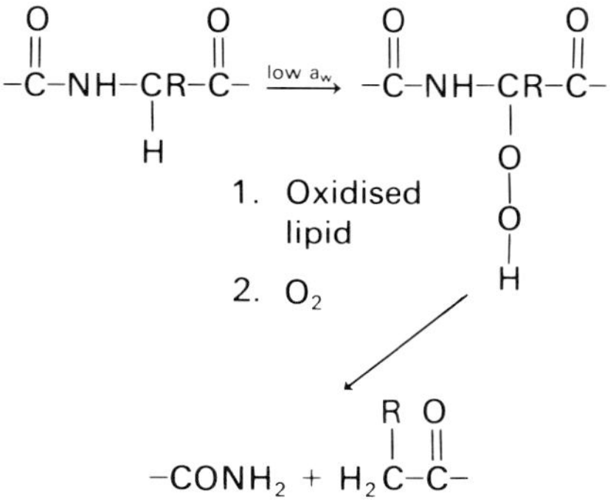

$$-\overset{\overset{\displaystyle O}{\|}}{C}-NH-\underset{\underset{\displaystyle H}{|}}{CR}-\overset{\overset{\displaystyle O}{\|}}{C}- \xrightarrow{\text{low } a_w} -\overset{\overset{\displaystyle O}{\|}}{C}-NH-\underset{\underset{\displaystyle O-O-H}{|}}{CR}-\overset{\overset{\displaystyle O}{\|}}{C}-$$

1. Oxidised lipid
2. O_2

$$\longrightarrow -CONH_2 + H_2\overset{\overset{\displaystyle R}{|}}{C}-\overset{\overset{\displaystyle O}{\|}}{C}-$$

Figure 14. Protein scission at low water activities (courtesy of Leatherhead Food RA).

resulted in polymerization of casein (80% m/m) [46], where none had existed at 0% humidity.

At equal concentrations of oxidizing lipid and protein, both casein and egg albumen polymerized and became insoluble. Matoba observed (unpublished data) that acylation of lysozyme reduced polymerization on exposure to oxidizing lipid. This was taken to mean that protein polymerization might occur via lysine residues, but Matoba was unsure whether arginine took part in the reaction.

A variety of polymerization reactions can occur depending on the conditions. If, for example, water activity is high ($a_w = 0.75$), polymerization as opposed to scission of protein is more likely to take place, in the presence of oxidized lipids.

2.4 Protein Scission and Amino Acid Damage

If the water activity is low, as may be found in a freeze-dried or dehydrated product, the interaction of protein and oxidized lipid may lead to the scission of the protein. For example, freeze-dried gelatin suffered a 4-fold decrease in molecular weight after exposure to peroxidizing methyl linoleate. However, the scission is inhibited by an increase in the water activity [47]. The decrease in molecular weight of the protein was associated with an increase in protein amide groups. To explain this Zirlin and Karel [47] suggested a mechanism involving the formation of a protein peroxide at the α-carbon atom. This unstable compound would cleave at the α-position, leading to the formation of an amide and a decrease in molecular weight, as shown in Figure 14.

However, at higher moisture contents, the protein radicals would be quenched by the water as it would exclude oxygen, preventing unstable peroxide formation and also provide hydrogen donors to react preferen-

Table 5. Damage caused to amino acids via reaction with oxidizing methyl linoleate at a relative humidity of 80%.

Amino acid	Protein	Method of hydrolysis	Percent loss of amino acid
Lysine	casein	acid	22
	casein	enzymic	41
Lysine	egg albumen	acid	22
	egg albumen	enzymic	37
Histidine	casein	acid	41
	casein	enzymic	49
Histidine	egg albumen	acid	31
	egg albumen	enzymic	42
Methionine	casein	acid	9
	casein	enzymic	94
Methionine	egg albumen	acid	1
	egg albumen	enzymic	88

tially with the protein radicals. The higher water activity could lead to protein radicals forming cross-links.

However, it has been observed that incubation of methyl linoleate with casein and egg albumen did not result in a reduction in molecular weight of either protein or produce smaller peptide fractions [46].

Protein scission can deleteriously affect the nutritive properties of food. Nevertheless, a more common problem is amino acid damage. The following amino acid residues are particularly susceptible to damage: cysteine/cystine, histidine, lysine, methionine, tyrosine and tryptophan. Matoba et al. [46] found that damage occurred to lysine, histidine and methionine, and no other amino acid residues, in protein that had been exposed to an equal weight of oxidizing methyl linoleate, at a relative humidity of 80% (see Table 5). The loss of methionine appeared to be very low after acid hydrolysis, compared with that after enzymic hydrolysis, indicating that methionine had been converted to a sulphoxide, which had reverted to methionine after acid hydrolysis (see Figure 15).

When casein was treated with oxidizing methyl linoleate [48] extensive damage occurred to the tryptophan, histidine and lysine residues, all being reduced by 40% after 10 days of incubation. As previously postulated by Tannenbaum et al. [49], methionine residues suffered most damage and were totally converted to the sulphoxide.

The destruction of tryptophan leads to the yellowing of wool, the production of cataracts in the human eye and inactivation of enzymes [42, 44]. Certain by-products of tryptophan degradation are thought to be animal carcinogens [43].

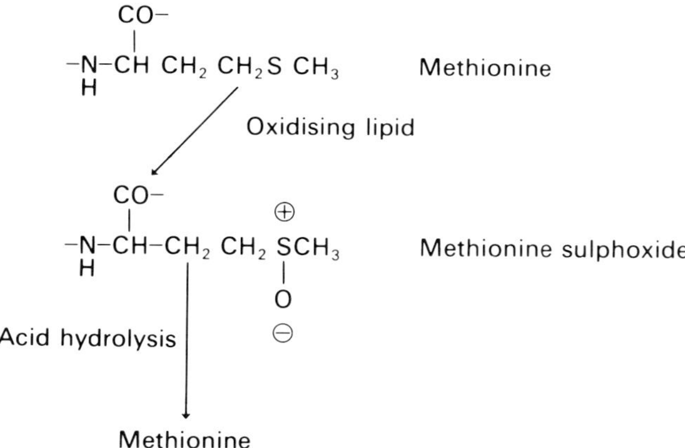

Figure 15. Reactions of methionine with oxidizing lipid and subsequent acid hydrolysis.

Oxidizing methyl linoleate degrades tryptophan to produce kynurenine and *N*-formyl kynurenine [50]. The rate of tryptophan oxidation was found to be dependent on a number of factors, the rate decreasing over the pH range 4.0–7.5, in agreement with Cuq and Cheftel [51] and Marcuse [52]. The rate of tryptophan oxidation increased with increasing buffer ionic strength, but fell when EDTA was added, to chelate and inactivate pro-oxidant transition metals. When EDTA (20 ppm) was added to a solution containing 5 ppm copper, the rate of tryptophan oxidation fell by 50%.

Tryptophan oxidation proceeded at a greater rate for residues chemically bound in a tripeptide than for the unbound amino acid, indicating that the latter is more resistant. Even in the absence of methyl linoleate (control experiments) oxidation of tryptophan was observed, although the rate of oxidation was lower. Control experiments consisted of the tripeptide in buffer with Tween 80, the latter containing linoleate and large quantities of oleate, both of which are liable to oxidation, resulting in potential damage to tryptophan.

After 72 h at room temperature, approximately 1–2% of the tryptophan had been converted to the products kynurenine and *N*-formyl kynurenine by the oxidizing methyl linoleate (Figure 16).

Similar losses were observed by Nielsen [53] but were considered of minor nutritional importance in comparison with the potential losses of lysine.

Finley and Lundin [41] demonstrated that considerable damage could occur to the cysteine and cystine residues in lysozyme and glutathione after exposure to linoleic acid hydroperoxide. In common with other observations it was found that cysteine and cystine residues were oxidized to sulphoxides, sulphones, sulphinic and sulphonic acids.

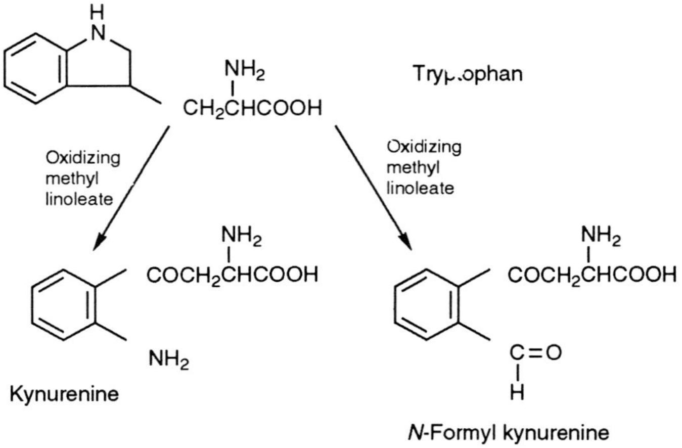

Tryptophan

$CH_2CH(NH_2)COOH$

Oxidizing methyl linoleate

Oxidizing methyl linoleate

$COCH_2CH(NH_2)COOH$

NH_2

Kynurenine

$COCH_2CH(NH_2)COOH$

$C(=O)H$

N-Formyl kynurenine

Figure 16. Products of tryptophan oxidation [50].

Mauron [54] had made a study of the influence of processing on protein quality, and made reference to the unpublished work of Hurrell and Nielson. The latter two researchers incubated whey protein with methyl linolenate under the conditions shown in Table 6.

Determination of total amino acids showed important losses for lysine, histidine and tryptophan only. When total methionine was measured there was no apparent change, but separte determination of methionine and methionine sulphoxide showed that large quantities of the former had been converted to the latter.

All amino acid losses showed the same pattern, with degradation being greatest at high water activity, excess oxygen and high temperature. It was found that methionine reacted more quickly than lysine, which was followed

Table 6. Storage conditions for the investigation of the influence of different parameters on lipid oxidation and reactions with proteins (0.5 g methyl linolenate/g whey protein) [54].

Conditions	Temperature (°C)	Water activity	Moisture content (%)	Oxygen: lipid (moles)
Basal conditions	37	0.84	17	4
Oxygen limitation	37	0.84	17	1
Low temperature	20	0.84	17	4
High temperature	55	0.84	17	4
Low water activity	37	0.33	2	4
Medium water activity	37	0.67	7	4

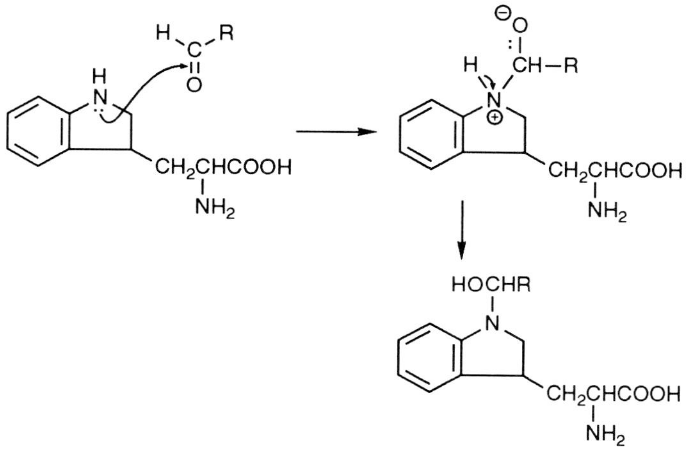

Figure 17. Reaction of tryptophan with carbonyl compounds formed from decomposition of lipid hydroperoxides.

by tryptophan. The differing rates of reaction were explained by suggesting that methionine reacted with the hydroperoxide, whereas lysine reacted with the secondary oxidation products, the carbonyl compounds. The very slow rate of reaction of tryptophan indicated that this amino acid only reacted with a very active secondary breakdown compound.

It was noted that lysine was lost from the system, which suggested that the reaction between this acid and carbonyl compounds progressed beyond the Schiff's base stage or that the lysine skeleton was itself broken. Available lysine measured by the direct fluorodinitrobenzene (FDNB) method was still lower than the total value, which indicated that some residues were blocked and were unable to complex with FDNB, but were regenerated during acid hydrolysis.

With regard to the reaction of tryptophan it was suggested that a carbonyl–amino reaction involvng the nitrogen of the indole ring was taking place (Figure 17).

Growth trials and nitrogen balance studies on rats confirmed the results obtained chemically on amino acid losses. These were that at low water activity ($a_w = 0.33$) very little reaction between oxidizing lipids and proteins occurred, whereas with increasing water activity and temperature, protein efficiency ratio, true digestibility and biological value all fell. In a number of samples, there was a large reduction in the protein digestibility; it was considered that this might have been due to the cross-linking formed between peptide chains during protein free-radical reactions or via lipid degradation products such as malondialdehyde.

Table 7. Amino acid analysis following the reaction of 13-oxo-9(*Z*)-11-(*E*)-octadecadienoic acid with GSH.

Compound	Cysteine	Glutamic acid	Glycine
Glutathione	100.0	100.0	100.0
Following reaction (1)	65.8	68.3	100.0
Following reduction of (1) with $NaBH_4$	53.4	71.4	100.0

The *in vivo* availability of the sulphur amino acids cysteine and methionine, as well as tryptophan and lysine, was determined using a slope ratio rat assay. Mauron [54] reported on unpublished work Nielsen and Liardon had undertaken, and stated that the bioavailability of methionine sulphoxide, in free or bound form, was high at 87 and 96%, respectively, while for cysteine it was markedly reduced. For tryptophan, the losses observed *in vivo* were larger than those determined chemically. This latter observation might explain the overall reduction in protein digestibility, as the chemically determined tryptophan values multiplied by true nitrogen digestibility were in good agreement with the rat assay values for available tryptophan.

With regard to lysine similar results as for tryptophan were found with the *in vivo* availability of lysine being much lower than that determined chemically (i.e. by FDNB). This appeared to be due to the low digestibility of the highly damaged samples since, again, if the FDNB-lysine values were multiplied by true protein digestibility they came close to the values of available lysine as determined in the rat assay.

Comparatively, the losses in bioavailability were least for tryptophan, intermediate for sulphur amino acids and highest for lysine.

While the above work by Nielson and Liardon is of great interest it should be noted that the experiments were undertaken on model systems under severe conditions not likely to be found in food processing. Mauron [54] points out that they might, however, occur in the processing of animal feed, especially fish meal.

Zamora et al. [55] undertook studies in which they reacted unsaturated oxoacids with glutathione (GSH) in order to investigate the formation of fluorescent compounds. It was found that the only oxoacid to induce fluorescence on reaction with GSH was oxidized octadecadienoic acid.

The reaction of GSH and the oxidized lipids destroyed large quantities of cysteine and glutamic acid while leaving glycine essentially unaffected (see Table 7). Ametani et al. [56] prepared an emulsion by homogenising soyabean oil with an α_{s1}-casein solution. The protein adsorbed to the oil/water in the emulsion was digested with trypsin and then analysed by reverse-phase HPLC. Two peaks were shown to have been formed, not present in the casein prior to emulsification. Both peaks were found to have the same amino acid sequence: ^{194}Thr–Thr–Met–Pro–Leu–Trp. Further analysis of these amino acid sequences revealed ^{196}Met of one of the peaks had been oxidized to methionine sulphoxide, while the other peptide remained unoxidized in the other peak. In effect, two peptides were formed, with the 196th, methionine, being oxidized in one species but not in the other. The two peaks were resolved, therefore, due to the inherent hydrophilicity of the sulphoxide in comparison with that of the methionine.

These results suggest that the peroxides formed in the soya-bean oil during storage oxidized ^{196}Met in the α_{s1}-casein molecules that were adsorbed to the emulsified oil droplets.

Further to the oxidation of methionine it was observed that the proline residue underwent *cis–trans* isomerization in casein during emulsification.

Ametani et al. [56] considered that the isomerization of the methionine–proline bond could not be attributed to denaturation of the protein during emulsification, but was due to the oxidation of ^{196}Met to its sulphoxide. The oxidation of this amino acid decreased the hydrophobicity of the peptides containing it. The peptide bond of ^{196}Met–Pro might be fixed in the *trans* form by some structural factor in native α_{s1}-casein. If this is the case, the oxidation of ^{196}Met destroys this factor, through which the *cis–trans* isomerization of ^{197}Pro may occur.

It appears that the amino acids that are readily lost after interaction with oxidized lipids are those that will form stable radicals. Indeed, Schaich and Karel [39] observed that, with the exception of methionine and tyrosine, the remaining sensitive amino acids all gave rise to electron paramagnetic resonance signals, which are associated with radicals.

The numerous products obtained from the interaction of oxidized lipids and proteins are summarized in Table 8.

Yong and Karel [57] suggested the mechanism shown in Figure 18 for the production of imidazole acetic and lactic acids from histidine.

A guide to enzyme inactivation, induced by amino acid damage is given in Table 9 [25].

In conclusion, amino acid residues are all potentially susceptible to attack by oxidized lipids. However, when chemically bound as in a protein it will be only those amino acid residues with reactive side chains, such as methionine and tryptophan, that will be able to react with oxidized lipids to readily form radicals.

Nevertheless, destruction of amino acid residues can cause reduction in the nutritive quality of food.

2.5 Reactions of Secondary Lipid Oxidation Products with Proteins

When unsaturated lipid material oxidizes, lipid hydroperoxides are formed. These are reactive compounds, liable to decompose, resulting in

Table 8. Products of the reactions between oxidized lipids and proteins. (Courtesy of Leatherhead Food RA.)

Amino acid	Products
Cysteine $HS-CH_2CH(NH_2)CO_2H$	Cysteine (−S−S−), cystine sulphoxide (−S−S(=O)−), Cystine sulphone (−S−S(=O)(=O)−), cysteine sulphinic acid ($-SO_2H$) and Cysteine sulponic acid ($-SO_3H$), alanine thiazolidine derivatives, hydrogen sulphide.
Methionine $CH_3SCH_2CH_2-CH(NH_2)-CO_2H$	Methionine sulpoxide, methionine sulphone, homocysteine.
Lysine $H_2N(CH_2)_4CH(NH_2)CO_2H$	Diaminopentane, aspartic acid, glycine, alanine, pipecolic acid, α-aminoadipic acid.
Histidine (imidazolyl)$-CH_2CH(NH_2)CO_2H$	Imidazole lactic acid, imidazole acetic acid, histamine, valine, aspartic acid, ethylamine.
Tryptopan (indolyl)$CH_2-CH(NH_2)CO_2H$	Kynurenine, *N*-formyl kynurenine.

products that include epoxides, ketones, aldehydes and alcohols. Such products are often referred to as secondary decomposition products [5].

It has, however, been difficult to contrast the relative importance of protein reactions with secondary products or with lipid hydroperoxides. Gamage and

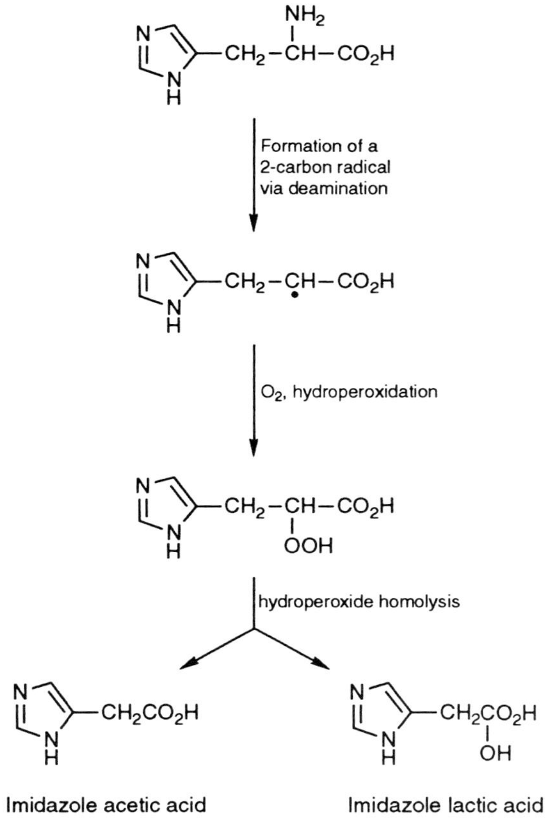
NH2
CH2-CH-CO2H
Formation of a
2-carbon radical
via deamination
CH2-CH-CO2H
O2, hydroperoxidation
CH2-CH-CO2H
OOH
hydroperoxide homolysis
CH2CO2H
CH2CO2H
OH
Imidazole acetic acid
Imidazole lactic acid

Figure 18. Radical-induced damage to histidine (courtesy of Leatherhead Food RA).

Table 9. Inactivation of enzymes by oxidized lipids.

Enzyme	Per cent Inhibition	Amino acids damaged
Ribonuclease	100	aspartic acid, tyrosine, methionine, lysine, histidine, threonine,
Trypsin	5	methionine, threonine, histidine
Pepsin	30	methionine, aspartic acid, tyrosine, arginine

Matsushita [58] demonstrated that pepsin was inactivated by linoleic acid hydroperoxide, but activated by secondary breakdown products. Malondialdehyde (MDA) is an important secondary breakdown product of oxidized fatty acids, containing three or more double bonds, and may be determined by the TBA test, an important method of detecting rancidity in food.

Kuusi et al. [59] found that as the TBA value of frozen herring increased, the quantity of free ε-amino groups (as measured by the ninhydrin test) decreased. Similar observations have been made by Crawford et al. [60] and Buttkus [9].

MDA can form cross-links between free amino groups of proteins because of its inherent bifunctionality. For example, Andrews [61] and Menzel [62] showed that mixing ribonuclease and MDA resulted in the production of dimers and higher oligomers. The cross-linking between MDA and protein groups was attributed to the formation of a Schiff's base [63] producing an *N,N′*-disubstituted 1-amino-3-imino-propene, and Dillard and Tappel [64] proposed the following to illustrate the point further:

$$\underset{\text{malondialdehyde}}{\text{O=CHCH=CHOH}} + \text{RNH}_2 \rightarrow \underset{\text{enamine}}{\text{O=CHCH=CHNHR}} + \text{H}_2\text{O}$$

$$\underset{\text{enamine}}{\text{O=CHCH=CHNHR}} + \text{RNH}_2 \rightarrow \text{RN=CHCH=CHNHR} + \text{H}_2\text{O}$$

Consequently, cross-linking could occur via the intermolecular reaction in which MDA reacts with two free amino groups, e.g.

$$\text{O=CHCH=CHOH} + 2\,(\text{protein–NH}_2) \rightarrow \text{protein–NHCH=CHCH=N–protein}$$

MDA Schiff's bases are fluorescent, with an excitation maxima of 370 nm and emission maxima of 450 nm.

Chio and Tappel [63] demonstrated the fluorescent properties of MDA-induced polymers in ribonuclease. However, intra-molecular reactions were observed because fluorescence was observed in ribonuclease monomer. The discovery that MDA-induced fluorescence existed led Tappel [65] to suggest that it was related to age pigments (ceroid or lipofusion pigments). This is expanded upon in Gardner's review [28].

MDA will react with various nitrogen groups, illustrated by the loss of lysine and arginine, which contain a free amino group, and also methionine and tyrosine, which do not [9].

Buttkus [66] suggested that MDA cross-linked proteins via reaction with

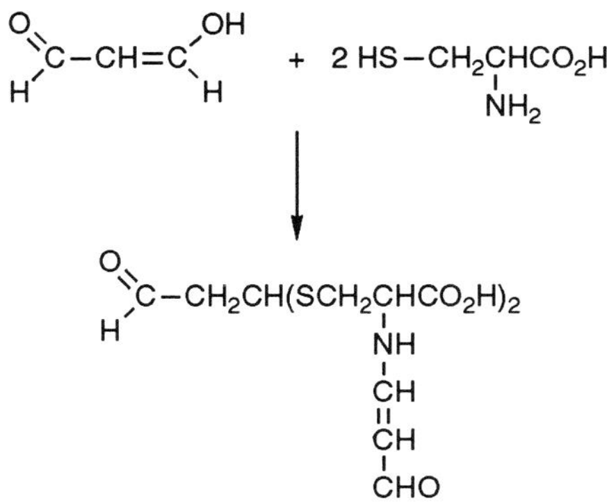

$$\mathrm{OHC{-}CH{=}CH(OH)} + 2\,\mathrm{HS{-}CH_2CH(NH_2)CO_2H} \longrightarrow \mathrm{OHC{-}CH_2CH(SCH_2CH(NH{-}CH{=}CH{-}CHO)CO_2H)_2}$$

Figure 19. Reaction of amino and sulphydryl groups of cysteine with MDA [66].

both the sulphydryl and amino groups, but the reaction products postulated did not include the 1-amino-3-imino-propene cross-link (see Figure 19).

Only compounds such as MDA that contain two reactive groups are able to form cross-links. Therefore, aldehydes will only do so if present in the dihydro or hydrated form shown in Figure 20.

The breakdown of lipid hydroperoxides leads to many compounds other than MDA that are potentially reactive towards proteins. Pokorny et al. [67] studied the effects of the autoxidation of hexanal in the presence of non-lipidic substances, reported as being rapidly autoxidized at 25°C. The formation of peroxides followed the kinetics of a first-order reaction with respect to hexanal, the rate being increased in the presence of casein or lysine-impregnated cellulose. Pokorny suggested that the autoxidation of aldehydes is a free radical chain reaction, initiation occurring via UV radiation, metal salts or interaction with radicals. The reaction scheme (Figure 21) postulates that R-C=O radicals react with oxygen to produce acylperoxy radicals, which abstract a proton from another aldehyde molecule, producing a chain reaction.

Since lipid hydroperoxides are precursors of aldehydes, their decomposition can be considered as the main initiation reaction of the above aldehyde oxidation. Pokorny et al. [67] suggested that the autoxidation of aldehydes is rapid when exposed to oxygen. Peroxides, although generally thought to be bland, were in this case detected by their odour, but the quantity could not be established because of their rapid decomposition.

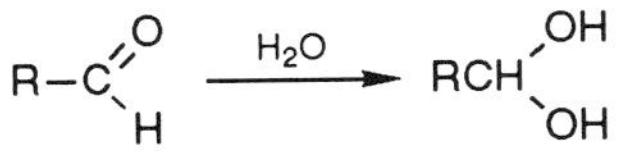

$$\mathrm{R{-}CHO} \xrightarrow{H_2O} \mathrm{RCH(OH)_2}$$

Figure 20. Production of the dihydro or hydrated form of aldehydes.

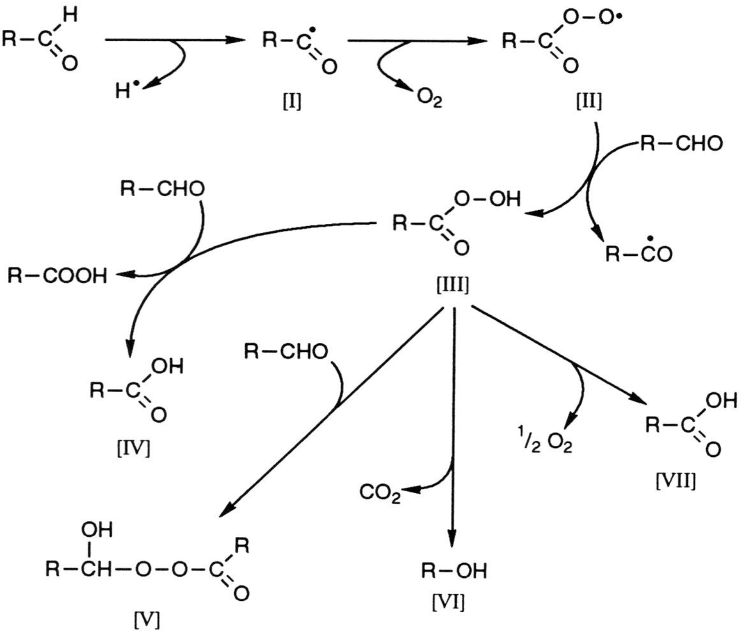

Figure 21. Mechanism of autoxidation of alkanals (From Porkorńy [68].).

Similar experimental work [68] suggested that propanal, butanal and hexanal were all partially oxidized and aldolized. The aldolization was catalysed by primary amino groups of proteins and amino acids. Reactions of aldehydes and amino groups resulted in the formation of Schiff's bases (i.e. $RCH_2CH = NHR^1$). Pokorny et al. [68] illustrated the variety of reactions that can occur (Figure 22).

The organoleptic properties of the mixture were altered as a result of the aldolization, with the result that fruity odours were formed. When lysine was added, the formation of weak odours similar to cheese and mushrooms were observed. When aldehydes were reacted with casein and egg albumen, similar odours were formed.

Non-enzymic browning of proteins has been observed after their reaction with aldehydes [69, 70] resulting in a change in appearance as well as flavour of foods. El-Zeany [71] showed that crotonaldehyde reacted with proteins and amino acids to give a rapid browning reaction, resulting in only liposoluble products with proteins, but both lipophilic and hydrophilic products with amino acids. The browning reactions were inhibited by blocking either the free amino or carbonyl groups. El-Zeany and El-Tarras [72] demonstrated that butylamine (containing an amino group similar to bound lysine) reacted with secondary breakdown products rapidly, and that glycine and lysine

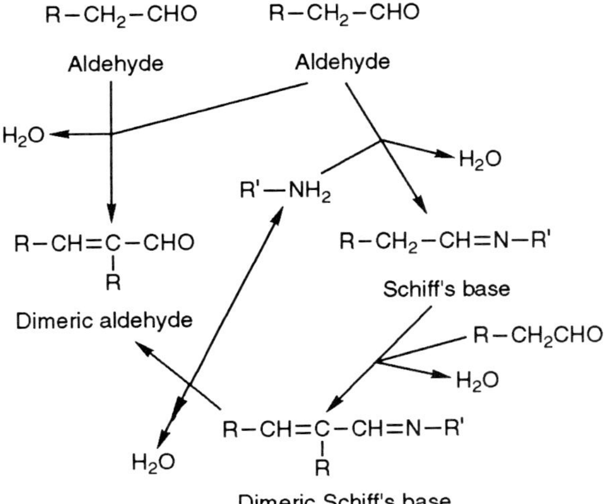

Figure 22. Interaction of aldehydes and amino groups. (From Porkorny [68].)

hydrochloride did not brown to the same extent, owing to the high volatility of butylamine, enabling it to come into close contact with lipid oxidation products more easily than the amino acids. El-Zeany and El-Tarras [72] observed that the products were liposoluble and suggested two mechanisms for their formation—firstly, the partial solubility of amino acid–aldehyde products in organic solvents and secondly, the release by lipid hydroperoxides of ammonia from α-amino acids, in a similar manner to the Strecker degradation (Figure 23). This may be followed by the reaction of ammonia with lipid hydroperoxides to produce liposoluble pigments. The reactivity of α- and ε-amino groups appeared to be very similar.

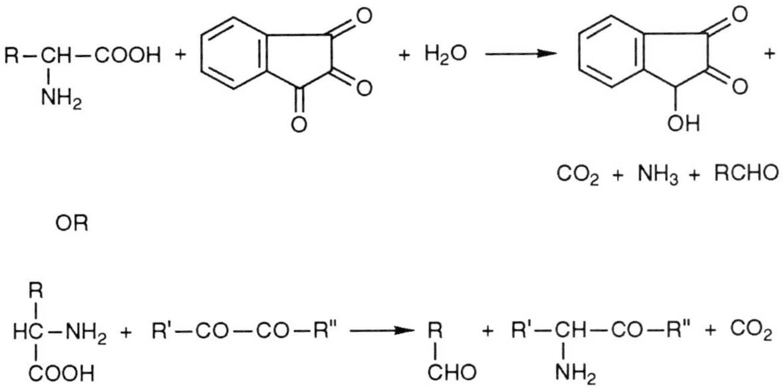

R−CH(NH2)−COOH + [ninhydrin] + H_2O ⟶ [hydrindantin, OH] + CO_2 + NH_3 + RCHO

OR

R−HC(NH2)−COOH + R'−CO−CO−R" ⟶ R−CHO + R'−CH(NH2)−CO−R" + CO_2

Figure 23. Strecker degradation.

The discolouration of white fish muscle is accompanied by changes in the sensory and nutrition values of the product. Stansby [73] attributed the brown discolouration of stored fish to oxidative changes within the unsaturated fish lipids. An emulsion of menhaden oil in an aqueous protein solution became brown on storage [74]. Three competitive reactions were assumed:

(1) the formation of brown lipid oxypolymers;
(2) the reaction of active groups of oxidized lipids and proteins producing brown pigments; and
(3) the oxidation of polyunsaturated fatty acids leading to a light brown discolouration.

Fujimoto et al. [75] and Fujimoto [76] separated the amines from a variety of fish into steam-volatile and non-volatile fractions, and reacted them with oxidized methyl linoleate. The volatile amines caused a greater degree of browning.

Pokorny et al. [77–79] suggested that in mixtures of polyunsaturated fatty esters with fish muscle homogenate, fish myosin or pure animal protein, the browning reaction proceeds in two steps—the formation of light-coloured intermediary products, such as Schiff's bases, and also the transformation of the intermediate products into polymeric brown pigments by aldolization reactions.

Phospholipids participate in browning reactions because they contain unsaturated lipid material as well as amines. Prithi [80] isolated phospholipids from butterfat and exposed them to heat for 16 h at 160°C, while determining their colour. The browning activity was highest with phosphatidyl ethanolamine and lowest with sphingo-lipids. The phosphatidyl choline experienced intermediate browning. In storage experiments at 30°C only the phosphatidyl ethanolamine caused discolouration.

Pokorny et al. [18] demonstrated that oxidation of the phosphatidyl ethanolamine fraction of egg phospholipids resulted in an increased browning, with a fall in the levels of polyunsaturated fatty acids and amine groups, but with an increase in the quantity of imine derivatives.

Tomioka and Kaneda [81] suggested that the browning of phosphatidyl choline was caused by 9-oxo-10,12-dienoic and 13-oxo-9,11-dienoic acids produced by autoxidation.

Browning may be desirable or unwanted but its rate can be controlled as shown in Table 10 [79]. Browning reactions increase with increasing lipid unsaturation, indicated by the fact that hydrogenation retards the process.

The extent of browning decreases with increasing antioxidant concentration or exclusion of oxygen but its total elimination cannot be achieved if small quantities of oxidation products (i.e. aldehydes) are already present in

Table 10. The control of browning reactions between oxidized lipids and proteins [79].

Factor	Stimulation	Inhibition
Temperature	increase above 40°C	cold or frozen storage
Atmosphere	oxygen	inert gas, nitrogen
pH	above 7	below 5
Catalysts	transition/heavy metals	antioxidants, vitamin E
Fatty-acid composition	polyunsaturated fatty acid	saturated or mono-unsaturated fatty acids—hydrogenation of oils
Composition of non-lipids	amines	blocking of amine and carbonyl groups
Condition of oil	use of used and oxidized oils, addition of aldehydes	use of fresh unoxidized oil

the oil. Additionally, antioxidants are poor at preventing the transformation of light-coloured intermediary products into stable brown products.

Flavour development, apart from browning, is also associated with alkanals and alkenals. Rancidity decreased when these aldehydes were stored in the presence of proteins. Pokorny [17] demonstrated that this was due to the formation of aldolization products and polymers, and also that brown lipid–protein products tasted bitter, especially in phospholipid rich materials.

The reduction in solubility, caused by peptide chains cross-linking, affects the texture of a protein [82].

Chang [83] observed that more than 220 volatile products were formed after fats had been heated under frying conditions, many being produced from the decomposition of oxidized lipids.

Flavours are also produced by the interaction of food and frying oil. For instance, amino acids, particularly methionine, interact with oxidized oils to produce both pleasant and unpleasant flavours. For example, it is claimed that carbonyl derivatives containing 4, 6 and 10–12 carbon atoms impart a pleasant fried food flavour, whilst homologues containing 3, 5 and 7 carbon atoms produce objectionable flavours [84].

Josephson and Lindsay [85] have suggested that certain secondary breakdown products of linoleic acid, of which potato lipids contain considerable amounts, contribute to the typical flavour of some potato products. At relatively high concentrations, greater than 0.7 ppb, *cis*-4-heptenal produced stale flavours in fresh, mashed and reconstituted dehydrated potatoes. However, when added at lower concentrations, 0.1–0.4 ppb, *cis*-4-heptenal produced earthy, potato-like flavours in freshly boiled potatoes, but still produced stale off-flavours in reconstituted dehydrated potatoes. This may be a consequence of the importance of water activity mentioned earlier.

Kamman and Labuza [86] suggested that at low water activities lipids may act as a solvent phase for sugars, flavours and hydrolysed vegetable protein and, by bringing these components together, cause a reduction in shelf-life of corresponding products due to non-enzymic browning.

It has been observed that free cysteine reacts with aldehydes [87] and to form a cyclic thiazolidine carboxylic acid derivative, with the production of meaty flavours (see Figure 24) [88].

The production of brown colours is not exclusive to reactions between aldehydes and amine groups. For example, 10-oxo-9-hydroxystearic and 9-oxo-10-hydroxystearic acids reacted with albumin to form a brown colour [89]. It was shown [90] that 9,10-epoxystearic acid or its methyl ester could bind to albumin. Although the structure was not identified, it has been suggested [27] that the nucleophilic substitution reaction shown in Figure 25 may take place.

Such reactions should not be discounted, because decomposition of lipid hydroperoxide can often lead to fatty epoxides being a major product. For

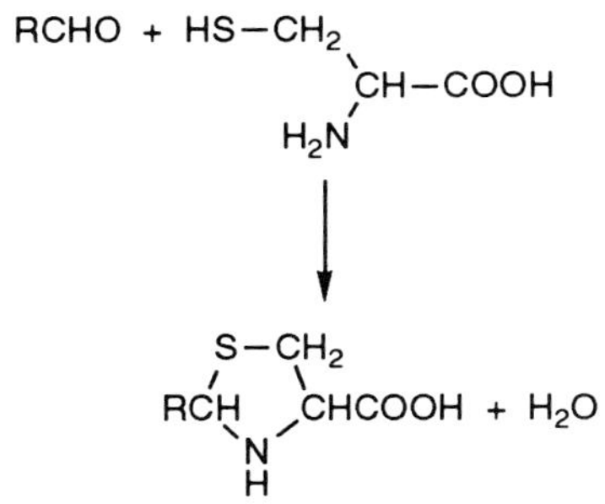

$$RCHO + HS{-}CH_2{-}CH(NH_2){-}COOH \longrightarrow \text{RCH(S–CH}_2\text{–CHCOOH–NH–)} + H_2O$$

Figure 24. Cyclization of cysteine. (From Gardner [28].)

example, Gardner et al. [29] showed that 9-oxo-*trans*-12,13-epoxy-*trans*-10-octadecenoic acid is a significant product of homolytic degradation of linoleic acid hydroperoxide.

2.6 Reactions of Oxidized Lipids in Food Systems

2.6.1 *Effect of Oxidized Lipid–Protein Interactions on the Nutritive Quality of Food*

The nutritive quality of food will, in general, be lower following the interaction of oxidized lipids with proteins. It is clear that a number of essential amino acids, such as tryptopan, lysine and methionine, will be destroyed; this loss will, of course, be in addition to destruction of the essential fatty acids. It is also possible that a number of vitamins might be damaged following interaction with oxidized fatty acids.

There will be a decrease in the digestibility of food due to the incomplete and slower enzymic hydrolysis of bound lipids and proteins.

It was shown, for example, that when egg albumin and carbonyl lipid oxidation products were digested by proteases both *in vitro*, by pepsin, and *in vivo* in mice and rats, there was a reduction in the net protein utilization (NPU). This was, in particular, manifested by a 32% lower calorific value being obtained from the protein. Furthermore, there was partial destruction and lower availability of essential amino acids, the quantity of available lysine being lowered considerably. In addition, the digestibility of the complex was 4% lower, as compared with untreated albumin [16].

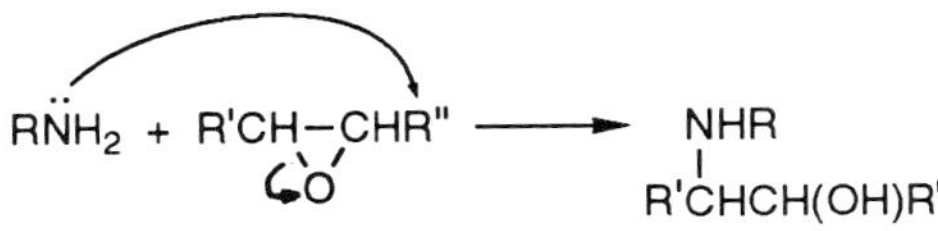

$$R\ddot{N}H_2 + R'CH{-}CHR''\ (\text{epoxide, O}) \longrightarrow R'CH(NHR)CH(OH)R''$$

Figure 25. Nucleophilic substitution of epoxides [28].

Horigome and Miura [91] reported similar findings. In the case of feeding proteins that had been stored with auto-oxidizing ethyl linoleate, the biological value and digestibility were reduced by almost 10%.

It was also observed that the protein efficiency ratios of milk proteins were impaired following incubation with oxidized lipids [92].

2.6.2 Alteration of Organoleptic Properties of Foods Following the Interaction of Oxidized Lipids with Proteins

Not only can the nutritive quality of food be lowered by the interaction of proteins with oxidized lipids, but the sensory quality can also be changed. There are three main ways in which these alterations manifest themselves.

(1) Changes to the colour owing to the interaction of lipidic carbonyl compounds with free amino groups, for example. This may be considered analogous to the Maillard reaction, which involves the formation of a Schiff's base following the reaction of sugar with an amino acid.
(2) Changes to the flavour due to the binding of off-flavour compounds into neutral or less active substances, or as a result of the production of new flavour-active compounds.
(3) Changes to the texture caused by the denaturation of proteins and by the cross-linking of polypeptide chains.

The discolouration of food was first noticed by Stansby [73] in frozen fish muscle. This is a particularly interesting system in which to study discolouration reactions, since fish lipids contain large quantities of highly unsaturated lipids and considerable amounts of protein rich in basic amino acids. The oxidation of lipid eventually leads to the production of carbonyl compounds, which readily react with the free amino groups and in doing so cause browning.

There is clear evidence that the rancid flavour that occurs in oils is due to the decomposition of lipid hydroperoxides to produce aldehydes and ketones. Furthermore, there is evidence that treatment of rancid oils with amino acids can eliminate off-flavours and odours. For example, Yamamoto and Kogure [93] succeeded in removing the rancid odour from fat following its treatment with lysine. Okumura and Kawai [94] also prepared odour-free oils by reaction with amino acids.

It has been observed [95] that in mixtures of the short-chain alkanals and proteins, the typical off-flavour odour characteristic of rancid fats gradually disappeared on storage and new odour compounds were formed, which produced stale, gluey or even fishy odour notes.

Changes to the partial odour notes of model systems containing ethyl linolenate and various non-lipidic substances during storage at 60°C are illustrated in Table 11. In general, changes in odour depend to a great

Table 11. Changes of selected odour notes during storage of mixtures containing ethyl linolenate and various non-lipidic substances. (From Pokorny [95].)

Description of odour notes	Intensity changes on storage with			
	casein	myosin	dehydrated meat	cellulose
Peroxidic	I	N	I	I
Varnish-like	R	N	N	R
Fishy	R	N	D	R
Algae, seaweed	I	N	N	N
Grassy	N	N	N	D
Muddy	I	I	N	N
Flowery	D	D	D	O
Camphory	D	N	N	D
Mushroom	M	N	N	N
Woody	D	D	D	D
Sulphuric	M	D	D	N
Spicy	N	D	N	O
Cheesy	N	M	N	N
Meaty	S	I	N	N
Rotten	N	D	D	N
Hydrogen sulphide	N	M	S	N
Oily	N	N	M	N
Rancid	N	N	I	M
Fried	N	N	N	D

Notes: Changes during storage: I, intensity increasing; R, rapidly increasing, later constant; S, slowly increasing; D, decreasing; B, decreasing in the beginning, later constant; M, decreasing moderately; O, increasing in the beginning, later decreasing; N, no pronounced change.

extent on the non-lipidic substrate with which the oxidized lipids react. During storage of lipid-free muscle mixed with phospholipid, flavour notes were produced that reminded panellists of stored meat, fish, sweat and stale flour.

Pokorny [95] undertook studies in which the lipid secondary oxidation product hexanal was stored with casein at 60°C, in systems that were described as dry and as containing 50% water. Changes in sensory profile were determined at various intervals of storage (Table 12). A number of the odour notes formed were clearly due to oxidation products (i.e. peroxidic, acidic); others, however, may be due to aldolization products, since this reaction is very rapid in the presence of protein.

The effect of the low lipid content on bread and bread-making properties has been studied. Hoseney et al. [96] stated that polar glycolipids are bound to gliadin by hydrophilic and to glutenin by hydrophobic bonds, respectively, and that this simultaneous binding may contribute structurally to the gas-retaining complexes in gluten.

Table 12. Changes of selected partial odour notes during storage of casein with hexanal. (From Pokorny [95].)

Description of odour note	Changes on storage
Rancid	increasing at later stages
Peroxidic	rapidly increasing
Acidic	increasing
Aldehydic	decreasing
Burnt	decreasing
Spicy	decreasing
Heavy	decreasing
Fruity	decreasing
Flowery	decreasing
Sweat-like	increasing
Grassy	increasing
Buttery	increasing
Stale	slowly increasing

A number of researchers [97, 98] have indicated that lipid- and sulphur-containing proteins affect the bread-making and rheological properties of doughs. Pomeranz and Chung [99] found that the removal of lipid from flour resulted in the subsequent loaf volume being reduced 3-fold, owing to enhanced protein aggregation. The loaf volume did not alter on the replacement of lipid, but it returned to the original size after addition of cysteine. Pomeranz suggested that this was due to cysteine binding lipids and thereby reducing protein aggregation.

Frazier et al. [100] showed that radio-labelled triolein formed a strong complex with a highly interactive protein, molecular weight 9000, on a 1 : 1 molar basis. The protein, later called ligolin, contained more cysteine than glutenin or purothionin, and was considered to be responsible for a very significant amount of lipid binding during dough formation.

Tsen and Hlynka [101, 102] found that dough mixed in air or oxygen contained lipid hydroperoxides, whose concentration was increased by addition of lipoxidase and decreased by the antioxidant, propyl gallate. Addition of sulphydryl-blocking or improving agents to the dough led to increased lipid oxidation, and it was suggested that flour lipids and sulphydryl groups compete for available oxygen.

Lipids and non-lipidic peroxides produced equal improvements in the structural relaxation constant of the dough. It was concluded that peroxides oxidize sulphydryl groups and promote an improving effect.

Loaf volume improved when dough making took place under oxygen and in the presence of soya lipoxygenase [103, 104]. There was simultaneous release of lipid, which was then oxidized to hydroperoxide.

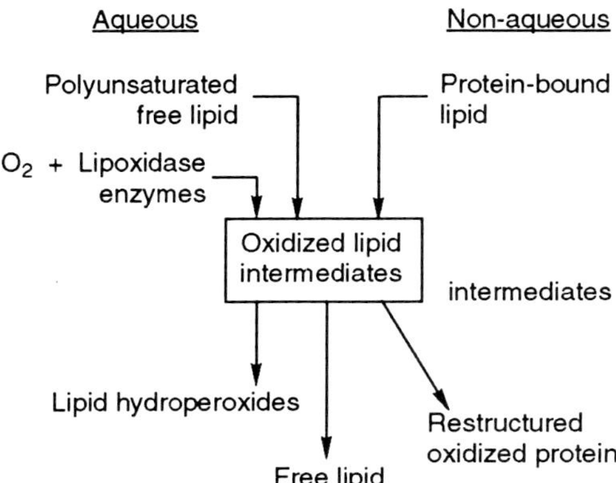

Figure 26. Proposed mechanism for the release of bound lipid during rough mixing in the presence of air [103].

Under nitrogen or in the absence of the enzyme, loaf volume was poor [105]. Removal of lipid and its replacement showed conclusively that for beneficial rheological effects to occur the dough required the presence of oxidizable, free lipid along with active lipoxygenase and oxygen. Addition of nordihydroguaiaretic acid, a lipoxygenase inhibitor, reduced hydroperoxide formation and also the rheological improvement.

Daniels et al. [103] postulated that when the dough is wetted, lipid becomes bound to protein. However, polyunsaturated fatty acids react with lipoxygenase and oxygen to produce an oxidized lipid intermediate, which breaks lipid–protein links, oxidizing the sulphydryl groups in the process. The resultng gluten network is such that it has the ability to retain gas and, therefore, to increase the loaf volume. Conversely, high levels of bound lipid greatly alter the gluten network and reduce its ability to retain gas, thereby reducing loaf volume (Figure 26).

If the structure of an oilseed is damaged, the oil may become prone to oxidation. Vegetable oils contain oleic, linoleic and linolenic acids, oxidation of which can result in an alteration of nutritional and functional properties via reaction with protein. Ory and St Angelo [106], St Angelo and Ory [107] and St Angelo and Graves [108] stored ground peanut samples (1) at 4°C in a sealed glass jar, (2) at 30°C in an open, gauze-covered jar and (3) at 30°C, mixed with 10% rancid vegetable oil, also in an open, gauze-covered jar.

The amounts of brine-extractable protein and brine-insoluble residue increased and fell, respectively, with increasing lipid peroxidation. Incorporation of lipid peroxides or their breakdown products into the protein may account for this.

The compounds, 9,12,13-trihydroxy-octadec-*trans*-10-enoic acid and

9,10,13-trihydroxy octadec-*trans*-11-enoic acid have been isolated from soya beans, and found to give rise to a bitter taste [109]. Gardner [110] presented results in which these trihydroxy compounds were formed from the decomposition of linoleic acid hydroperoxidation by cysteine and Fe^{3+}.

Fish can be regarded as similar to some oilseeds in that they contain high levels of polyunsaturated fatty acids and proteins. Castell, Bishop and Neal [111] suggested that fish protein became water-insoluble when oxidized lipid free radicals promoted polymerization of the myofibrillar protein. Connell [112], however, observed that formation of insoluble myofibrillar proteins took place in cod even when the lipids remained unoxidized. Castell [7] suggested that the lack of oxidized lipid material might be due to rapid reaction with the fish proteins forming insoluble polymers. Oxidized rather than unoxidized fatty acids were more effective at reducing the solubility of fish proteins [113]. Malondialdehyde cross-linked tuna myosin [114] and trout myosin [9], in the latter case with loss of the 5-amino acid residues, presumably via a similar reaction. Jarenbäck and Liljemark [115] demonstrated that the incubation of 0.8 and 8 μmol of linoleic acid hydroperoxide and linoleic acid, respectively, per gram of myofibril resulted in a 90% reduction in the amount of protein soluble in aqueous potassium chloride. It was also found that no hydroperoxide remained, indicating either its breakdown or complete combination with the protein.

Misleading answers may be obtained in the analytical determination of peroxide, TBA and anisidine values in fish products if there is very rapid interaction of lipid hydroperoxides with the proteins.

The reduction of *in vitro* digestibility of soya-bean leaf protein concentrate, after 12 weeks of storage at 27°C, was associated with the reaction of oxidized linoleic acid and protein [116]. Similar leaf protein concentrates showed a reduction in protein quality when stored at increasing temperatures at high moisture levels (12%). The reduction in protein quality followed the increase in lipid oxidation [117]. Whitfield and Shipton [118] showed that off-flavours in frozen peas are due to at least 12 lipid oxidation products, such as *n*-alkanals, 2-alkanals, 2,4-alkandienals and alkan-2-ones. However, Haydar and Hadziyei [119] showed that the interaction between oxidized pea lipids and proteins was small, and that albumins retarded the lipid oxidation. Cationic, but not anionic, glutathione S-transferase and glutathione peroxidase decompose hydroperoxides derived from microsomes or lipoprotein [120] suggesting that the level of hydroperoxide, which has a high toxicity, could be controlled by such compounds.

Examples of proteins producing lipid hydroperoxides have been mentioned in relation to lipoxygenase and bread making. In addition, a protein fraction from mushrooms oxidized linoleic acid through the intermediate 10-hydroperoxy-*trans*-8-*cis*, *cis*-12,15 octadecatrienoic acid to the products 1-*cis*-5-octadien-3-ol, 2-*cis*, 5-*cis*-octadien-1-ol and 10-oxo-*trans*-8-decenoic acid [121].

It has been well established that, in model systems, oxidizing fatty acids catalyse the oxidation of myoglobin (MbO_2), initially to form the brown, unpleasant metmyoglobin (met MbO_2). It is the presence of MbO_2 that gives fresh meat its very bright red appearance [122, 123].

Verma et al. [124] showed that the rapid oxidation of MbO_2 in refined lard emulsions was indicative of catalysis by peroxidizing lipids. It must be pointed out, however, that as in most studies reported in the literature the lipid material was oxidizing far more rapidly, as measured by the increase in thiobarbituric acid (TBA) number, than would be normally expected in meat products. Consequently, the practical implications of this work may be regarded as difficult to evaluate.

Studies involving meat emulsions made with rancid and fresh fat and comparison of the kinetics of formation in different meat systems led Verma to suggest that the catalytic effect of peroxidizing lipids on metmyoglobin formation in meat is minimal. It was further suggested by Verma that since in non-meat emulsions very rapid formation of met MbO_2 does take place, some protective system might operate in meat that prevents the intermediates formed during the peroxidation of the unsaturated lipids from reacting with the ferrous haematin.

3. CONCLUSIONS

From the literature surveyed above, it can be seen that there are a considerable number of reactions that can occur between proteins (or amino acids) and oxidizing or oxidized lipid material and their breakdown compounds.

In most living systems the lipid material is carefully protected from contact with oxygen, e.g. by cell walls. However, in many food systems, the processing techniques involved, such as grinding, heat and exposure to light, break down the cell structure, and in so doing enable the oxidation of unsaturated lipid material. This may, in turn, react with protein as a hydroperoxide or in the form of aldehydes and other carbonyl compounds. If hydroperoxides react with proteins, it generally results in oxidation of amino acid residues, with the lipid being reduced. Consequently, it may be assumed that the lipid is acting as a catalyst in the oxidation of a protein. The major interaction between the two species may be summarized as follows.

(1) Formation of hydrophobic and hydrogen bonds with protein and oxidized lipid, i.e. non-covalently bound complexes.
(2) Formation of protein–lipid complexes, which tend to be rare, due to the instability of the hydroperoxide.
(3) Damage, generally oxidation, of the amino acid residues, by the lipid hydroperoxides. Cysteine, for example, may react to form a disulphide,

sulphoxide or a sulphinic acid. The lipid may be regenerated and in this case acts as a catalyst. In systems of this type a low level of polyunsaturated lipid may promote considerable protein damage.

(4) Protein–protein complexes are formed via polymerization, involving the decomposition of a hydroperoxide and formation of protein radical.

(5) Carbonyl compounds, the secondary decomposition products of lipids, react with proteins, particularly those containing free amino groups, and in some cases form cross-links between proteins, if the carbonyl compound is bifunctional, such as malondialdehyde.

It is clear that one of the most important processes in the formation of many of the above products is that of radical transfer, from the lipid to protein. It appears that at least two factors are critical in the process—the presence of oxidized lipid and contact between the reactants. Schaich [25] has proposed an overall mechanisms (Figure 27) in which, although contact is critical for direct hydrogen abstraction from proteins by lipid alkoxy and peroxy radicals, the rate-limiting step is the formation of complexes between lipid hydroperoxides and nitrogen or sulphur centres of reactive amino acids. Radical transfer then occurs either as the hydroperoxide and complex homolyse simultaneously, or subseuqent to decomposition of the complex as the protein dismutates in the immediate vicinity of the protein sites. Lysine, histidine, tryptophan and cysteine form charge complexes with Cu^{2+}, and it may be that such species are involved in the production of lipid hydroperoxide–protein complexes (see Figure 27).

In such a system the maximum radical transfer occurs at weight ratios of 1 : 2 of lipid : protein. This occurs [25] because the number of lipid molecules (designated '○' in Figure 28) is sufficient to form a monolayer around the protein and is in a small excess of the number of reactive sites (+) on the surface of the protein. If more lipid were available, other reactions would take place in the lipid phase, rather than with reactive protein sites. If more protein is present, lipid molecule separation limits radical chain length, and the number of transfer sites exceeds the number of radicals available for transfer.

From Table 13, it is clear that there are both striking simularities as well as differences between the effects of ionizing the UV radiation and lipid oxidation on protein material.

Many of the reactions discussed above cause changes in the nutritive quality of proteins, affect their solubility and texture, and can produce changes in the organoleptic quality of food products. Enzymes are particularly susceptible to biological inactivation following interaction with an oxidized lipid.

The foregoing text attempts to illustrate the effects of oxidized lipids on

Initiation

$$LH \rightarrow L^{\cdot}$$
$$LOOH \rightarrow LO^{\cdot} + {}^{\cdot}OH$$
$$2LOOH \rightarrow LOO^{\cdot} + LO^{\cdot} + H_2O$$

$$^{*}\ LOOH + PH \rightarrow [LOOH \ldots HP]$$

$$\swarrow \qquad\qquad \searrow$$

$$LO^{\cdot} + P^{\cdot} + H_2O \qquad LO^{\cdot} + {}^{\cdot}OH + PH$$
$$PH + {}^{\cdot}OL \rightarrow P^{\cdot} + LOH$$

Propagation

$$LO_2^{\cdot} + LH \rightarrow LOOH + L^{\cdot}$$
$$LO^{\cdot} + LH \rightarrow LOH + L^{\cdot}$$
$$L^{\cdot} + O_2 \rightarrow LO_2^{\cdot}$$
$$^{**}\ LO_2^{\cdot} + PH \rightarrow P^{\cdot} + LOOH$$
$$^{**}\ LO^{\cdot} + PH \rightarrow P^{\cdot} + LOH$$

Termination

$$\left.\begin{array}{l} LO_2^{\cdot} + LO_2^{\cdot} \\ LO_2^{\cdot} + LO^{\cdot} \\ P^{\cdot} + P^{\cdot} \\ P^{\cdot} + LO_2^{\cdot} \\ P^{\cdot} + LO^{\cdot} \end{array}\right\} \text{products}$$

where LH = lipid
PH = protein

* = rate-limiting step
** = contact required

Figure 27. Proposed overall reaction scheme for free radical interactions of proteins with peroxidizing lipids [25].

proteins. In many cases there is very considerable destruction of the protein. However, studies have been undertaken, both *in vitro* and *in vivo*, that show a limited or in certain cases zero effect. For example, Okiy and Oke [125] fed thermally degraded palm oil to rats. Following a 10-day trial period, the rats were killed and the protein was examined. The above researchers came to the

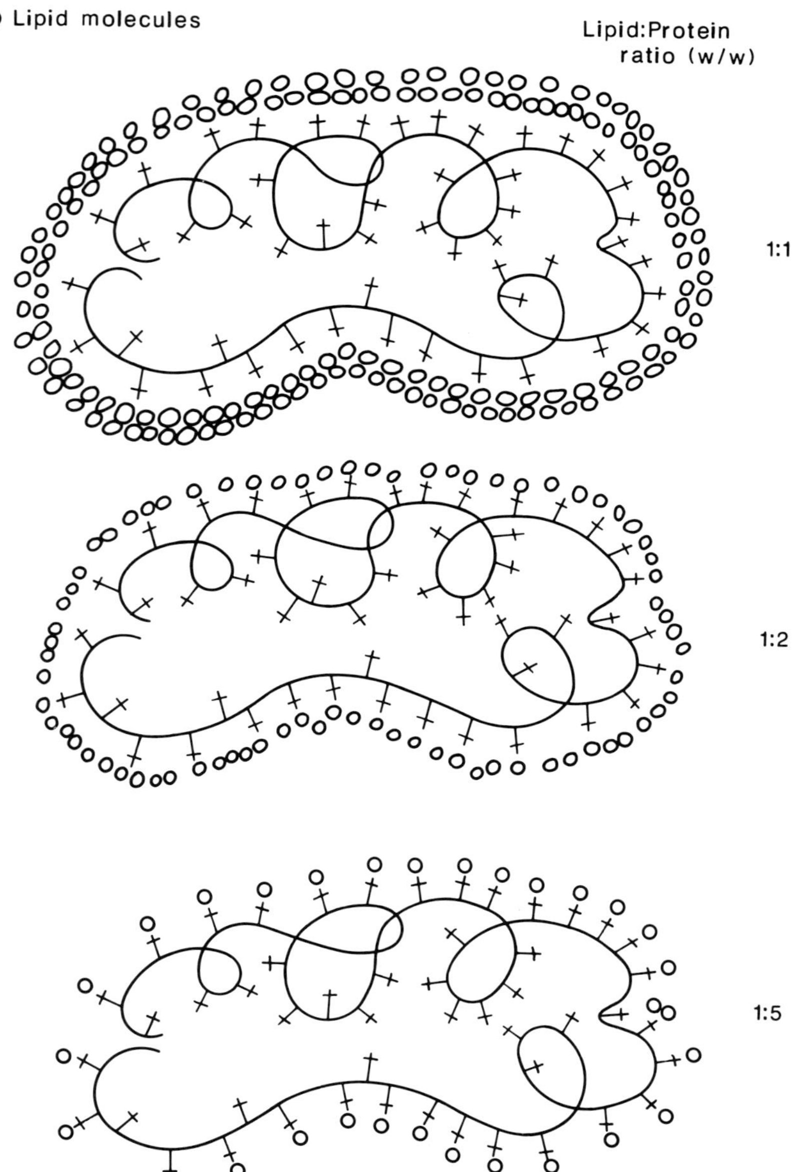

Figure 28. Schematic diagram illustrating the molecular arrangements in lyophilized emulsions of protein and lipid, in varying concentrations [25].

Table 13. Comparison of radical production in proteins and amino acids by ionization, UV light and oxidized lipids [25].

	Ionizing radiation	UV	Peroxidizing lipid free radicals
Primary process	physical (ionization)	physical (excitation)	chemical
Secondary process	electron transfer and hydrogen abstraction		hydrogen abstraction
Propagating species	$H^{\bullet}$, $HO^{\bullet}$, e^{-}_{aq}, $R^{\bullet}$		$ROO^{\bullet}$, $RO^{\bullet}$, $^{\bullet}OH$
Time scale	seconds–minutes	minutes–hours	hours–days
Amino acids most affected	cystine, cys, his, asp, arg, trp, phe, tyr	met, his, sulphur amino acids, aromatic amino acids	trp, lys, his, cys, arg
Localization of protein free radicals	throughout	surface	surface
Damage in proteins	peptide scission, polymerization decreased solubility loss of enzyme activity	fragmentation altered conformation and configuration decreased solubility	peptide scission polymerization decreased solubility loss of enzyme activity

conclusion that probably no changes had occurred to the protein of the animals.

In conclusion, it can be said that although certain reactions between oxidized lipids and proteins are favourable, as they impart attractive flavours or improve loaf volume, in general such reactions are probably detrimental.

ACKNOWLEDGMENTS

The authors wish to thank the management of the Leatherhead Food RA for permission to publish this work. They are also indebted to Ann Pernet, Alison Brodrick and Lee Worsfold, as well as the researchers whose work is described in the preceeding text.

APPENDIX 1: AMINO ACIDS OF MAJOR RELEVANCE

Glycine $H{-}CH(NH_2)CO_2H$

Methionine $CH_3SCH_2CH_2{-}CH(NH_2)CO_2H$

Tryptophan $CH_2{-}CH(NH_2)CO_2H$ (attached to indole ring; N, H)

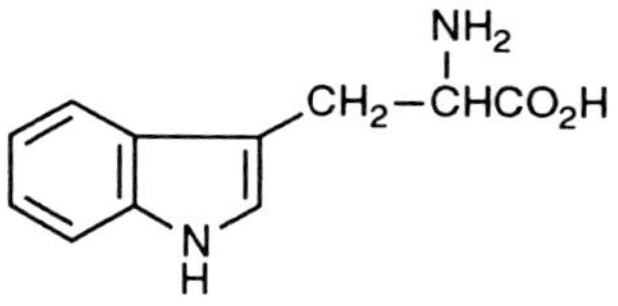

Cysteine $HS{-}CH_2{-}CH(NH_2)CO_2H$

Lysine $H_2NCH_2CH_2CH_2CH_2{-}CH(NH_2)CO_2H$

Arginine $H_2NC(=NH)NHCH_2CH_2CH_2{-}CH(NH_2)CO_2H$

Histidine $CH_2{-}CH(NH_2)CO_2H$ (attached to imidazole ring; N, N, H)

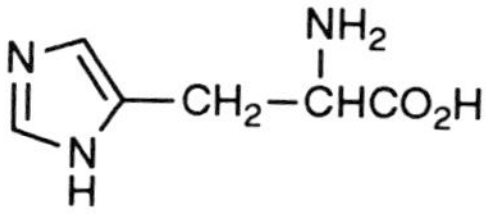

APPENDIX 2:
NOMENCLATURE OF OXYGEN-CONTAINING SULPHUR COMPOUNDS

Sulphonic acid	$-SO_3H$ or $-\overset{\overset{O}{\|}}{\underset{\underset{O}{\|}}{S}}OH$
Sulphinic acid	$-SO_2H$ or $-S(=O)OH$
Sulphone	$-\overset{\overset{O}{\|}}{\underset{\underset{O}{\|}}{S}}-$
Sulphoxide	$-\overset{\overset{O}{\|}}{S}-$
Disulphoxide	$-\overset{\overset{O}{\|}}{S}-\overset{\overset{O}{\|}}{S}-$

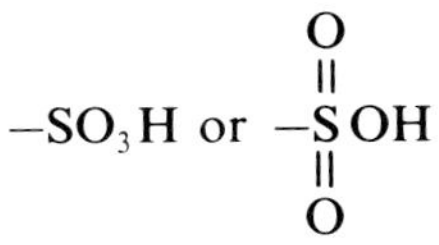

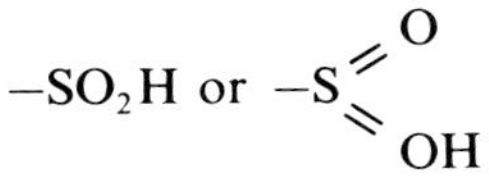

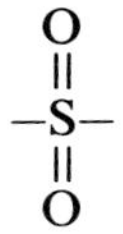

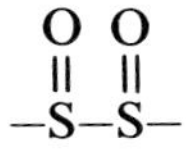

APPENDIX 3:
ABBREVIATIONS OF MOST COMMONLY USED TERMS

PV: peroxide value
TBA: thiobarbituric acid
MDA: malondialdehyde
–SH: sulphydryl group
NMR: nuclear magnetic resonance
ESR: electron spin resonance
a_w: water activity
LH: lipid
LOOH: lipid hydroperoxide
PH: protein
EDTA: ethylenediamine tetra acetic acid

APPENDIX 4: OXIDATION OF OILS AND FATS

Examination of the literature reviewed in this survey demonstrates that the oxidation of lipids and their subsequent interaction with proteins have been associated with the formation of pleasant and rancid flavours, and the deterioration of membrane structure, and are also linked with the ageing process.

What follows is a very brief introduction to the initial stage of the interaction between oxidized lipids and proteins, namely the oxidation of unsaturated fatty acids.

The oxidation of unsaturated lipid material (LH) follows a free radical pathway and, as a consequence, can be considered to have three main steps.

Initiation

$$\underset{\text{unsaturated lipid}}{LH} + O_2 \longrightarrow \underset{\text{lipid radicals}}{\begin{matrix} L^{\cdot} \\ LO^{\cdot} \end{matrix} + \begin{matrix} {}^{\cdot}OH \\ H_2O^{\cdot} \end{matrix}}$$

Propagation

$$\underset{\text{lipid radical}}{L^{\cdot}} + O_2 \longrightarrow \underset{\text{lipid peroxy free radical}}{LO_2^{\cdot}}$$

$$LO_2^{\cdot} + LH \longrightarrow \underset{\text{lipid hydroperoxide}}{LOOH} + \underset{\text{lipid free radical}}{L^{\cdot}}$$

Termination

$$2\,LO_2^{\cdot} \longrightarrow LOOL + O_2$$

$$2\,L^{\cdot} \longrightarrow L{-}L$$

The initiation step can be catalysed by heat, light, transition metal ions and haem pigments, for example. As can be seen from the propagation step, oxidized lipid intermediates are autocatalytic. The oxidation of methyl oleate is shown below:

$$-CH_2CH{=}CHCH_2-$$

$$\downarrow -H^{\cdot}$$

$$-CH_2CH{=}CH\dot{C}H + -\dot{C}HCH{=}CHCH_2-$$

$$\updownarrow \qquad\qquad\qquad \updownarrow$$

$$-CH_2\dot{C}HCH{=}CH- \quad -CH{=}CH\dot{C}HCH_2-$$

Consequently, the production of a free radical on any of four carbon atoms can occur, with the result that four hydroperoxy compounds may be formed. Such hydroperoxy compounds are unstable and are liable to break down to produce a number of reactive ketones, aldehydes and alcohols. As will be shown, all of the above chemical species can react with protein.

REFERENCES

[1] Leibowitz, M. E. and Johnson, M. C. *J. Lipid Res.*, 12 (1971) 662.
[2] McKnight, R. C. and Hunter, F. E. *J. Biol. Chem.*, 241 (1966) 2757.
[3] Victoria, E. J. and Barber, A. A. *Lipids*, 4 (1969) 582.
[4] Roubal, W. T. and Tappel, A. L. *Archs Biochem. Biophys*, 133 (1966) 5.
[5] Kochhar, S. P. and Meara, M. L. Leatherhead Fd RA Sci. Tech. Surv. No. 87 (1975).
[6] Toyamizu, M., Matsumura, Y. and Tomiyosu, Y. *Bull. Jap. Soc. Scient. Fish*, 29 (1963) 854.
[7] Castell, C. H. *J. Am. Oil Chem. Soc.*, 48 (1971) 645.
[8] Desai, I. D. and Tappel, A. L. *Lipid Res.*, 4 (1963) 204.
[9] Buttkus, H. J. *Food Sci.*, 32 (1967) 432.
[10] Fukuzumi. Oléagineux 36 (1981) 251.
[11] Horowitz, I. and Hartroft, W. S. *J. Nutr.*, 101 (1971) 959.
[12] Esterbauer, H. Aldehydic products of lipid peroxidation. In *Free Radicals, Lipid Peroxiation and Cancer*. McBrien, D. C. H. and Slater, T. F. (eds), Academic Press, New York.
[13] Reiss, U. and Tappel, A. L. *Lipids*, 8 (1973) 199.
[14] Narayan, K. A. and Kummerow, F. A. *J. Am. Oil Chem. Soc.*, 35 (1958) 52.
[15] Narayan, K. A. and Kummerow, F. A. *J. Am. Oil Chem. Soc.*, 40 (1963) 339.
[16] Narayan, K. A., Sugai, M. and Kummerow, F. A. *J. Am. Oil Chem. Soc.*, 41 (1964) 254.
[17] Pokorny, J. *Fette Seifen AnstrMittel.*, 65 (1963) 278.
[18] Pokorny, J., Kocourek, V. and Zajic, *J. Nahrung*, 20 (1976) 707.
[19] Kanner, J. and Karel, M. J. *J. Agric. Chem.*, 24 (1976) 468.
[20] Harmuth-Hoene, A. E. and Delincée, H. Int. *J. Vitamin Nutr. Res.*, 48 (1978) 62.
[21] Matsushita, S. and Kobayashi, M. *Agric. Biol. Chem.*, 34 (1970) 825.
[22] Matsushita, S., Kobayashi, M. and Nitta, Y. *Agric. Biol. Chem.*, 34 (1970) 817.
[23] Franzen, K. L. and Kinsella, J. E. *J. Agric. Food Chem.*, 22 (1974) 675.
[24] Beyeler, M. and Solms, J. *Lebensmittel.-Wiss. Technol.*, 7 (1974) 217.
[25] Schaich, K. M. CRC Crit. *Rev. Food Sci. Nutr.*, November (1980) 189.
[26] Gardner, H. W., Weisleder, D. and Kleiman, R. *Lipids*, 11 (1976) 127.
[27] Gardner, H. W., Kleiman, R., Weisleder, D. and Inglett, G. E. *Lipids*, 12 (1977) 655.
[28] Gardner, H. W. *J. Agric. Food Chem.*, 27 (1979) 220.
[29] Gardner, H. W., Kleiman, R. and Weisleder, D. *Lipids*, 9 (1974) 696.
[30] Kanawaza, K., Danno, G. and Natake, M. *J. Nutr. Sci. Vitaminol.*, 21 (1975) 373.
[31] Kanawaza, K., Danno, G. and Natake, M. *Agric. Biol. Chem.*, 47 (1983) 2035.
[32] Kanawaza, K., Ashida, H. and Natake, M. *J. Food. Sci.*, 52 (1987) 475.
[33] Wren, J. J. *Nature*, 185 (1960) 295.
[34] Pokorny, J., Davidek, J., Tran Ha Chi, Valentova, H., Matejicek, J. and Dlaskova, *Z. Nahrung*, 32 (1988) 343.
[35] Roubal, W. T. *J. Am. Oil Chem. Soc.*, 47 (1970) 141.
[36] Pokorny, J., Phan-Throng Tai, Nguyen-Thein Lang and Janicek, G. *Nahrung*, 17 (1973) 621.
[37] Frankel, E. N. *Prog. Lipid. Res.*, 22 (1982) 1.
[38] Roubal, W. T. *Lipids*, 6 (1971) 62.
[39] Schaich, K. M. and Karel, M. *Lipids*, 11 (1976) 392.

[40] Finley, J. W., Walker, H. G. and Wheeler, E. L. *Proceedings of the Conference on Autoxidation of Food Biological Systems: Joint Meeting of the AOCS and JOCS*, San Francisco, 29 April–3 May. Plenum, New York, 1979.
[41] Finley, J. W. and Lundin, R. E. *Proceedings of the Conference on Autoxidation of Food Biological Systems: Joint Meeting of the AOCS and JOCS*, San Francisco, 29 April–3 May. Plenum, New York, 1979.
[42] Leake, L. and Karel, M. *J. Food Sci.*, 47 (1982) 737.
[43] Schaich, K. M. and Karel, M. *J. Food Sci.*, 40 (1975) 456.
[44] Funes, J. A., Weiss, U. and Karel, M. *J. Agric. Food Chem.*, 30 (1982) 1204.
[45] Delincée, H. and Paul, P. *J. Food Proc. Preserv.*, 5 (1981) 145.
[46] Matoba, T., Yoshida, H. and Yonezawa, D. *Agric. Biol. Chem.*, 46 (1982) 979.
[47] Zirlin, A. and Karel, M. *J. Food Sci.*, 34 (1969) 160.
[48] Matoba, T., Yonezawa, D., Nair, B. M. and Kito, M. *J. Food Sci.*, 49 (1984) 1082.
[49] Tannenbaum, S. R., Barth, H. and Le Roux, J. P. *J. Food Sci.*, 17 (1969) 1353.
[50] Krogull, M. K. and Fennema, O. *J. Agric. Food Chem.*, 35 (1987) 66.
[51] Cuq, J. C. and Cheftel, J. C. *Food Chem.*, 12 (1983) 1.
[52] Marcuse, R. *J. Am. Oil Chem. Soc.*, 39 (1962) 97.
[53] Nielson, H. K., De Weck, D., Finot, P. A., Liardon, R. and Hurrell, R. F. *Br. J. Nutr.*, 53 (1985) 281.
[54] Mauron, J. Biblthia, *Nutr. Dieta*, 34 (1985) 56.
[55] Zamora, R., Hidalgo, F. J., Alaiz, M., Millan, F., Maza, M. P. and Vioque, E. *Grasas y Aceites*, 38 (1987) 318.
[56] Ametani, A., Shimizu, M., Kaminogawa, S., Yamauchi, K. R., Takahashi, S. *Agric. Biol. Chem.*, 51 (1987) 477.
[57] Yong, S. H. and Karel, M. *J. Am. Oil Chem. Soc.*, 55 (1978) 352.
[58] Gamage, P. T. and Matsushita, S. *Agric. Biol. Chem.*, 37 (1973) 1.
[59] Kuusi, T., Nikkila, O. E. and Savolainen, K. *Z. Lebensmittelunters. u.-Forsch.*, 159 (1975) 285.
[60] Crawford, D. L., Yu, T. C. and Sinnhuber, R. O. *J. Food Sci.*, 32 (1967) 332.
[61] Andrews, F., Bjorksten, J., Trenk, F. B., Henick, A. S. and Koch, R. B. *J. Am. Oil Chem. Soc.*, 42 (1965) 779.
[62] Menzel, D. B. *Lipids*, 2 (1967) 83.
[63] Chio, K. S. and Tappel, A. L. *Biochemistry*, 8 (1969) 2821.
[64] Dillard, C. J. and Tappel, A. L. *Lipids*, 6 (1971) 715.
[65] Tappel, A. L. *Fedn Proc. FASEB*, 32 (1973) 1870.
[66] Buttkus, K. *J. Am. Oil Chem. Soc.*, 46 (1969) 88.
[67] Pokorny, J., Kminek, M., Janitz, W., Novotna, E. and Davidek, J. *Nahrung*, 29 (1985) 459.
[68] Pokorny, J., Janitz, W., Viden, I., Velisek, J., Valentova, H. and Davidek, J. *Nahrung*, 31 (1987) 63.
[69] Montgomery, M. W. and Day, E. A. *J. Food Sci.*, 30 (1965) 828.
[70] Davidek, J. and Jirousova, J. *Z. Lebensmittelunters. u.-Forsch.*, 159 (1975) 153.
[71] El-Zeany, B. A. Egypt. *J. Food Sci.*, 3 (1975) 81.
[72] El-Zeany, B. A. and El-Tarras, M. F. *Riv. Ital. Sost. Grasse*, 53 (1976) 175.
[73] Stansby, M. E. *Comml. Fish. Rev.*, 19 (1957) 24.
[74] Venolia, A. W., Tappel, A. L. and Stansby, M. E. *Comml. Fish. Rev.*, 19 (1957) 32.
[75] Fujimoto, K., Mariyama, M. and Kaneda, T. *Nippon Suisan Gakkaishi*, 34 (1968) 519.
[76] Fujimoto, K. *Nippon Suisan Gakkasishi*, 36 (1970) 850.
[77] Pokorny, J., El-Zeany, B. A. Luzan, N. T. and Janicek, G. *Z. Lebensmittelunters. u-Forsch.*, 161 (1974) 271.
[78] Pokorny, J., El-Zeany, B. A. and Janicek, G. *Proceedings of Fourth International Congress of Food Science Technology*, 1974, Vol. 1, pp. 271–283.

[79] Pokorny, J. J. *Prog. Ed. Nutr. Sci.*, 5 (1981) 421.

[80] Prithi, T. D., Narayan, K. M. and Bhalerao, V. R. *Indian J. Dairy Sci.*, 24 (1972) 185.

[81] Tomioka, F. and Kaneda, T. *Yukagaki*, 23 (1974) 782.

[82] Davidkova, E. and Svadlenka, I. *Z. Lebensmittelunters. u.-Forsch.*, 158 (1975) 279.

[83] Chang, S. S., Peterson, R. J. and Ho, C. T. *J. Am. Oil Chem. Soc.*, 55 (1978) 718.

[84] Blumenthal, M. M., Trout, J. R. and Chang, S. S. *J. Am. Oil Chem. Soc.*, 53 (1976) 496.

[85] Josephson, D. B. and Lindsay, R. C. *J. Food Sci.*, 52 (1987) 328.

[86] Kaman, J. F. and Labuza, T. P. *J. Food Proc. Preserv.*, 9 (1985) 217.

[87] Schmolka, I. R. and Spoem, P. E. *J. Org. Chem.*, 22 (1957) 943.

[88] Hall, G. M. *Food Sci. Technol. Today*, 1 (1987) 155.

[89] Pokorny, J., Klein, S. and Zelinkova, M. *Nahrung*, 11 (1967) 121.

[90] Pokorny, J., Klein, S. and Koren, J. *Nahrung*, 10 (1966) 321.

[91] Horigome, T. and Miura, M. *Nippon Nogei Kagaku Kaishi*, 48 (1974) 437.

[92] Lohrey, E. E., Hughes, I. R. and Gray, I. K. *J. Am. Oil Chem. Soc.*, 61 (1978) 104.

[93] Yamamoto, A. and Kogure, M. *Nippon Shokuhia Kogyo Gukkaishi*, 16 (1969) 44.

[94] Okumura, S. and Kawai, H. Japanese Patent 70 39 622 (1970).

[95] Pokorny, J. In Paolini, R., Jacini, G. and Porcellati, R. (eds), *Lipids*. Raven Press, New York, 1976, Vol. 2.

[96] Hoseney, R. C., Finney, K. F. and Pomeranz, Y. *Cereal Chem.*, 47 (1970) 135.

[97] Zawistowska, U., Bietz, J. A. and Bushuk, W. *Cereal Chem.*, 63 (1986) 414.

[98] Bloksma, A. H. *Cereal Chem.*, 49 (1972) 105.

[99] Pomeranz, Y. and Chung, O. K. *Proceedings of Seventh World Cereal and Bread Congress*, Prague, 1982.

[100] Frazier, P. J., Daniels, N. W. R. and Russell Eggitt, P. W. *J. Sci. Food Agric.*, 32 (1981) 877.

[101] Tsen, C. C. and Hlynka, I. *Cereal Chem.*, 39 (1962) 209.

[102] Tsen, C. C. and Hlynka, I. *Cereal Chem.*, 40 (1963) 145.

[103] Daniels, N. W. R., Wood, P. S., Russell Eggitt, P. W. and Coppock, J. B. M. *J. Sci. Food Agric.*, 21 (1970) 377.

[104] Frazier, P. J., Leigh-Dugmore, F. A., Daniels, N. W. R., Russell Eggitt, P. W. and Coppock, J. B. M. *J. Sci. Food Agric.*, 24 (1973) 421.

[105] Frazier, P. J., Brimblecombe, F. A. and Daniels, N. W. R. *Proceedings of Fourth International Congress of Food Science Technology*, Madrid, 1974, Vol. 1, pp. 127.

[106] Ory, R. L. and St Angelo, A. J. In Cherry, J. P. (ed.), *Food Protein Deterioration*. Americal Chemical Society, Washington, DC.

[107] St Angelo, A. J. and Ory, R. L. *J. Agric. Food Chem.*, 23 (1975) 141.

[108] St Angelo, A. J. and Graves, E. E. *J. Agric. Food Chem.*, 34 (1986) 643.

[109] Moll, C., Biermann, U. and Grosch, W. *J. Agric. Food Chem.*, 27 (1979) 239.

[110] Gardner, H. W. *J. Agric. Food Chem.*, 23 (1975) 129.

[111] Castell, C. H., Bishop, D. M. and Neal, W. E. *J. Fish. Res. Bd Can.*, 25 (1968) 921.

[112] Connell, J. J. In Hawthorn, J. and Rolfe, E. J. (eds), *Low Temperature Biology of Foodstuffs*. Pergamon Press, Oxford, 1968, p. 353.

[113] Ota, F. and Nishimoto, J. *Progress in Frigeration Science and Technology*. Pergamon Press, Oxford, 1963, Vol. 2, p. 911.

[114] Kwon, T. W., Menzel, D. B. and Olcott, H. S. *J. Food Sci.*, 30 (1965) 808.

[115] Jarenbäck, L. and Liljemark, A. *J. Food Technol.*, 10 (1975) 437.

[116] Betschart, A. A. and Kinsella, J. E. *J. Food Sci.*, 40 (1975) 271.

[117] Hudson, B. J. F. and Karis, I. G. *J. Sci. Food Agric.*, 25 (1974) 1041.

[118] Whitfield, F. B. and Shipton, L. *J. Food Sci.*, 31 (1966) 328.

[119] Haydar, M. and Hadziyei, H. *J. Food Sci.*, 38 (1973) 772.
[120] Miwa, T., Adaichi, T., Hirano, K. and Sigiura, M. *J. Pharm. Agric.*, 6 (1983) 459.
[121] Wurzenberger, M. and Grosch, W. *Lipids*, 21 (1986) 261.
[122] Kendrick, J. and Watts, B. M. *Lipids*, 1 (1969) 146.
[123] Ledward, D. *Food Sci. Technol. Today*, 1 (1987) 153.
[124] Verma, M. M., Paranjape, V. and Ledward, D. A. *Meat Sci.*, 14 (1985) 91.
[125] Okiy, D. A. and Oke, O. L. *Oléagineux* 41 (1986) 77.

SYNTHESIS OF LIPIDS IN YEASTS
BIOCHEMISTRY, PHYSIOLOGY, PRODUCTION

R. Julian Davies and Jane E. Holdsworth

OUTLINE

Advances in Applied Lipid Research, Volume 1, pages 119–159

ISBN: 1-55938-317-8

1. HISTORICAL INTRODUCTION

Lipid accumulation in microorganisms was first recognized over a century ago by Nageli and Loew [1] who observed the occurrence of lipid bodies in yeast cells. The first commercial production of microbial lipid was considered approximately 40 years later by Lindner [2] whose pilot scale work was prompted by the World War I blockade of Germany and the consequent fat and oil shortages. During the intervening war years considerable effort was put into research directed at understanding the process of oil accumulation; however, no commercial process was established. Following World War II, normal trading resumed and interest in microbially-produced lipids declined until Woodbine [3] produced a review outlining the industrial potential of microbial lipids. During the 1970s and 1980s there was a resurgence of interest in the commercial potential of microbial oil production which can be attributed to two major factors. Firstly, technological advances in large-scale continuous fermentation had demonstrated the feasibility of such processes. Secondly, world prices of oils and fats had been variable and increasing. The volatility of the market made regular production of a microbial oil product with a stable price an attractive proposition.

This resurgence of interest has included study of a wide range of potentially useful microorganisms including, yeasts, algae, moulds and bacteria. The scope of the advances is broad and in this chapter we will concentrate only on microbial oil synthesis in yeasts. The reader is referred to excellent articles by Ratledge [4, 5] who has covered all other microbial lipid sources in depth.

2. THE PROCESS OF LIPID ACCUMULATION

Lipid accumulation by oleaginous microorganisms is dependent on cultural conditions. The intrinsic requirement is an excess of carbon over some other

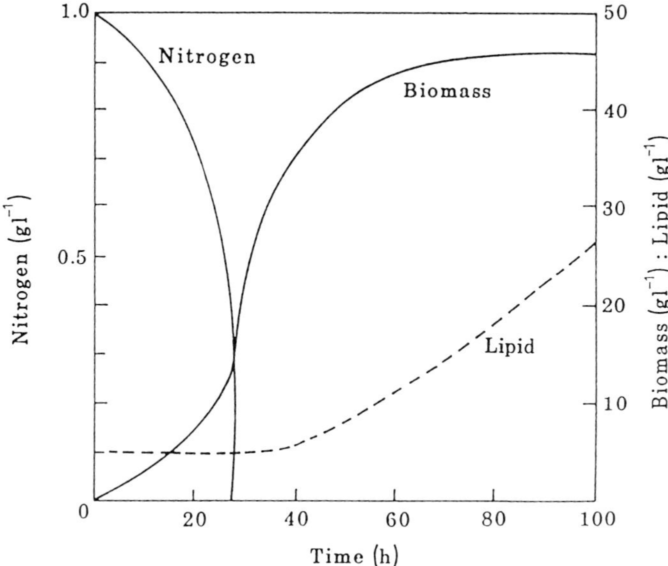

Figure 1. Pattern of lipid accumulation in oleaginous yeasts grown in batch culture.

limiting nutrient. Therefore, in batch culture there will be a period of cell proliferation followed by an increase in lipid levels initiated by the exhaustion of the limiting nutrient, usually nitrogen. Other limiting nutrients such as phosphate [6], sulphate [7], iron [8] and magnesium [9] have all been used effectively to promote lipid accumulation. However, low biomass yields relative to those with nitrogen limitation were obtained and therefore lower lipid contents were also observed.

On exhaustion of the limiting nutrient within the cell, the organism can no longer grow and divide and thus it utilizes the excess carbon source to accumulate lipid as a carbon store. Should conditions of available nitrogen source recur, the organism is capable of mobilizing the lipid store and utilizing it as a carbon source for growth and maintenance [10]. A typical pattern of lipid accumulation in batch culture is shown in Figure 1.

The concentration of the limiting nutrient will determine the quantity of biomass obtained, whilst the concentration of the excess carbon source will largely determine the amount of oil accumulated. The relative proportions of carbon and nitrogen within a medium thus play a key role in determining the oil content and biomass density of the organism. Oleaginous microorganisms do not generally begin to accumulate appreciable quantities of oil unless the C:N (carbon:nitrogen) ratio in the media is greater than 20. Optimal con-

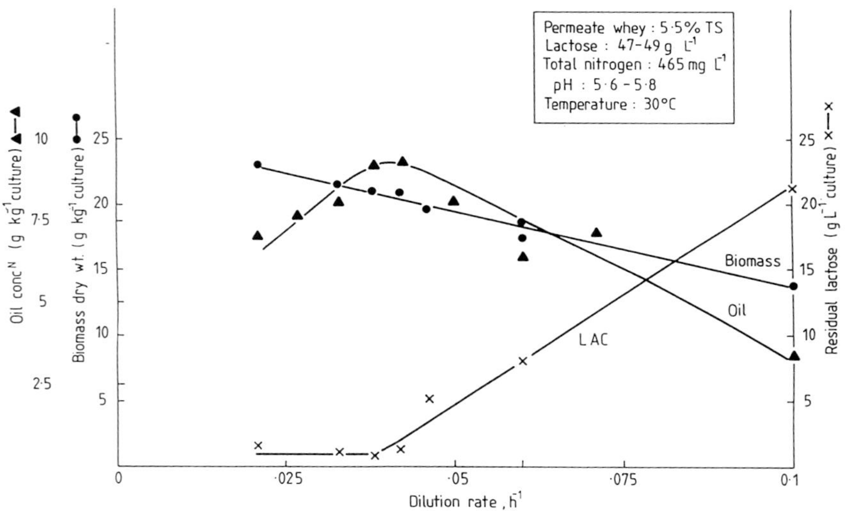

Figure 2. Effect of dilution rate on oil production under nitrogen limited conditions.

ditions vary from yeast to yeast but are generally between C:N ratios of 30 : 1 and 80 : 1, the most common value used being approximately 40 : 1. A notable exception was reported by Pederson [11, 12] who observed lipid accumulation in *Cryptococcus terricolus* even with low C:N ratios.

It should be noted that during lipid accumulation the actual rate of lipid synthesis does not increase above rates observed during cell proliferation, rather the lipid accumulates due to the fact that synthesis of other nitrogenous cellular compounds ceases.

Early studies of lipid accumulation were accomplished either in batch culture or in two-stage continuous culture. It was generally believed that lipid concentrations equal to those attained in batch culture could not occur in a single-stage chemostat. However, the work of Gill *et al.* [9], Hall and Ratledge [13] and Ratledge and Hall [14] demonstrated that providing dilution rates of 0.03–0.15 h^{-1} were used and the medium had a high C:N ratio, oleaginous yeasts could attain lipid concentrations equal to those found in batch culture. Therefore lipid accumulation occurred because cell proliferation was restricted by nitrogen limitation and the residence time of any yeast in the chemostat was such that excess carbon in the media could be converted to lipid. Figure 2 shows the effect of dilution rate on oil production by *Apiotrichum curvatum* ATCC 20509 in continuous culture grown on cheese whey. The results are in agreement with the findings of other researchers [15, 16] and optimal oil yields were obtained at a dilution rate (D) of 0.04 h^{-1}.

Table 1. Oleaginous yeasts.

Organism	Lipid content (% w/w)	Reference
Candida curvata D	58	[21]
Candida lipolytica	36	[22]
Candida sp. 107	42	[23]
Cryptococcus terricolus	55–65	[11, 12]
Hansenula saturnus	28	[24]
Lipomyces lipofer	37	[25]
Lipomyces starkeyi	37	[26]
Rhodosporidium toruloides	56	[27]
Rhodotorula gracilis	57	[28]
Rhodotorula graminis	36	[29]
Rhodotorula glutinis	59	[30]
Trichosporon cutaneum	45	[21]

3. THE BIOCHEMISTRY OF LIPID ACCUMULATION

An oleaginous microorganism was defined by Ratledge [4] as one that is capable of producing 20% or more of its biomass as lipid, some examples are shown in Table 1. The figure of 20% is somewhat arbitrary and a better description of such organisms can be defined in terms of a biochemical test notably the detection of the enzyme ATP:citrate lyase [17]. This enzyme cleaves citrate into acetyl-CoA and oxaloacetate and hence produces the building blocks for fatty acid and lipid biosynthesis. It should be noted, however, that a few strains of yeast exist which have ATP:citrate lyase but contain only low amounts of lipid [18].

Although lipid accumulation has been recognized for many years, studies of the biochemical aspects of the process have largely been confined to the non-oleaginous yeasts *Saccharomyces cerevisiae* and *Candida utilis*. Such yeasts generally contain less than 10% (w/w) lipid. In the mid-1970s Ratledge and co-workers began examining the pathways of carbohydrate metabolism in the oleaginous yeast, *Candida* sp. 107. There followed a series of investigations in which the possible reasons for oleaginicity were successively eliminated and it was eventually hypothesized that the phenomenon was due to the presence of the enzyme ATP:citrate lyase. This work continued and expanded and after 15 years of research a biochemical picture of the process of lipid accumulation in yeasts has now emerged. This is shown in Figure 3.

Fatty acid biosynthesis has been established as a cytosolic process [19, 20] employing acetyl CoA as the basic building block [31]. The study of the mechanism of lipid accumulation can therefore be divided into

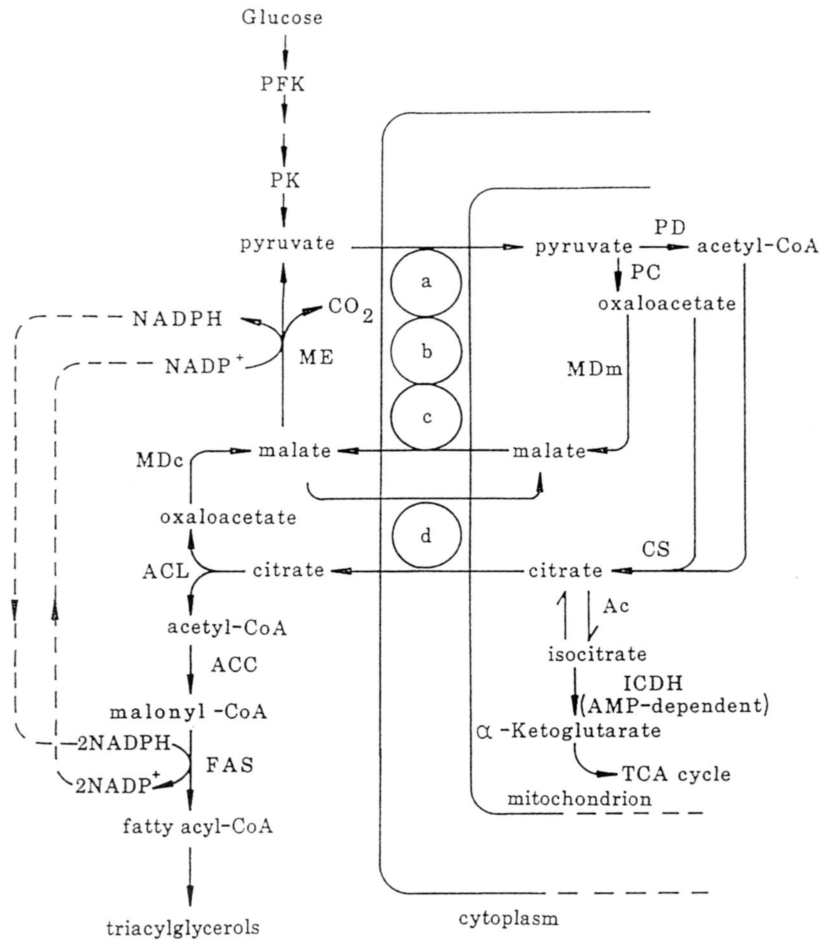

Figure 3. Intermediary metabolism as linked to fatty acid biosynthesis in oleaginous microorganisms. Mitochondrial transport process: a, b, c, interlinked pyruvate-malate translocase systems; d, citrate–malate translocase. Enzymes: ACC, acetyl-CoA carboxylase; Ac, aconitase; ACL, ATP:citrate lyase; CS, citrate synthase; FAS, fatty acid synthetase complex; ICDH, isocitrate dehydrogenase; MDc, malate dehydrogenase (cytosolic); MDm, malate dehydrogenase (mitochondrial); ME, malic enzyme; PC, pyruvate carboxylase; PD, pyruvate dehydrogenase; PFK, phosphofructokinase; PK, pyruvate kinase. Based on material given in Ratledge [5].

two basic areas: (1) intermediate metabolism leading to the production of cytosolic acetyl-CoA and (2) fatty acid biosynthesis and triacylglycerol production.

3.1 Intermediate Metabolism

The net result of the operation of glycolysis and the pentose phosphate pathway is the production of pyruvate in the cytosol which then passes into the mitochondrion. Metabolism of this intermediate to acetyl-CoA occurs by the action of pyruvate dehydrogenase, an enzyme which cannot be detected in the cytosol [32]. The acetyl-CoA thus formed will either enter the tricarboxylic acid cycle or be transported back into the cytosol for fatty acid synthesis. The mitochondrial membrane is impermeable to acetyl-CoA and transfer of this intermediate across it probably occurs in the form of acetylcarnitine mediated by the enzyme carnitine acetyl transferase [33]. The primary role of this enzyme is, however, to transfer the acetyl-CoA to the mitochondrion from the cytosol for lipid degradation. The key to lipid accumulation lies mainly in the change in intercellular concentrations of various cellular metabolites. Immediately following nitrogen exhaustion in the medium, there is a rapid decrease in the AMP concentration of the yeast [34]. This is due to the activation of AMP deaminase [35] which converts AMP into IMP and NH_3, thus providing a nitrogen source to the cell. The resultant drop in AMP concentration in the cell leads to a decreased NAD^+: isocitrate dehydrogenase activity due to the fact that this enzyme is completely dependent on the presence of AMP for its activity [17]. Loss of this activity leads to build up of isocitrate which then equilibrates with citrate via aconitase thus enabling the citrate to accumulate in the mitochondrion.

Transport of citrate into the cytosol is thought to be mediated by a specific tricarboxylate carrier in mammalian tissue. It is situated in the inner mitochondrial membrane [36] and catalyses the stoichiometric exchange of tricarboxylate anion and L-malate. Similarly in the oleaginous yeast the citrate is thought to exchange for malate which previously entered the cytosol in exchange for pyruvate [37]. The malate within the cytosol is then thought to be converted back to pyruvate via the action of malic enzyme thereby generating NADPH for use in fatty acid synthesis.

As previously mentioned, the enzyme which is the key to oleaginicity is ATP:citrate lyase. It is a cytosolic enzyme and its action yields the majority of the cytosolic pool of acetyl-CoA. The activity of this enzyme does not mutually relate to the actual lipid content of the cell; however, its activity can be correlated with the specific rate of lipid synthesis, i.e. g lipid free biomass produced per unit time. ATP:citrate lyase is also inhibited by long-chain fatty acyl-CoA esters (FACEs).

Points of control of the supply of acetyl-CoA units also lie in the glycolysis

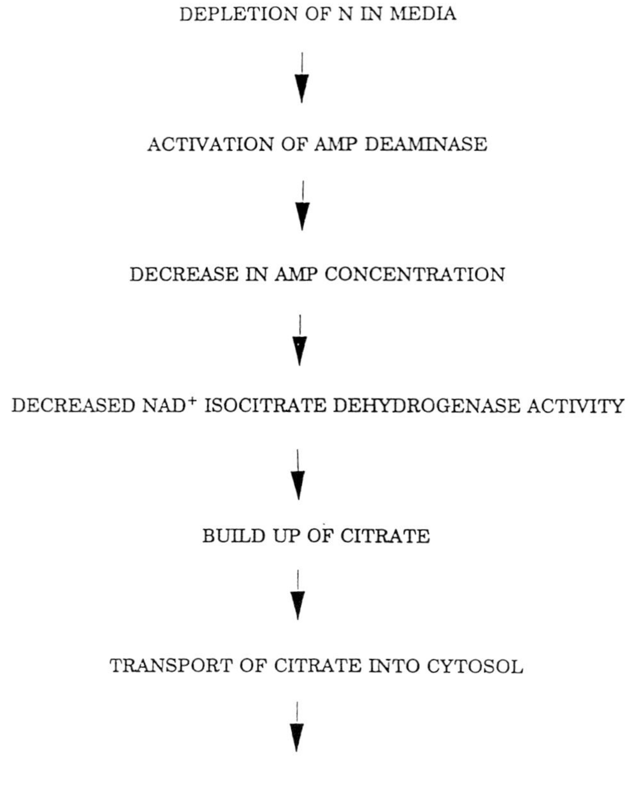

Figure 4. Steps leading to lipid accumulation.

chain, notably phosphofructokinase (PFK) and pyruvate kinase (PK) are under regulatory control. PFK and to a lesser extent PK are both inhibited by citrate, but inhibition can be relieved by a high concentration of NH_4^+ within the cell.

A schematic representation of the control steps leading to lipid accumulation is shown in Figure 4.

3.2 Fatty Acid Biosynthesis

The priming reaction of fatty acid synthesis is catalysed by acetyl-CoA carboxylase; it is biotin dependent and its action occurs in two stages which can be summarized as follows:

$$\text{acetyl-CoA} + HCO_3^- + \text{ATP} \dashrightarrow \text{malonyl-CoA} + \text{Pi}$$

The enzyme is cytosolic and is thought to be the rate limiting step in fatty acid biosynthesis [38], this is probably also true for microorganisms not possessing ATP:citrate lyase. The enzyme is activated by citrate in the oleaginous yeast *Candida* sp. 107 but not in the non-oleaginous yeast *C. utilis* [39, 40]. It is also inhibited by FACEs, but only at exceptionally high concentrations (up to 200 μM). The latter inhibition may be due to the detergent qualities of FACEs. The importance of this enzyme in control of lipogenesis is still uncertain.

The remainder of fatty acid synthesis involves the action of the fatty acid synthetase multi-enzyme complex, the reactions of which can be summarised as:

$$\text{acetyl-CoA} + 7\,\text{malonyl-CoA} + 14\,\text{NADPH} + 14\,\text{H}^+ \dashrightarrow$$

$$\text{palmitoyl-CoA} + 7\,\text{CO}_2 + 14\,\text{NADP}^+ + 7\,\text{CoASH} + 6\,\text{H}_2\text{O}$$

The final reaction of this enzyme in yeasts differs from that of mammalian systems in that there is a terminal acylation to produce FACEs instead of free fatty acids [41]. This enzyme has not been studied in oleaginous yeasts other than by Gill and Ratledge [39], who found it to be repressible with no other indication of a significant regulatory role.

3.3 Triacylglycerol Biosynthesis

Triacylglycerols are the principal form in which lipid is accumulated in yeasts. Other lipid classes are also biosynthesised, e.g. the phospholipids, however, they comprise only a small amount (less than 10%) of the oil produced by the oleaginous yeast and hence their biosynthetic routes will not be discussed in this chapter.

The most common route for triacylglycerol biosynthesis in plants and eukaryotes involves the stepwise acylation of α-glycerol phosphate to form L-α phosphatidic acid. It is known as the α-glycerol phosphate pathway and is shown in Figure 5. The enzyme phosphatidic acid phosphatase is known to be the rate limiting step of this pathway [42, 43] and it catalyses the conversion of phosphatidic acid into diacylglycerols. The distribution of fatty acids in triacylglycerols suggests that the stage involving the conversion of α-glycerol phosphate to phosphatidic acid involves two enzymes. The first would have an absolute specificity for unsaturated fatty acids and would esterify at the two position whilst the second would be less specific and esterify at the one position. These two enzymes have been identified in both bacterial and animal cells [44, 45], the first enzyme is glycerophosphate acyl transferase, the second monoacylglycerol acyl transferase. The biosynthesis of yeast lipids has not been investigated to any significant extent and the majority of studies have been with non-oleaginous yeasts [46–50]. These

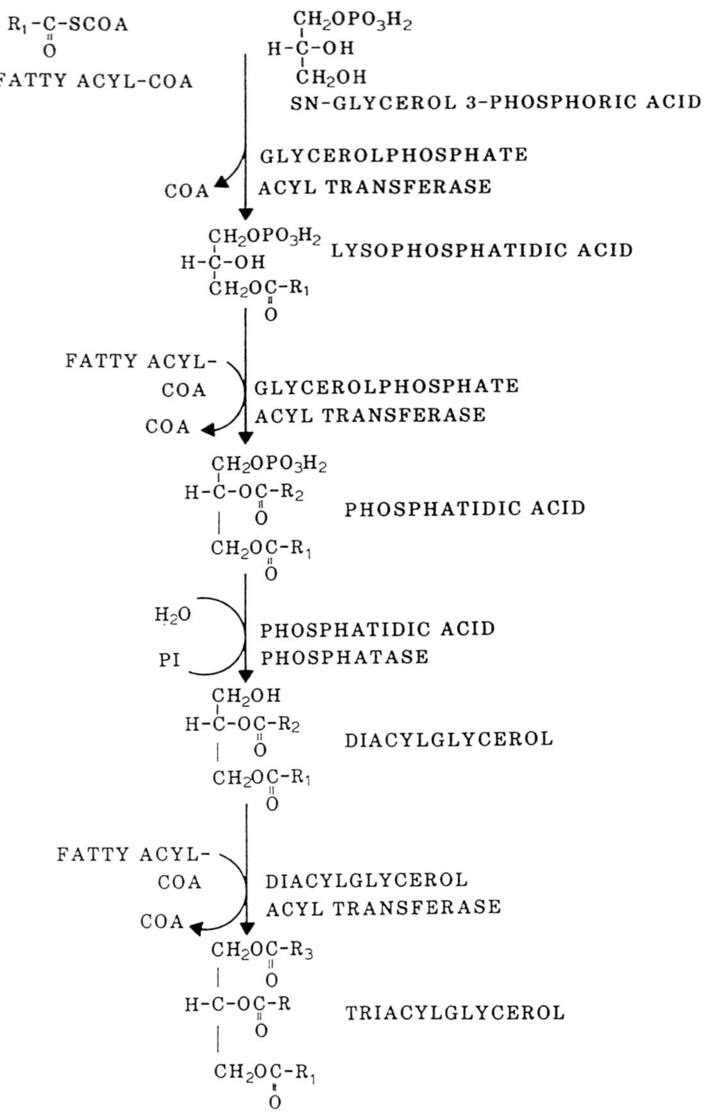

Figure 5. Biosynthesis of lipids via the α-glycerol phosphate pathway.

indicate that the α-glycerol phosphate pathway is an active route for triacylglycerol biosynthesis in the non-oleaginous yeasts and similar findings have been reported for the oleaginous yeasts *Candida curvata* D and *Lipomyces starkeyi* CBS 1809 [51].

4. PHYSIOLOGY OF LIPID ACCUMULATION

4.1 Carbon Sources for Lipid Accumulation

A number of carbon sources have been considered for growth of oleaginous yeasts. These organisms can utilize most commonly available fermentation substrates with the notable exceptions of methanol and cellulose. The biochemical explanation for the inability of the yeasts to utilize methanol for oil accumulation is unclear but the possibility of genetic manipulation of an organism to facilitate this may be a viable option.

The simple carbohydrates, e.g. glucose, are assimulated rapidly and efficiently by most oleaginous yeasts. The yields of lipid (w/w) from carbohydrates are generally 20–22% with the theoretical value being approximately 35% [52]. More complex carbohydrates, e.g. starch, can also be utilized by some oleaginous yeasts provided that the material has been previously hydrolysed by acid or amylases. *L. starkeyi* was reported to grow on starch and produce a lipid yield of 24% of its biomass [53], and a symbiotic mixture of *Saccharomycopsis fibuliger* and *Rhodosporidium toruloides* was shown to grow on starch with a maximum lipid productivity of $0.28\,g\,l^{-1}\,h^{-1}$ [54].

Higher alkanes and crude oil fractions have been utilized successfully as carbon sources for microbial lipid production. Use of such substrates does however have a marked effect on the fatty acid composition of the yeast [55] and cellular lipid compositions reflect the substrate fatty acid being used.

Ethanol has been considered as a carbon source for lipid production by a number of workers. It has the advantage that it is an extremely clean substrate with no residue disposal problems. A theoretical yield of lipid from ethanol would be 54% w/w [56]; however, actual yields observed in fermentations are much lower. Values of 31% for *Cryptococcus albidus* [57] and 21% for *L. starkeyi* [58] have been recorded. These yields represent only about 50% of the theoretical value suggesting that there is scope for improvement. An essential requirement of using ethanol as a substrate is to maintain the concentration below the toxic threshold of the organism, thus making the fed-batch mode of fermentation preferable.

Whey has been considered by many workers as an ideal substrate for oil accumulation [15, 52, 59, 60]. It contains approximately 5% sugar in the form of lactose as well as nitrogen and various mineral components. It naturally has a high C:N ratio making it appropriate for use as a substrate for oil accumulation without further additions. Disadvantages of the use of whey lie in the fact that it is a weak source of sugar and transportation of the large bulk to a central processing facility would be expensive.

Molasses which contains approximately 50% fermentable sugars (principally sucrose) is a common fermentation feedstock. It was first identified as a potential substrate for a single cell oil process by Woodbine [3]. Oil yields with this substrate are, however, slightly lower than those obtained when using sucrose as substrate due to presence of non-fermentable components and nitrogenous compounds [61].

Other substrates considered for a yeast oil process have included domestic waste water [62]; peat hydrolysates [63]; various fruit wastes [64] and xylans [65].

4.2 Nitrogen Sources for Lipid Accumulation

The effect of nitrogen source on the amount of oil accumulated by oleaginous yeasts has also attracted considerable attention from researchers. Blinc and Hocevar [66] examined the different lipid yields of *Rhodotorula gracilis* when using uric acid, urea, asparagine, ammonium sulphate and aspartic acid as nitrogen sources. They concluded that aspartic acid was the best source of nitrogen followed by asparagine. Evans and Ratledge [67] compared asparagine, L-glutamate and ammonium chloride as nitrogen sources in shake flask trials with a number of oleaginous yeasts. Some of the organisms examined showed stimulated oil accumulation when utilizing an organic nitrogen source rather than an inorganic one. The most marked observation was with the yeast *Rhodosporidium toruloides* which showed an increased oil content from 18 to 51% of dry biomass. A similar effect was observed by Moreton [68] in shake flask culture using *R. toruloides*; however, when studying the phenomenon in fermenter culture no effect of nitrogen source was observed. Moreton [68] postulates a possible explanation for the shake flask observations might be that the organic nitrogen sources provide additional buffering capacity, thus maintaining the pH optimal for growth in shake flask culture.

4.3 Effect of Growth Temperature on Lipid Composition and Production

Growth of organisms and plants at low temperatures is known to have an effect on their oil composition. An increase in unsaturated fatty acid content is generally observed, making the lipid softer hence facilitating maintenance of membrane fluidity [69]. Ferrante *et al.* [70] demonstrated that there was an increase in the C18:2 fatty acid content of *C. lipolytica* when grown at 10°C relative to that at 25°C which corresponded to an increase in the 18:1 CoA desaturase activity of the cells. Moon and Hammond [66] demonstrated that the fatty acids produced by three oleaginous yeasts grown at 15°C were slightly more unsaturated than at the optimum temperature (30°C).

Effect of temperature on oil production by *Apiotrichum curvatum* was examined in continuous culture by Davies [71] and the results are shown in Table 2. Above 32°C a drastic decrease in biomass produced and lipid content occurred with no significant change in the fatty acid composition of the yeast. At 35°C it was also noted that the morphology of the yeast was altered and pseudomycelia lacking in oil coexisted with the usual single cell form. This would indicate that with *A. curvatum* there is no advantage in using elevated growth temperatures.

4.4 Effect of pH on Lipid Composition and Production

Most oleaginous microorganisms are able to grow over a wide range of pH without any change in the lipid composition or production rate. This was demonstrated with *Candida* sp. 107 by Ratledge *et al.* [72] over the pH range 3.5–7.5. Davies [71] showed similar results between pH 3.45 and 5.7 for *A. curvatum*; however, at pH 2.7 a marked decrease both in growth and lipid production was reported.

4.5 Effect of Oxygen on Lipid Composition and Production

Various workers have examined the effect of oxygen levels on the fatty acid composition of yeasts. Notably, Choi *et al.* [73] reported that with the obligate aerobe *Rhodotorula gracilis* growing in continuous culture under nitrogen limitation, significant alterations to the fatty acid profile occurred with decreasing oxygen concentration. Both stearate and oleate increased whilst the content of polyunsaturated fatty acids (linoleate and linolenate) decreased. A decrease in palmitate proportional to the increase in stearate was also noted.

Hall and Ratledge [13] working with *Candida* sp. 107, however, found no significant alteration to the fatty acid profile at low aeration rates (0.05 vvm) in a fermenter (vvm ≡ volumes of air per volume fermenter per minute).

Recent work performed in the authors' laboratories has extensively examined the effect of oxygen uptake rate (OUR) on the oleaginous yeast *A. curvatum* [74]. At oxygen uptake rates lower than $7\,\mathrm{mmol\,O_2\,L^{-1}\,h^{-1}}$ applied during the oil accumulating phase of fermentation, a decrease in total unsaturated fatty acids was observed. It was possible to decrease the unsaturated fatty acids markedly (oleate from 55 to 41% and linoleate from 9 to 3%) by limiting the OUR of the culture to less than $3\,\mathrm{mmol\,O_2\,L^{-1}\,h^{-1}}$. This is illustrated in Figure 6. Disadvantages of operating under low aeration conditions are the decrease in oil coefficient observed at low oxygen uptake rate and the increased fermentation time. A practical limit for *A. curvatum* below which adverse effects on oil yields and increased fermentation times occured was $5\,\mathrm{mmol\,O_2\,L^{-1}\,h^{-1}}$. At this oxygen uptake rate the accumulated oil contained approximately 45% oleate and 5% linoleate.

Table 2. Effect of temperature on the biomass, oil concentration and fatty acid profile of *A. curvatum* at constant dilution rate in continuous culture.

Cultivation temperature (°C)	Biomass ($g\,kg^{-1}$)	Residual lactose ($g\,L^{-1}$)	Oil ($g\,kg^{-1}$)	Fatty acid in oil (%)							
				C14:0	C16:0	C18:0	C18:1	C18:2	C18:3	C20:0	C24:0
30	20.8	0.50	9.21	0.4	25.9	11.9	53.0	5.1	1.1	0.8	1.8
30	24.8	2.50	9.01	—	—	—	—	—	—	—	—
30	24.8	0.75	9.36	—	—	—	—	—	—	—	—
32	21.0	1.10	9.23	0.5	25.8	11.7	53.2	5.1	1.1	0.7	1.9
35	17.0	12.50	5.41	—	—	—	—	—	—	—	—
35	16.5	11.00	6.64	0.5	25.9	11.6	53.3	5.0	1.2	0.7	1.9

Dilution rate $D = 0.04\,h^{-1}$; pH 5.5; total *N*: $465\,mg\,L^{-1}$, lactose $48\,g\,L^{-1}$.

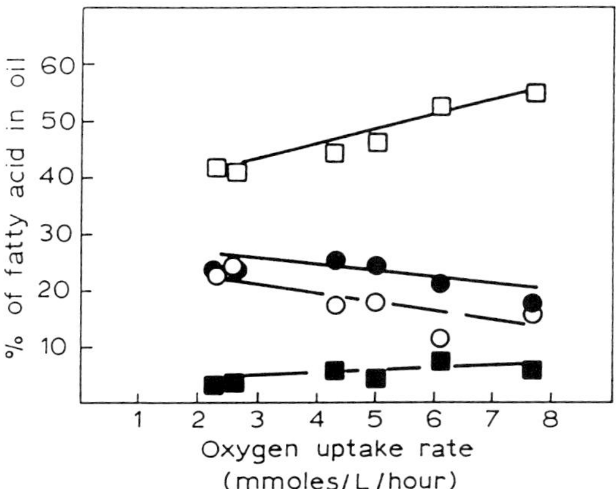

Figure 6. Effect of varying oxygen uptake rate on the fatty acid profile of oil accumulated by *A. curvatum* ATCC 20509: □, oleic acid; ■, linoleic acid; ○, palmitic acid; ●, stearic acid. From Davies *et al.* [74].

Effects of oxygen levels on yeast fatty acid profiles appear to be strain dependent. Much lower oxygen levels were required to demonstrate fatty acid alteration in *A. curvatum* than those reported by Choi *et al.* [73] for *R. gracilis*.

The enzyme involved in desaturating fatty acids is a specific monoxygenase system which requires one molecule of oxygen to desaturate one molecule of stearic acid. The fact that limiting oxygen availability to the organism has an effect on the fatty acid profile of the yeast is therefore not surprising. Oxygen is also required by the yeast for other cellular functions and therefore an adverse effect on these would also be predictable if the oxygen concentration became too low.

5. EXTRACTION AND ANALYSIS OF SINGLE CELL OILS

Accurate analysis of lipids produced by oleaginous yeasts is often difficult and the action of lipases and phospholipases within the cell can lead to erroneous results. Many lipases are able to function in the organic solvents which are generally employed for lipid extraction and hence this compounds the problem. Reports of substantial quantities of free fatty acids within yeast lipids are almost certainly artifacts produced as a result of the extraction technique or the action of lipases.

To prevent this, Hammond and Glatz [78] recommend thermal processing of cells prior to extraction or the adjustment of the pH during extraction to a value at which lipases are inactive.

In order to extract oil from a yeast either mechanical or chemical disruption of the cell wall is necessary. Chemical agents used for this purpose have included acid [76] and alkali [76]. The work of Moon and Hammond [21] with *C. curvata* details direct extraction of oil from the undisrupted yeast with a variety of organic solvents. They consider the hydration of the membranes around the oil bodies a necessary prerequisite to the extraction and use an alcohol, e.g. ethanol or methanol, for this purpose followed by a second non-polar solvent, e.g. benzene or chloroform, to remove the lipid. Moon and Hammond [21] consider that due to the expense and toxicity of such solvents as benzene and chloroform a better approach is to use ethanol/hexane (1:1 v/v) followed by a wash with hexane, as they report this to be almost as effective as other approaches. Hammond *et al.* [77] reported that if after dehydrating wet cells with ethanol, the yeast was treated with hexane (possibly to remove residual ethanol) prior to benzene or chloroform extraction a more efficient extraction was achieved. The general applicability of a direct solvent extraction procedure to all undisrupted oleaginous yeasts has not been shown and verification of a solvent method should be carried out with each new yeast examined.

Mechanical disruption methods for yeasts include ultrasonic treatment, ball milling, passage through a French press or a Hughes press and freeze drying. Enzymic digestion of the cell wall has also been employed. Following any of these methods solvent extraction is necessary to effect removal of lipid from the cell debris. A combination of solvents is normally used with chloroform:methanol, the most popular mixture employed at the laboratory scale.

The commercial extraction of a yeast oil for food product use would inevitably impose restrictions on the type of solvent available for use in processing. This is discussed later within this chapter (yeast cell disruption and oil recovery).

Once extracted, analysis of lipids from oleaginous yeasts can easily be effected by conventional methods which will not be discussed in detail here. The reader is recommended to Hammond and Glatz [78] for a summary of the techniques generally used.

6. TRIACYLGLYCEROL PRODUCTS FROM YEASTS

Lipid is stored within oleaginous yeasts in the form of discrete droplets which are clearly visible under the light microscope. Oil globules have been isolated from both *L. starkeyi* [79] and *C. curvata* [51], and a typical scanning electron micrograph is shown in Figure 7. Such globules extracted from the oleaginous yeasts are predominantly composed of triacylglycerols which can account for greater than 90% of their total. With *A. curvatum* contents of

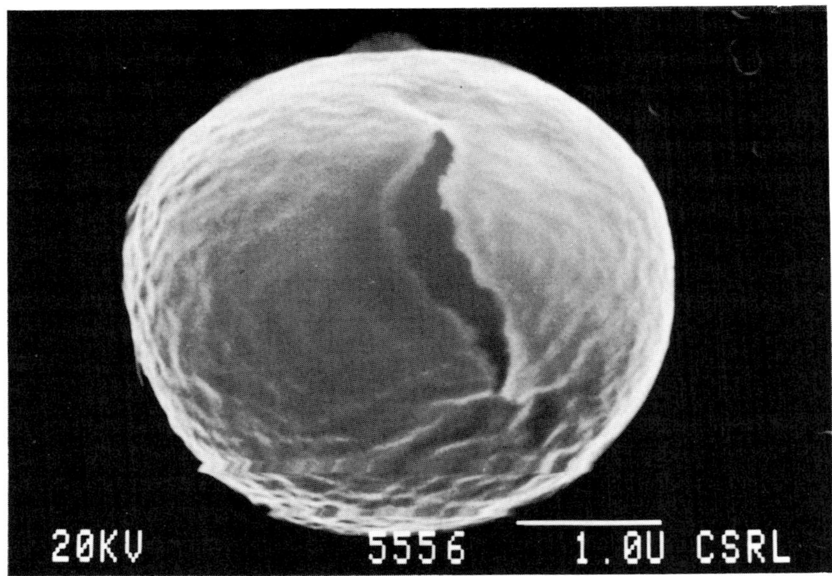

Figure 7. Scanning electron micrograph of lipid globules isolated from *C. curvata* D. (From Holdsworth [51]).

triacylglycerol as high as 95% have frequently been found [71] and values up to 92% have been reported for *C. terricolus* [11]. Phospholipids are usually the second largest component of the lipid with phosphatidylcholine, phosphatidylethanolamine and phosphatidylinositol usually predominating in this class. Sterols, both in the free form and esterified to fatty acids, are also present in yeast oils and can comprise between 3 and 10% of the total lipid content [4].

Mono- and di-acylglycerols as well as free fatty acids have also been reported as components of accumulated oil. All these products are, however, breakdown components of triacylglycerols and phospholipids which may have occurred as an artifact of extraction and hence reporting of their presence should be viewed with care.

The fatty acids of triacylglycerols in oleaginous yeasts are unexceptional in their nature and relative proportions. The predominant fatty acid is oleic (C18:1), followed by palmitic acid (C16:0) with linoleic acid (C18:2) usually being the third most abundant. The stearic acid content of triacyglycerols of oleaginous yeasts is low with contents generally below 15%. Other minor components generally include linolenic acid (C18:3), palmitoleic acid (16:1), myristic acid (C14:0) and some longer-chain acids (C20:0, C22:0 and C24:0). The fatty acids within the triacylglycerol molecule in

Table 3. Similarity of palm olein to yeast oil (*A. curvatum*).

	Palm olein (PORIM)[a]	Yeast oil
Fatty acid composition (% w/w)		
C12:0	0.2	0.1
C14:0	1.0	0.5
C16:0	39.8	25–30
C18:0	4.4	10–15
C18:1	42.5	45–50
C18:2	11.2	5
C18:3	0.4	0.5
C20:0	0.4	0.6
Triglyceride composition (carbon number)		
C46	0.4	0.7
C48	2.3	1.2
C50	42.0	31.0
C52	45.7	50.3
C54	9.9	12.9
	21.6	20–23
Melting point (°C) and solid fat content (%)		
5°C	51.1	59.1
10°C	37.0	—
15°C	19.2	25.1
20°C	5.9	15.2
30°C	0	0.2
40°C	0	0.0
Saponification value	194–202	206
Iodine value	58.0	50–55

[a]Palm Oil Research Institute, Malaysia.

oils extracted from oleaginous yeasts bear a similarity with plant derived oils [4].

The most abundant triacylglycerols in yeasts are of the type SUS and SUU where S = saturated acid and U = unsaturated acid. Yeast triacylglycerols show a further similarity to plant derived oils in that there is a notable lack of saturated fatty acids in the *sn*-2 position of the glycerol molecule.

If yeast lipids are to be produced commercially then they will inevitably have to compete with similar products derived from plant oils. The similarity between palm olein (a lower melting point fraction from palm oil) and yeast oil from *A. curvatum* is shown in Table 3 [71]. Misra *et al.* [80] also noted that oil from *R. glutinis* bears a close resemblance to palm oil.

The current price of palm olein is US $300 per tonne and, bearing in mind the economics of production of such a microbial oil (described later in this chapter), it is difficult to envisage a process producing a product at less than or equal to this price. Interest has thus turned to ways of altering oil accumulated by oleaginous yeasts in order to manufacture a higher value plant oil equivalent, e.g. cocoa butter.

7. MODIFICATION OF THE LIPID COMPOSITION OF OLEAGINOUS YEASTS

The pursuit for a cocoa butter substitute using yeast technology has been active ever since Tatsumi *et al.* [27] filed the first Fuji patent in this area. All the yeast species examined in this patent, gave low stearate contents in the yeast oils, i.e. less than 12.5%. The SUS type triglyceride content of all the oils was less than 35%, well below the level of such triglycerides within cocoa butter. There are five approaches one can adopt to increase the stearate content of yeast oils in the hope of simulating cocoa butter.

The first approach was reported in another Fuji Oil patent by Matsuo *et al.* [81], who described work with similar yeasts grown in the presence of methyl esters of palmitate and stearate. With *Torulopsis* sp. ATCC 20507, an oil with a fatty acid composition of 18.3% palmitate, 32.2% stearate, 43.0% oleate and 3.6% linoleate was claimed. On fractionation of this crude oil, a medium melting point fraction (54% of crude oil) with a fatty acid composition of 20.2% palmitate, 36.5% stearate, 38.5% oleate, 2% linoleate was produced. This fraction gave a similar cooling curve to that for cocoa butter. Two patents filed by CPC International [82, 83] describe similar processes with *R. toruloides* in which rather than using esters, the fatty acids—stearic and palmitic emulsified with Tween 20—were incorporated into the medium.

A second approach in order to increase the stearic acid content of oleaginous yeasts is to limit the oxygen uptake rate of the yeast during the oil accumulation phase, as previously described in an earlier section. This method has the potential for tailoring fats to suit individual requirements within a certain range. There is a limit to which one can push the oleate : stearate ratio by limiting oxygen and the oil analysis shown in Table 4 is the limit to which *A. curvatum* can be altered using this technique without adversely affecting yeast growth.

Both oleate and linoleate contents have decreased considerably in the oil as a result of using the low oxygen technique and the solid fat content has improved compared to the oil extracted from *A. curvatum* grown under well aerated conditions. The melting point has increased and the iodine value has decreased (*cf.* Tables 3 and 4). However, the yeast oil produced by this technique has a softening effect when blended with cocoa butter which is reflected in the solid fats NMR results. This is probably due to the high content of unsaturated fatty acids and the high ratio of di and tri/mono-unsaturated triglycerides.

Table 4. Analysis of Crude Oil from *A. curvatum* ATCC 20509 grown in a pilot scale fermenter under low oxygen uptake rate and compared to cocoa butter.

Analysis	100% yeast oil	50% yeast oil in cocoa butter	20% yeast oil in cocoa butter	Cocoa butter
Free fatty acids (M = 282)	0.84%			
Peroxide value	1.0			
Refractive index (40°C)	1.45088			
Iodine value	45.3			33–36
Mettler dropping point	29.1°C			
Solid fat by NMR (%)				
15°C	47.5	65.5	70.5	75
20°C	36	62	66	71
25°C	25	44	54	61
27.5°C	17	37	47	54.5
30°C	3.5	26	35.5	40
32.5°C	—	6	11.5	17
35°C	—	—	—	—
Fatty acid composition				
C12:0	0.3			
C14:0	1.1			
C16:0	26.1			23–30
C16:1	0.2			
C17:0	0.1			
C18:0	22.6			32–37
C18:1	42.8			30–37
C18:2	2.4			2–4
C18:3 + C20:0	0.4			
C20:0				
C22:0	0.5			
C24:0	2.3			
Triglycerides HPLC				
SSS	0.7			0.5
SSU/SUS	61.4			82.2
SUU/SUS	28.1			10.5
UUU	3.5			0.7

A third approach to increase the stearate content of yeast oils in order to mimic cocoa butter is to employ a desaturase enzyme inhibitor in the fermentation broth. Gierhart [82, 83] cited the use of sterculic acid to prevent desaturation of stearic acid to oleic acid. Moreton [84] showed how the Δ9 desaturase inhibitor present in sterculia oil and kapok oil could be used to increase the stearate content of oil in three species of yeast from 3–5% up to greater than 40% with a concomitant reduction in the oleate content. The authors have utilised sterculia oil in fermentation experiments both at 200 l pilot scale and 10 m^3 plant scale with *A. curvatum*. Table 5 shows that the oil

Table 5. Oleochemical data on crude oil samples of *A. curvatum* ATCC 20509 grown at 200 l pilot and 10 m^3 plant scale in the presence of 100 ppm sterculia oil compared to cocoa butter.

Analysis	200 l pilot scale oil	10 m^3 plant scale oil	cocoa butter	10 m^3 plant scale oil fractionated	
				stearin	olein
Iodine value	41.1	45.6	33–36	31.0	68.4
Free fatty acid (%)	—	0.5	—	—	—
Slip point (°C)	38.9	37.1	32–37	39.2	16.5
NMR British Standard 684 (% solids) Method 2					
20°C	65.1	50.4	70–75	89.0	—
25°C	58.7	38.8	65–70	79.2	—
30°C	46.3	20.0	45–50	45.8	—
35°C	10.2	4.6	5–10	9.5	—
Yield (%)				61.6	38.4
Fatty acid composition					
C12 : 0	0.1	0.1		—	0.1
C14 : 0	0.4	0.5		0.3	0.8
C16 : 0	14.3	17.6	23–30	18.9	15.0
C18 : 0	43.4	34.7	32–37	41.9	22.6
C18 : 1	30.1	34.7	30–37	30.5	42.5
C18 : 2	5.7	5.7	2–4	1.5	13.1
C18 : 3	0.5	0.7		0.1	1.5
C20 : 0	1.5	1.4		1.6	0.8
C22 : 0	1.0	1.0		1.2	0.7
C24 : 0	2.9	3.5		3.9	2.6

sample from the 200 l pilot experiment possesses characteristics approaching those of cocoa butter. On scale up to 10 m^3 bubble column fermenters operating at a dairy site, the effect could not be exactly reproduced and the iodine value was higher whilst the fat was softer. However, on fractionation, which yielded 61% stearin, the hard fraction possessed excellent cocoa butter characteristics apart from a high slip point. The oleate and linoleate rich triglycerides concentrate within the olein fraction which leaves an acceptably low content of linoleate containing triglycerides in the stearin fraction.

The drawbacks to this technique are that sterculia oil, extracted from the nuts of the tropical tree, *Sterculia foetida*, is in short supply. Chemical synthesis of sterculic acid is technically achievable but is not economically viable. Sterculic acid is also rumoured to be a mild carcinogen but it is destroyed in the deodourization stages during refining.

A fourth approach is that advocated by Nijkamp's group in Vrije Uni-

Table 6. Fermentation time (h), cell dry weight yield, lipid yield (as g/g lactose) and fatty acid compositions of lipids extracted with chloroform from revertants cultivated in diluted whey permeate (containing 10 g lactose l^{-1}) with an effective C : N ratio of 30, compared with results of *A. curvatum* wild-type (cultivated in the same medium) and cocoa butter (fatty acids are expressed as % (w/w) of total lipids).

Strain	R22.72	R25.75	R26.17	Wild-type	Cocoa butter
Fermentation time[a]	67	48	80	44	—
Dry weight yield	0.30	0.35	0.33	0.44	—
Lipid yield	0.14	0.16	0.15	0.12	—
Fatty acids					
C16 : 0	26.4	23.8	22.4	28.4	23–30
C18 : 0	36.2	29.5	53.7	15.0	32–37
C18 : 1	20.4	28.4	12.3	42.7	30–37
C18 : 2	8.7	7.7	6.3	7.6	2–4
C18 : 3	1.8	0.9	—	1.0	—
C24 : 0	2.7	2.1	2.9	1.7	—
% SFA[b]	68.6	62.7	80.2	47.2	60–64

[a]Fermentation time = time from inoculation to complete lactose consumption.
[b]Percentage saturated fatty acids (as w/w of total lipid).

versity, Amsterdam. Unsaturated fatty acid (Ufa) mutants of the oleaginous yeast *A. curvatum* have been produced, which are blocked in the conversion of stearic acid to oleic acid [85]. Stable revertants of these Ufa mutants have been isolated on media without oleic acid supplementation [86]. Some of these revertants show comparable fatty acid compositions and differential scanning calorimetry plots to cocoa butter. Table 6 is taken from Ykema *et al.* [86] and shows characteristics of the revertants isolated.

Lipids from revertants R22.72 and R25.75 closely resemble cocoa butter and differential scanning calorimetric plots of these lipids show similar congelation characteristics to cocoa butter, although a minor fraction of triglyceride (probably 1, 2, 3 trisaturated triacylglycerols) with a high congelation temperature was present. Of all the methods reviewed for creating high stearate yeast oils, this method holds the most promise for the economic production of a cocoa butter equivalent using yeast technology. The techniques described using saturated fatty acid supplementation and sterculic acid would substantially increase costs of raw materials for the process and significantly affect process economics.

The last approach for obtaining a cocoa butter equivalent type oil from an oleaginous yeast is by an extensive isolation and screening programme of organisms. The authors have conducted some work in this area and 80 oil accumulating yeast strains have been isolated. Table 7 shows fatty acid

Table 7. Biomass lipid content and fatty acid profiles of oil accumulating yeasts isolated from New Zealand dairy drains.

Yeast isolate code number	Dry biomass yield on whey ($g\,kg^{-1}$)	Lipid content (%)	Fatty acid in oil (%)										
			C14:0	C16:0	C16:1	C18:0	C18:1	C18:2	C18:3	C20:0	C22:0	C22:1	C24:0
K11-7	3.1	15	2.2	19.2	0.6	24.2	41.6	7.8	1.6	0.2	0.6	0	1.5
W10-6	2.5	10	2.7	21.9	2.7	21.4	38.7	7.4	1.45	0	0.38	0	1.45
Rep1-1	19.5	16	0.82	22.7	3.1	23.4	40.6	6.8	0.95	0	0.14	0	0.88
K7-1	21.0	13	0.85	21.0	9.9	10.6	46.1	8.8	0.76	0	0.37	0	0.78
K7-2	19.0	18	0.92	25.6	3.7	23.7	37.7	6.2	0.74	0	0.15	0	0.55
K7-4	18.3	20	0.69	20.1	2.7	23.8	40.1	7.2	1.5	0.57	0	0.69	1.5
K11-6	18.0	15	0.81	19.6	8.9	11.0	46.1	10.1	1.35	0	0.29	0.19	0.83
TA4-5	18.0	20	1.0	24.6	8.6	11.4	43.0	9.0	0.75	0	0	0	0.74
K11-8	2.0	10	1.9	21.0	0.75	25.3	40.5	0.78	1.5	0	0.1	0	0.65

compositions of some of these isolates which appeared promising in shake flask culture, however all exhibited poor oil yields in fermenters compared with *A. curvatum*. A claim has been made by a Canadian company, Diversified Research Laboratories Ltd of Toronto, to have developed a fermentation process for the production of a superior cocoa butter equivalent using their own yeast isolate. Unfortunately no data has been published to substantiate these claims.

8. COMMERCIAL PRODUCTION OF YEAST LIPIDS

8.1 Introduction

The technical and economic aspects of single cell oil production using yeast technology have recently been addressed by Moreton [87], Davies [71], Slater [88] and Floetenmeyer *et al.* [16]. Slater [88] examined the economic aspects of lipid production by fermentation as an alternative to the agricultural route. Four groups of lipids which could be produced by fermentation were reported along with the market values of these products.

(1) Single cell triglycerides
(2) Fatty acids from hydrolised fats
(3) Wax esters C32–C40
(4) Other lipids such as:
 monoglycerides
 biosurfactant glycolipids
 highly polyunsaturated fatty acids (PUFA)

All of these products are relatively low value (less than US $\$10\,kg^{-1}$) with the exception of highly polyunsaturated fatty acids (PUFA) such as γ-linolenic acid (GLA) and eicosapentaenoic acid (EPA). However, GLA is produced by certain phycomycete fungi and EPA by marine bacteria, and not by yeasts. No product niche has yet been found for a yeast lipid with a high market value.

Fermentation processes are sophisticated, utilizing complex technology, and they therefore have high capital costs. They all operate in dilute aqueous solutions with maximal final broth concentrations of product of the order of $100\,g\,l^{-1}$. Dilute solutions require large reactor volumes leading to high operating costs associated with water removal. In order to manufacture low value products, a high conversion efficiency and low processing costs are essential to achieve an acceptable return on investment. In view of the above restraints it is therefore not surprising that there have been few commercial scale examples of yeast lipid products. To date there are only four of note.

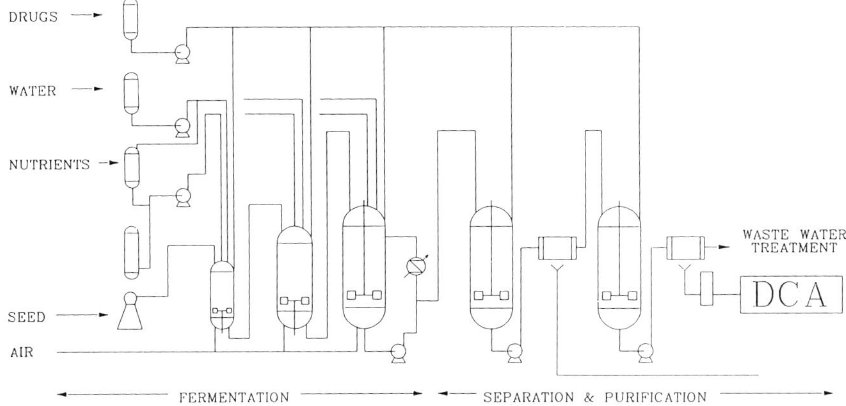

Figure 8. Flow diagram of the dicarboxylic acid process.

Detailed accounts of production data relating to commercial processes for the first three products described below are understandably unavailable. Therefore we have concentrated to a greater extent on the fourth product —single cell oil.

8.2 Production of the Dicarboxylic Acid Brassylic Acid $HOOC(CH_2)_{11}COOH$ by *Candida tropicalis*

Dicarboxylic acid-13 (DC-13) is now being commercially produced at a rate of 150 t $year^{-1}$ in Japan by fermenting together two carbon substrates, one a sugar (glucose, sucrose, molasses) together with *n*-tridecane (*n*-C13) [89, 90]. In a 20 m^3 fermenter, yields of 140 kg m^{-3} have been reported after 120 h cultivation. The process shown in Figure 8 consists of a fermentation section and a dicarboxylic acid separation and recovery section. As the substrate *n*-tridecane is almost completely utilized by the end of the fermentation, dicarboxylic acid product recovery is considerably simplified. It is to be noted also that dicarboxylic acids with various carbon chain lengths can be manufactured by simply changing the fermentation alkane substrate.

Brassylic acid is used to produce ethylene brassylate, a macrocyclic musk compound. The purity of the fermented product is much higher than the presently used product from the ozonization of rapeseed oil. This technology represents the first petroleum fermentation process for manufacturing chemicals on a commerical scale. Other dicarboxylic acids with long carbon chain lengths can be used as raw materials for the preparation of plasticizers, lubricants, polyurethanes, polyamides and perfumes.

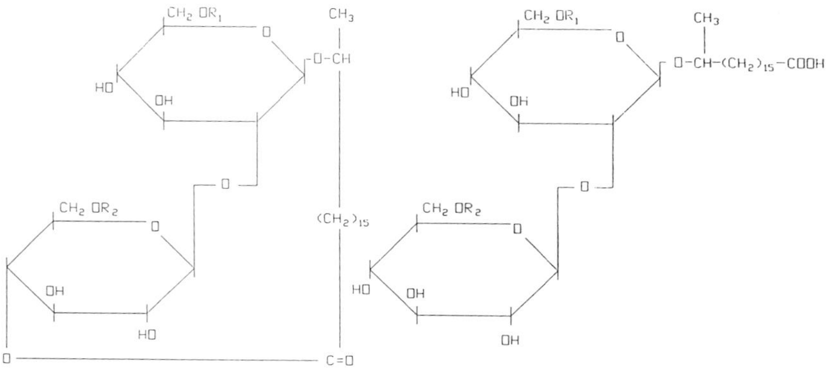

i $R_1 = R_2 = COCH_3$
ii $R_1 = COCH_3 . R_2 = H$
iii $R_1 = H . R_2 = COCH_3$
iv $R_1 = R_2 = H$
v isomeric Lactones

vi $R_1 = R_2 = COCH_3$
vii $R_1 = COCH_3 . R_2 = H$
viii $R_1 = H . R_2 = COCH_3$
ix $R_1 = R_2 = H$

Figure 9. Structure of sophorolipids produced by *Torulopsis*.

8.3 Production of the Biosurfactant Sophorolipid by Torulopsis bombicola

Inoue [91] reported the commercialization by Kao Corporation of Japan of a yeast glycolipid product for use in the cosmetic industry. The structure of the sophorolipids produced by *Torulopsis bombicola* is shown in Figure 9. As in the previous example, the fermentation is based on two carbon substrates—glucose and palm oil. From a medium containing 10% glucose and 14% palm oil a yield of $120\,g\,l^{-1}$ of sophorolipid was obtained after a week's cultivation. The method of recovery of the sophorolipid from the spent broth was not discussed. Sophorolipids have found commercial use as skin moisturizers.

8.4 Astaxanthin

Phaffia rhodozyma was first described in 1976 [92] and shown to contain the carotenoid (3,3′ dihydroxy-8,8′ carotene-4-4′-dione) as its principal pigment [93]. Recently, there has been considerable commercial interest in using *P. rhodozyma* as a pigment source for aquacultured animals particularly pen-reared salmon. Red colouration of salmonoid flesh occurs in the wild when the animals' diet includes crustacea. Farm rearing of salmon is expected to reach more than 200 000 tonnes in 1990 and, hence, there is a large demand for an effective red–pink pigment. Various workers have grown *Phaffia* in simple cheap fermentation media but the yield of astaxanthin is generally less than $0.5\,mg\,g^{-1}$ dry yeast [94, 95]. Recently, however, Gil-hwan *et al.* [96]

have produced mutants of *P. rhodozyma* with increased astaxanthin content (greater than 1 mg g^{-1}) and state that Danisco Biotechnology of Denmark have reported yields greater than 2.5 mg g^{-1}. One commercial outcome of this extensive research is that Lefersa Alimentos of Chile, a salmon farmer and yeast producer, and Igene Biotechnology US will jointly produce and sell astaxanthin diet supplement for salmon farms. Perceived global sales are expected to reach $75–100 million per annum. The 'Astaxin' will be produced by plants to be built in Europe and the USA [97]). Other companies such as Microbio Resources of California are following similar development paths.

8.5 Single Cell Oil (Triacylglycerols)

Over the past 10 years there has been extensive research and development into single cell oil production using yeast technology—Ratledge's group in Hull [4], Karanth's group in Mysore [98], Nijkamp's group in Amsterdam [99], Diversified Research Laboratories Ltd, Toronto (J. Fein, personal communication, 1989), Yamauchi in Nagoya [58], Hammond's group in Iowa [77], Eroshin's group in the Soviet Union [57], Moreton in the UK [87] and in our laboratory here in New Zealand [59, 71]. Yet, so far as is known, no yeast single cell oil production process is operating commercially. Here in New Zealand we conducted trials at a dairy site during the 1986/87 and 1987/88 seasons using a 200 m^3 bubble column fermenter and casein whey as a substrate for yeast oil production. These were feasibility trials to produce sufficient quantity of oil for market assessment and it was sent to a number of notable CBE manufacturers for analysis.

The reason for a lack of success in this product area stems from process economics. If fermentation is to compete with agriculturally produced oils, then product costs must be comparable. Table 8 shows the market values of a range of commercially interesting oils.

To produce a single cell oil product to compete with these low value commodity oils, the substrate costs for the fermentation must be very low. Various carbon sources could be used for the production of single cell oil.

Table 9 lists typical fermentation substrates and costs to produce 1000 tonnes of single cell oil. Other wastes could possibly be used for economic production of single cell oil e.g. fruit wastes [100] and xylans from wood waste [65]. These are excluded from the analysis through lack of economic data on the waste source. It can be seen that of the substrates listed only whey lactose could possibly be used to produce such an oil selling at US$1000 per tonne. In order to use such substrates as raw sucrose sugar, cane molasses or ethanol for yeast lipid production, the selling price for the product must be of the order of greater than $10 000 per tonne to command an attractive return on investment in capital plant.

Using an ultrafiltered whey permeate as substrate, Moon and Hammond

Table 8. Current costs of commodity oils.

Crude vegetable oils	US$ per tonne[a]
Coconut	500
Corn	500
Linseed	840
Palm	300
Peanut	900
Soya bean	425
Cocoa butter	3000 (£2500)
Olive oil (pure spanish)	2400 m^{-3}
Tallow	300

[a] *Chemical Marketing Reporter,* December 1989.

[101] put forward a single cell oil process for producing a palm oil-type lipid using a newly isolated yeast, *C. curvata* D (*A. curvatum* ATCC 20509). This work set the scene for a number of groups who have been investigating the possibility of producing a cocoa butter substitute/equivalent, the most highly priced commodity fat (J. Fein, personal communication 1989 [74, 85, 86]). The substrate for the fermentation in all three cases is some form of whey, a byproduct of the dairy industry. Active commercial research has also been pursued into producing a cocoa butter substitute using yeast technology with other substrates, e.g. mixed carbon sources consisting of fatty acids, fatty acid esters and various sugars. This work has previously been described in this chapter and resulted in two patents each filed by both Fuji Oil [27, 81] and by CPC International [82, 83]. Although the patents show that acceptable cocoa butter substitutes can be produced, the costs of the fatty acid derivatives would undoubtedly make the process uneconomical.

Two process economic studies for a single cell oil process have recently been reported [71, 88]. Slater [88] in a general economic survey put forward

Table 9. Raw material costs for single cell oil processes.

Raw substrate	Substrate cost (US$ per tonne)	Substrate to SCO coefficient	Tonnes substrate required to produce 1000 tonnes of SCO	Cost of substrate for producing 1000 tonnes of SCO ($US)
Whey lactose (New Zealand)	15	0.17	6000	90 000
Ethanol (95%)	500 m^{-3}	0.25	4000	2×10^6
Tallow	300	0.5	2000	600 000
Cane molasses (50% fermentables)	150	0.17	12 000	1.8×10^6
Sucrose raw sugar	500	0.17	6000	3×10^6

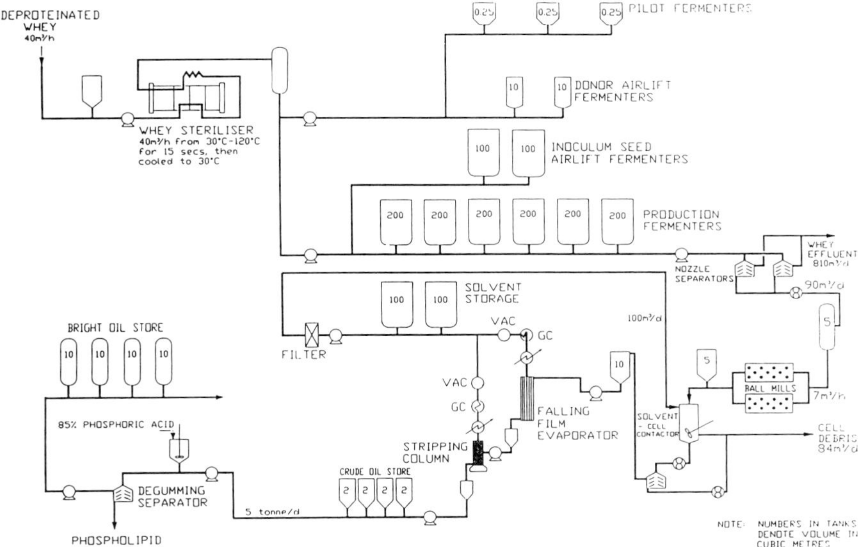

Figure 10. Yeast oil process.

an optimistic production cost of US$5000–6000 per tonne of single cell oil based on a molasses feedstock for a 10 000 tonne per annum production facility. At this production cost, a single cell oil could not compete with any of the plant produced oils set out in Table 8. Davies [71] put forward a detailed process based on the continuous fermentation of whole whey to produce 1000 tonnes per annum of yeast oil. Although the economics were based on 1985 costs, an acceptable return on investment of greater than 20% was shown provided the oil sold for more than US$1000 per tonne. Therefore using cheese whey as a feedstock, it is possible to manufacture a single cell oil that could compete with the plant derived oils.

8.5.1 Yeast Oil Process

The yeast oil process can be conveniently split into two steps: (1) the production of the yeast biomass containing oil by fermentation and (2) downstream processing of the yeast concentrate to yield oil product. A tentative equipment layout and flowsheet for the production of 1500 tonnes of crude yeast oil per annum is shown in Figure 10. The annual production level of 1500 tonnes of oil is the limit that any one dairy site in New Zealand could manage. A dairy site with a daily input of 1000–2000 m^3 of milk is a large processing facility and this sets the upper limit for a single cell oil production facility based on whey. Transport of further whey from other sites again adversely effects overall process economics. The design parameters for

Table 10. The influence of cultivation mode on the lipid production rate.

Cultivation mode	Experimentally reported lipid production rate ($kg\,m^{-3}\,h^{-1}$)	Reference
Batch	0.12	[71]
Continuous	0.4	[71]
Fed batch	0.4	[99]
Continuous with cell recycle	1	[99]

this yeast oil process have been obtained from pilot and large scale trials utilising *A. curvatum* ATCC 20509. Isolated by Moon *et al.* [60], it was originally known as *C. curvata* D but the present consensus of opinion is that it is a *Trichosporon* species (either cutaneum or beigelii).

Production of the yeast biomass containing oil by fermentation. The whey, which has been previously deproteinated either by heat treatment or ultrafiltration, is heat treated in a four-stage heat exchanger equipped for efficient heat regeneration (plate exchanger or spiraflo exchanger, Alfa Laval Co.) The whey is maintained at 120°C for 15 s, then cooled to 30°C before passing to the respective fermenter for yeast inoculation.

The fermentation stage can be operated in any number of modes—batch, fed batch, continuous and continuous with cell recycle. The cultivation mode will have a strong influence on the lipid production rate of the fermenter. Moreton [68] showed that the maximum possible productivity under industrial conditions of a lipid fermentation would be of the order of 1–2 $kg\,m^{-3}\,h^{-1}$. The continuous fermentation process advocated by Davies [71] is designed to operate at a productivity of 1 $kg\,m^{-3}\,h^{-1}$ of yeast oil (i.e. 200 $kg\,h^{-1}$ of oil from a 200 m^3 working capacity fermenter). The fermenter is operated at a high cell density (50 kg dry wt yeast m^{-3}) by recycling 67% of the yeast cells back to the fermenter using a nozzle separator. A similar recycling mode has been operated and successfully proved by Ykema *et al.* [99] on a laboratory scale fermenter using 0.1 μm membrane filters to return the cells to the fermenter. A continuous process although desirable in terms of economy and productivity, requires a constant assured supply of feedstock to maintain steady state growth conditions. The influence of cultivation mode on the lipid production rate is shown in Table 10. In an industrial dairy situation, the whey supply can vary in quality and quantity and is time dependant following upstream cheese or casein processing. A batch system is the most likely commercial option and this mode has been selected for this process design.

Figure 10 process flowsheet shows a number of fermenters of increasing capacity. The pilot fermenters (250 l) and 10 m^3 airlift fermenters function as a yeast inoculum preparation facility. These serve inoculi to the two 100 m^3

Table 11. Size and density of yeasts.

	S. cerevisiae	*A. curvatum*
Size (μm)	8	6.5 by 3
Density (kg m^{-3})	1200–1250	1030–1080

seed fermenters. Then, 25 m^3 inoculum aliquots serve to seed the 125 m^3 of fresh whey in each of the production fermenters. The inoculum and seed fermenters utilize 0.1% urea as a suplementary nitrogen source to produce high biomass levels for the production fermentations. The production fermenters do not include urea as the whey naturally has a high C:N ratio, favouring oil accumulation. The 20% inoculum ensures that rapid fermentation occurs. The first 6–8 h is a period of yeast growth (μ 0.2–0.3 h^{-1}) until the assimilable nitrogen is exhausted. Peak oxygen demand is 15 mol O_2 m^{-3} h^{-1} which can easily be supplied by an air sparged bubble column design fermenter. Within 24 h the expected yeilds are 17–18 kg m^{-3} dry yeast containing 30–35% oil from whey containing 35–40 kg m^{-3} lactose. Fermenters are harvested over a 4 h cycle leaving 25 m^3 as inoculum for another fresh 125 m^3 whey.

Downstream processing of the yeast concentrate to yield oil product. There are two challenging technical problems to overcome in the yeast oil process:

(1) Concentrating the yeast cells, containing the desired oil product, after fermentation.
(2) Economic extraction of the oil from the yeast cell concentrate.

Cell recovery. Yeasts are commonly recovered in the brewing and baker's yeast industry using nozzle disc separators. The efficiency of separation for oleaginous yeasts will not be as great as for *Saccharomyces cerevisiae* (bakers yeast, brewing yeast) owing to the lower density of the cell due to the presence of internal oil globules. Bell and Davies [102] cited that the separation of yeast cells can be described by the Stokes settling velocity:

$$v = \frac{d^2 \Delta_\rho}{18\mu} \omega^2 r$$

where d is the equivalent spherical particle diameter, Δ_ρ is the difference in density between the particle and the fluid, μ is the fluid viscosity, $\omega^2 r$ is the centrifugal force, and v is the settling velocity.

Table 11 shows the size and density for bakers/brewers yeast compared to that of *A. curvatum,* an efficient oleaginous yeast employed at present by most workers in the microbial oil field. It can be seen that $d^2 \Delta_\rho$ factor for oleaginous yeast cells would contribute to a significant drop in settling velocity and hence affect centrifugal separation efficiency, compared to *Saccharomyces* cells. Acceptable separation efficiencies can only be achieved at reduced centrifuge throughputs. Extensive trials with pilot scale and large scale nozzle separators have confirmed that good separation can be achieved with oleaginous yeasts. Table 12 shows results from our pilot scale laboratory and from large scale trials conducted at a New Zealand dairy factory.

Table 12 clearly shows that the oil content of the yeast cells significantly affects cell recovery. Trials 1 and 2 on the pilot scale were conducted with similar yeast feed characteristics (apart from different oil contents). Significant carry over of yeast cells into the supernatant occurred with trial 2. Likewise, on the larger scale, a similar effect was recorded. A practical limit for the Alfa Laval machine cited would be an approximate feed flow rate of $20\,m^3\,h^{-1}$ with a yeast cell concentration of $2\text{–}4 \times 10^8\,ml^{-1}$ containing around 30–35% oil by weight. Above this oil content, one would have to resort to cross flow filtration. Although 100% cell recovery with high cell concentration (up to 28% w/w) can be achieved using this technique [71] the major drawback is that low flux and membrane fouling can necessitate large membrane areas to achieve satisfactory throughputs. The flux needs to be greater than $100\,L\,m^{-2}\,h^{-1}$ to be competitive with conventional centrifugal separators. However, with the intensive research and development currently progressing in membrane technology, it can only be a matter of time before an economic cross flow filter unit is marketed which would solve oleaginous yeast separation problems.

Yeast cell distruption and oil recovery. A great deal has been reported on laboratory scale extraction of lipids from yeasts and the reader is referred to Hammond and Glatz's paper [78] on extraction and analysis of single cell oils for an account of this area. However, very little has been published on oil extraction from microorganisms on the pilot or large scale. Workers at Simon Rosedowns Ltd filed a patent in 1977 claiming a successful method for extracting oil from yeast powder, presumably after spray drying [103]. The method comprised of partially rehydrating the power to approximately 20% moisture content, forming pellets of yeast powder of approximately 3 mm, flaking the pellets and finally subjecting the flakes to a conventional *n*-hexane extraction percolation procedure. This method was an attempt to convert yeast cells into entities similar to oilseeds, thus making extraction a possibility in a conventional oilseed percolator.

In our laboratory a far simpler extraction method has been developed, using a continuous bead mill (Dynomill KDL bead mill, Willy Bachofen AG, Basle, Switzerland). In the Dynomill, grinding beads of 0.5–0.75 mm diameter are rotated at high speed by means of a number of agitator discs fitted symmetrically to a fast rotating shaft. The kinetic energy transferred from the agitator discs to the grinding beads creates impact and shear forces between individual beads, which efficiently disrupts yeast cells. By pumping spray

Table 12. Separation of the oleaginous yeast *A. curvatum* ATCC 20509 by continuous nozzle disc centrifuge.

	Feed condition	Cell concentrate	Supernantant	% recovery (count basis)
Pilot scale centrifuge	flow rate 0.42 $m^3 h^{-1}$			
Westfalia	yeast count 2.4 × 10^8 $m l^{-1}$		4.7 × 10^6 $m l^{-1}$	98
Separator SKOG205	yeast dry weight —	160 $kg m^{-3}$		
Trial 1	yeast oil content 39% w/w			
Pilot scale centrifuge	flow rate 0.39 $m^3 h^{-1}$			
Westfalia	yeast count 3 × 10^8 $m l^{-1}$		3.3 × 10^7 $m l^{-1}$	89
Separator SKOG205	yeast dry weight —	210 $kg m^{-3}$		
Trial 2	yeast oil content 46%			
Large scale centrifuge	yeast count 3 × 10^8 $m l^{-1}$			
Alfa Laval	yeast oil content 16% w/w			
FEUX 512U-31C	flow rate 15 $m^3 h^{-1}$	2.2 × 10^9 $m l^{-1}$	2.2 × 10^5 $m l^{-1}$	99.9
Trial 3	20 $m^3 h^{-1}$	4.9 × 10^9 $m l^{-1}$	1.1 × 10^6 $m l^{-1}$	99.7
	25 $m^3 h^{-1}$	2.8 × 10^9 $m l^{-1}$	1.5 × 10^6 $m l^{-1}$	99.6
	30 $m^3 h^{-1}$	4.5 × 10^9 $m l^{-1}$	5.6 × 10^6 $m l^{-1}$	98.3
Large scale centrifuge	flow rate 28 $m^3 h^{-1}$			
Alfa Laval	yeast count 4.4 × 10^8 $m l^{-1}$	5 × 10^9 $m l^{-1}$	8 × 10^7 $m l^{-1}$	82
FEUX 512U-31C	yeast dry weight —	200 $kg m^{-3}$		
Trial 4	yeast oil content 23% w/w			

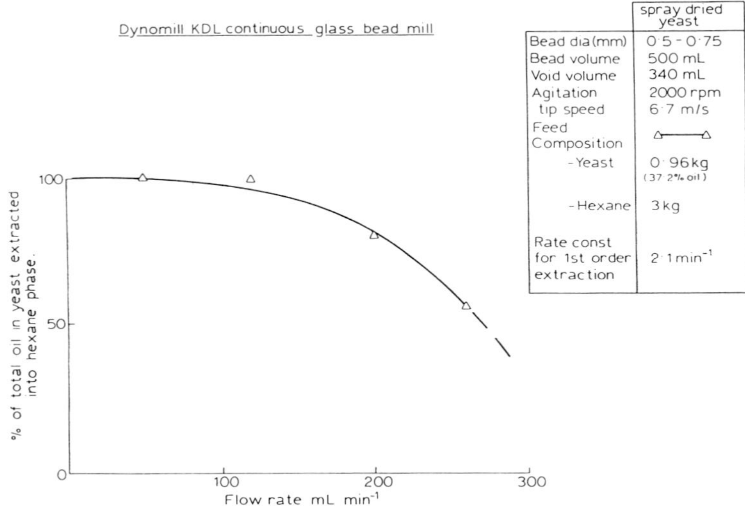

Figure 11. Distruption and oil extraction into hexane phase from spray dried *T. cutaneum* with Dynomill KDL bead mill.

dried yeast powder suspended in hexane through the Dynomill, excellent extraction of oil into the hexane phase can be achieved [71]. It can be seen from Figure 11 that 100% extraction can be achieved with 25% (w/w) yeast in hexane at a flow rate of 0.1 l min^{-1}. The disruptate separates easily on standing into a yellow oil:hexane top phase and a bottom yeast cell debris. Bead mills are available from the same manufacturer up to 250 l chamber capacity which could be used to process 500 kg dry yeast h^{-1}.

There are only two drawbacks with the dry yeast/milling method:

(1) There are reasonably high costs associated with the evaporation and drying of yeast cream suspensions from 20 to 95% (w/w).
(2) Careful control of temperature and particle residence time during drying is required to minimize oxidation of the yeast oil. Higher peroxide values than 5 meq kg^{-1} are not acceptable.

Bearing in mind the above two points, it would appear sensible and more economical to extract oil from wet yeast creams. Hammond *et al.* [77] compared the use of various solvent sequences for extraction of hexane-soluble oil from wet *C. curvata*. After dehydration with ethanol, they achieved efficient extraction by a sequential extraction first with hexane then with either chloroform or benzene. As a laboratory oil extraction routine, the authors regularly utilize this procedure (ethanol–hexane–chloroform) in the laboratory for all yeast lipid extraction work. However, a commercial extraction process for yeast oil intended for human consumption, would be restricted to ethanol, isopropyl alcohol, acetone and hexane as solvents.

In our laboratories we have worked extensively on oil extraction from wet yeast cell creams. Owing to the confidential nature of this work, we are unfortunately unable to divulge scientific data or operational details from pilot scale trials.

One must of course bear in mind the overall economics of a single cell oil process. It may appear economically advantageous to extract oil from yeast in the wet state rather than in the dry state, but one has to dispose of a large volume of wet cell debris on using the former method. Further processing to yield yeast polysaccharides or single cell protein are possible but inevitably down the line, a large disposal problem is still evident. A possible solution is to add the cell debris to the whey effluent and use as a land fertilizer. Fermented whey effluent from whey alcohol factory sites is routinely used as a pasture fertilizer in New Zealand.

The solvent is evaporated from the oil:solvent process stream in a falling film calandria evaporator. The last traces of solvent are separated in a steam sparged stripping column, with the oil leaving the column with a portion of condensed steam, which serves to hydrate the phosphatides. These are removed by a degumming oil separator to produce a bright crude oil.

Yeast oil destined for human consumption in any form would have to be refined. Oils from *A. curvatum* produced in our pilot plant and at a commercial dairy site have been successfully refined using conventional techniques by Abels Ltd, Auckland, New Zealand and by Friwessa BV in Holland. Providing careful attention is applied during handling through downstream processing operations, crude yeast oils can be manufactured with free fatty acids levels of less than 0.5% and at a peroxide level of less than 1 meq kg^{-1}. After physical refining (acid degumming, bleaching and deodourization) refined oils with less than 0.05% free fatty acid and phosphorus less than 1 ppm are achievable. With low free fatty acid levels in the crude yeast oil, a loss during refining of no greater than 5% can be expected.

8.5.2 Economics of a Yeast Cocoa Butter Equivalent Process

Lastly, some thoughts on the economics of a yeast cocoa butter equivalent (CBE) production process. Cocoa butter, at present, is selling for around US$4000 per tonne (1989) and is forecasted to stay at this low price for the next few years. For a cocoa butter equivalent to command the top price of 80% of the cocoa butter price, i.e. US$3200 per tonne, the fat must have over 70% of the symmetrical triglycerides POS and SOS. The linoleate and linolenate content should be less than 4%. Asymmetrical and trisaturated triglycerides and diglycerides should be absent. Provided:

(1) these standards can be met for a yeast CBE;

(2) that the various regulatory authorities have lodged no objections to a novel microbial oil being incorporated into human foods; and
(3) the yeast CBE compares favourably on performance criteria to competitive products based on palm oil, sheanut oil, illipe and other tropical sources,

there seems no reason why a yeast CBE should not command a price of US$2400 per tonne. However, if the yeast fat contains:

(1) trisaturated triglycerides as has been reported for the revertant yeasts [86];
(2) long-chain fatty acids such as C24 in high proportions within the triglycerides; or
(3) the linoleate content is greater than 4%,

then much lower prices would prevail.

Having established a tentative price for a yeast CBE of US$2400 per tonne, a process economic study can then be attempted. On considering a new fermentation process, two options are available. Firstly, one can establish a totally new greenfield facility or alternatively one could retrofit or modify an existing fermentation plant for manufacture of a new product. Table 13 shows an estimated operating budget for a yeast oil process (as depicted in Figure 10) using data from pilot scale experiments and actual costings from a currently operating commercial whey ethanol production facility. The eight large fermenters ($6 \times 200\,m^3$, $2 \times 100\,m^3$) are modified dairy milk storage tanks and are not aseptic, pressure rated fermenters. Operating the process in such vessels is feasible and considerably reduces the capital required for such a facility. From Table 13 one can see that an injection of US$5.4 M as plant capital is the maximum outlay that the process could bear. Under these conditions the profit/loss is neutral and the internal rate of return over a 10 year period is only 16%. Present estimates for a process sited in New Zealand reveal that a capital outlay of about US$9 M would be required for a greenfield facility. Therefore the process is only economically feasible whilst considering the retrofit option.

In the above analysis no credits are assumed for by products. No doubt on further research and development possible returns from sales of yeast phospholipids, cell wall polysacchrides or low value SCP would also improve process economics. Furthermore the process can also be looked at in terms of whey pollution abatement and, in a real situation, cost savings should be taken into account.

It seems probable that for a yeast to exactly produce POS or SOS type triglycerides is too much to expect and that fractionation will be necessary to produce the 'real' thing. The added cost of fractionation and the drop in

Table 13. Yeast CBE process: estimated operating budget.

		US$ M
Basis		
available whey:	200 000 m^3 per annum	
whey lactose content:	39 kg m^{-3}	
whey substrate cost:	US$ 0.5 m^{-3}	
actual CBE recovery:	0.16 kg kg^{-1} lactose	
CBE production	1250 tonnes per annum	
value of CBE:	US$ 2400 per tonne	
CBE sales value:	US$ 3 M	
by products:	—	
Direct manufacturing costs		1.0
manufacturing wages		
steam		
electricity		
consumables		
water		
culture costs		
cleaning and chemicals		
whey value		
oil recovery		
Manufacturing overheads		0.46
laboratory wages		
laboratory expenses		
overhead salaries		
personnel related expenses		
site recharges		
effluent disposal		
other materials		
travel/communication		
maintenance		
insurance and rates		
service overheads		
Financial and selling (distribution and freight, research and development)		0.3
Plant depreciation (over 10 years) for capital of US$5.4 M		0.54
Interest at 12%		0.65
	Profit/loss	− 0.05
5 year internal rate of return		2%
10 year internal rate of return		16%

actual yield of CBE from the yeast fat will all add further to the process costs and decrease the return on investment. The present drop in cocoa prices could not have come at a worse time for this technology but then one takes a risk in advancing products in the commodities market.

9. CONCLUSION

The most actively investigated area of microbial lipid technology is in yeast lipid accumulation/production. Further opportunities into economic methods for producing fatty acids and alcohols, partial and triglyceride esters, wax esters, free and esterified sterols and triterpenoids, terpenes, carotenoids, dicarboxylic acids and biosurfactants using yeast technology will present themselves in the future. CPC International recently filed a patent application on the production of wax esters by a yeast [104], even though these are not normally found in yeast lipids. Therefore in the future we can expect biotechnology researchers utilizing the potential of genetic engineering to produce an even wider range of lipids in yeasts.

ACKNOWLEDGMENTS

We would like to thank Diana Fenton of Abels Ltd, Auckland, New Zealand; Teun Biere and Rinus Heemskerk of Friwessa BV, Zaandijk, Holland; and Bent Andersen of Aarhus Oliefabrik A/S, Denmark for their very valuable comments and advice on oils and fats over the past few years. Also we thank Colin Ratledge for his continued interest and advice in our attempts to commercialize a single cell oil process. Finally, we acknowledge the New Zealand Cooperative Dairy Company, Hamilton for their continued financial support in the single cell oil field.

REFERENCES

[1] Nageli, C. and Loew, O. *Liebigs. Am. Chem.*, 193 (1878) 322–348.
[2] Lindner, P. *Angew. Chem.*, 35 (1922) 110–114.
[3] Woodbine, M. *Progr. Ind. Microbiol.*, 1 (1959) 179–245.
[4] Ratledge, C. *Progr. Ind. Microbiol.* 16 (1982) 119–205.
[5] Ratledge, C. *Microbial Lipids*, 2 (1989) 567–668.
[6] Nielson, N. and Nilsson, N. G. *Acta Chem. Scand.*, 7 (1953) 984–989.
[7] Maas-Forster, M. *Archiv. Mikrobiol.*, 22 (1955) 115–144.
[8] Nielson, N. and Rojowski, P. *Acta Chem. Scand.*, 4 (1950) 1309–1311.
[9] Gill, C. O., Hall, M. J. and Ratledge, C. *Appl. Environ. Microbiol.*, 33 (1977) 231–239.
[10] Holdsworth, J. E. and Ratledge, C. *J. Gen. Microbiol.*, 134 (1988) 339–346.
[11] Pederson, T. A. *Acta Chem. Scand.*, 16 (1962a) 1015–1026.
[12] Pederson, T. A. *Acta Chem. Scand.*, 16 (1962b) 359–373.
[13] Hall, M. J. and Ratledge, C. *Appl. Environ. Microbiol.*, 33 (1977) 577–584.
[14] Ratledge, C. and Hall, M. J. *Biotech. Lett.*, 1 (1979) 115–120.
[15] Evans, C. T. and Ratledge, C. *Lipids*, 18 (1983) 623–639.
[16] Floetenmeyer, M. D., Glatz, B. A. and Hammond, E. G. *J. Dairy Sci.*, 68 (1985) 633–637.
[17] Botham, P. A. and Ratledge, C. *J. Gen. Microbiol.*, 114 (1979) 361–375.

[18] Boulton, C. A. and Ratledge, C. *J. Gen. Microbiol.*, 127 (1981) 169–176.
[19] Van Baalan, J. and Gurin, S. *J. Biol. Chem.*, 205 (1953) 303–308.
[20] Popjak, G and Tietz, A. *Biochem. J.*, 60 (1955) 147–155.
[21] Moon, N. J. and Hammond, E. G. *J. Am. Oil. Chem. Soc.*, 55 (1978) 683–688.
[22] Zvyagint Seva, I. S. and Pitryuk, I. A. *Mikrobiologiya*, 45 (1976) 470–474.
[23] Thorpe, R. F. and Ratledge, C. *J. Gen. Microbiol.*, 72 (1972) 151–163.
[24] Fahmy, T. K., Hopton, J. W. and Woodbine, M. *J. Appl. Bacteriol.*, 25 (1962) 202–212.
[25] McElroy, F. A. and Stewart, H. B. *Can. J. Biochem.*, 45 (1967) 171–178.
[26] Roy, M. K., Vadalkar, K., Baruch, B., Misra, U., Bhadgat, S. D. and Barvah, J. N. *Indian J. Exp. Biol.*, 16 (1978) 511–512.
[27] Tatsumi, C., Hashimoto, Y., Terashima, M. and Matsuo, T. US Patent 4 032 405 (1977).
[28] Enebo, L., Anderson, L. G. and Lundin, H. *Arch. Biochem.*, 11 (1946) 383.
[29] Fuji Oil Co. Ltd. US Patent 1 501 355 (1978).
[30] Pidoplichko, G. A. and Zalashko, M. V. *Mikrobiol Zh. (Kiev)*, 39 (1977) 471–472.
[31] Wakil, S. J. *J. Lipid Res.*, 2 (1961) 1–24.
[32] Boulton, C. A. PhD Thesis, University of Hull, UK, 1982.
[33] Ratledge, C. and Gilbert, S. C. *FEMS Microbiol. Lett.*, 27 (1985) 273–275.
[34] Evans, C. T. and Ratledge, C. *Can. J. Microbiol.*, 31 (1985) 1000–1005.
[35] Evans, C. T., Scragg, A. H. and Ratledge, C. *Eur. J. Biochem.*, 132 (1983a) 609–615.
[36] Klingenberg, M. *Essays Biochem.*, 6 (1970) 119–159.
[37] Evans, C. T., Scragg, A. H. and Ratledge, C. *Eur. J. Biochem.*, 130 (1983b) 195–204.
[38] Volpe, J. J. and Vagelos, P. R. *Physiol. Rev.*, 56 (1976) 339–417.
[39] Gill, C. O. and Ratledge, C. *J. Gen. Microbiol.*, 78 (1973) 337–347.
[40] Botham, P. A., PhD Thesis, University of Hull, England 1978.
[41] Bloch, K. and Vance, D. *Annu. Rev. Biochem.*, 46 (1977) 263–298.
[42] Fallon, H. J., Lamb, R. G. and Jamdar, S. C. *Biochem. Soc. Trans.*, 5 (1977) 37–40.
[43] Brindley, D. N. *Int. J. Obesity*, 2 (1978) 7–16.
[44] Cronen, J. E. and Vagelos, P. R. *Biochim. Biophys. Acta*, 256 (1972) 25–60.
[45] Yamashita, S. and Numa, S. *Eur. J. Biochem.*, 31 (1972) 565–573.
[46] Kuhn, N. J. and Lynen, F. *Biochem. J.*, 94 (1965) 240–246.
[47] Steiner, M. R. and Lester, R. L. *Biochim. Biophys. Acta*, 260 (1972) 222–243.
[48] Schlossman, D. M. and Bell, R. M. *J. Bacteriol.*, 133 (1978) 1368–1376.
[49] Christiansen, K. *Biochim. Biophys. Acta*, 530 (1978) 78–90.
[50] Belov, A. P. and Davidova, E. G. *Mikrobiologiya*, 51 (1982) 253–258.
[51] Holdsworth, J. E. PhD Thesis, University of Hull, England 1986.
[52] Ykema, A., Verbree, E. C., van Verseveld, H. W. and Smit, H. Antonie van Leeuwenhoek, 52 (1986) 491–506.
[53] Guerzoni, M. E., Lambertini, P., Lercker, G. and Marchetti, R. *Starch*, 37 (1985) 52–57.
[54] Dostalek, M. *Appl. Microbiol. Biotechnol.*, 24 (1986) 19–23.
[55] Glatz, B. A., Hammond, E. G., Hsu, K. H., Bachman, L., Bati, N., Bernarski, W., Brown, D. and Floetenmeyer, M. In Ratledge, C., Dawson, P. and Rattray, S., (eds) *Biotechnology for the Oils and Fats Industry, American Oil Chemists Society Monograph no. 11.* AOCS, Champaign, Ill, 1986, pp. 163–176.
[56] Ratledge, C., In Moreton R. S. (ed), *Single Cell Oil* Longmans, London, 1988, pp. 33–70.
[57] Eroshin, V. K. and Krylova, N. I. *Biotechnol. Bioeng.*, 325 (1983) 1693–1700.
[58] Yamauchi, H., Mori, H., Kobayashi, T. and Shimizu, S. *J. Ferment. Technol.*, 61 (1983) 275–2780.
[59] Davies, J. Food Technology in New Zealand, June (1984) 33–37.
[60] Moon, N. J., Hammond, E. G. and Glatz, B. A. *J. Dairy Sci.*, 61 (1978) 1537–1547.
[61] Atamanyuk, D. I. and Vakar, L. I. In Garkavenko, A. I. (ed.), *Lipidy Gribov.* 'Shtiintsa': Kishinev, USSR, 1975, pp. 98–102.
[62] Fustier, P. and Simard, R. E. *Cand. Inst. Food Sci. Technol.*, 9 (1976) 182–185.

[63] Evdokimova, G. A., Raitsina, G. I., Kostyukevich, L. I., Lyakh, V. V., Zatashko, M. V., Gurinovich, E. S., Bogdanovskaya, Z. N. and Obraztsova, N. V. *Prikl. Biokhim. Mikrobiol.*, 10 (1974) 780–785.
[64] Glatz, B. A., Floetenmeyer, M. D. and Hammond, E. G. *J. Food Protection*, 48 (1985) 574–577.
[65] Fall, R., Phelps, P. and Spindler, D. *Appl. Environ. Microbiol.*, 47 (1984) 1130–1134.
[66] Blinc, M. and Hocevar, B. *Mh. Chem.*, 84 (1953) 674.
[67] Evans, C. T. and Ratledge, C. *J. Gen. Microbiol.*, 130 (1984) 1693–1704.
[68] Moreton, R. S., In Moreton, R. S. (ed.) *Single Cell Oil.* Longmans, London, 1988, pp. 1–32.
[69] Rattray, J. B. M., Schibeci, A. and Kidley, D. K. *Bacteriol. Rev.*, 39 (1975) 1971.
[70] Ferrante, G. and Kates, M., *Biochim. Biophys. Acta*, 876 (1983) 429–437.
[71] Davies, R. J. In Moreton R. S. (ed.), *Single Cell Oil.* Longmans, London, 1988, pp. 99–145.
[72] Ratledge, C., Boulton, C. A. and Evans, C. T. In Dean, A. C. R. *et al.* (eds), *Continuous Culture 8.* Ellis Horwood, Chichester, 1984.
[73] Choi, S. Y., Ryu, D. D. W. and Rhee, J. S. *Biotechnol. Bioeng*, 24 (1982) 1165–1172.
[74] Davies, R. J., Holdsworth, J. E. and Reader, S. L. *Appl. Microbiol. Biotechnol.*, 33 (1990) 569–573.
[75] Letters, R. *Biochim. Biophys. Acta*, 116 (1966) 489–499.
[76] Sobus, M. T. and Holmund, C. E. *Lipids*, 11 (1976) 341–348.
[77] Hammond, E. G., Glatz, B. A., Choi, Y. and Teasdale, M. In Pryde, E. H. *et al.* (eds), *New Sources of Fats and Oils.* AOCS, Champaign, Ill, 1981, pp. 171–187.
[78] Hammond, E. G. and Glatz, B. A. In Moreton, R. S. (ed.), *Single Cell Oil.* Longmans, London, 1988, pp. 147–165.
[79] Uzuka, Y., Takeshi, K., Koga, T., Tanaka, K. and Nagamuma, T. *J. Gen. Appl. Microbiol.*, 21 (1975) 157–168.
[80] Misra, S., Gosh, A. and Dutta, J. *J. Sci. Food Agric.*, 35 (1984) 59–65.
[81] Matsuo, T., Terashima, M., Hashimoto, Y. and Hasida, W. US Patent 4 308 350 (1981).
[82] Gierhart, D. L. UK Patent 2 091 285 A (1982).
[83] Gierhart, D. L. UK Patent 2 091 286 A (1982).
[84] Moreton, R. S. *Appl. Microbiol. Biotechnol.*, 22 (1985) 41–45.
[85] Ykema, A., Verbree, E. C., Nijkamp, H. J. J. and Smit, H. *Appl. Microbiol. Biotechnol.*, 32 (1989) 76–84.
[86] Ykema, A., Verbree, E. C., Verwoert, I. G. S., Van der Linden, K. H., Mijkamp, H. J. J. and Smit, H., in press (1989).
[87] Moreton, R. S. In Applewhite, T. H. (ed.), *Proceedings of the World Conference on Biotechnology for the Fats and Oils Industry.* AOCS, Champaign, Ill, 1988, pp. 102–109.
[88] Slater, N. K. H. In Applewhite, T. H. (ed.), *Proceedings of the World Conference on Biotechnology for the Fats and Oils Industry.* AOCS, Champaign, Ill, 1988, pp. 238–243.
[89] Uemura, N., Taoka, A. and Takagi, M. In Applewhite, T. H. (ed.), *Proceedings of the World Conference on Biotechnology for the Fats and Oils Industry.* AOCS, Champaign, Ill, 1988, pp. 148–152.
[90] Kato, K. and Uemura, N. US Patent 4 339 536 (1982).
[91] Inoue, S. In Applewhite, T. H. (ed.), *Proceedings of the World Conference on Biotechnology for the Fats and Oils Industry.* AOCS, Champaign, Ill, 1988, pp. 206–210.
[92] Miller, M. W., Yoneyama, M. and Soneda, M. *Int. J. Syst. Bacteriol.*, 26 (1976) 286–291.
[93] Andrewes, A. G., Phaff, M. J. and Starr, M. P. *Phytochemistry*, 15 (1976) 1003–1007.
[94] Johnson, E. A. and Lewis, M. J. *J. Gen. Microbiol.*, 115 (1979) 173–183.
[95] Haard, N. F. *Biotechnol. Lett.*, 10 (1988) 609–614.
[96] Gil-Hwan An., Schuman, D. B. and Johnson, E. A. *Appl. Environ. Microbiol.*, 55 (1989) 116–24.

[97] Food Technology, May (1988) 60.
[98] Sattur, A. P. and Karanth, N. G. *Biotech. Bioeng.*, 34 (1989) 863–874.
[99] Ykema, A., Verbree, E. C., Kater, M. M. and Smith, H. *Appl. Microbiol. Biotechnol.*, 29 (1988) 211–218.
[100] Vega, E. Z., Glatz, B. A. and Hammond, E. G. *Appl. Environ. Microbiol.*, 54 (1988) 748–752.
[101] Moon, N. J. and Hammond, E. G. US Patent 4 235 933 (1980).
[102] Bell, D. J. and Davies, R. J. *Biotechnol. Bioeng.*, 29 (1987) 1176–1178.
[103] Alexander, D. G., Forster, A. and Farmery, D. W. UK Patent 1 466 853 (1977).
[104] Sekula, B. C. *J. Am. Oil. Chem. Soc.*, 65 (1988) 1584–1585.

LONG-CHAIN POLYUNSATURATED FATTY ACIDS

SOURCES, BIOCHEMISTRY AND NUTRITIONAL/CLINICAL APPLICATIONS

Robert G. Ackman and Stephen C. Cunnane

OUTLINE

Advances in Applied Lipid Research, Volume 1, pages 161–215

ISBN: 1-55938-317-8

1. INTRODUCTION

Polyunsaturated fatty acids (PUFAs) comprise two series of 18–22 carbon fatty acids, $\omega 6$ ($n-6$) and $\omega 3$ ($n-3$), differentiated by the location of the double bonds in relation to the methyl group (see Figure 1). These two series of fatty acids are present in all complex organisms. Two PUFAs are designated as *dietarily* essential because they cannot be synthesized *de novo* by mammals; linoleic acid (18:2 $n-6$) and α-linolenic acid (18:3 $n-3$). Excellent reviews of the biochemistry and essentiality of 18:2 $n-6$ and 18:3 $n-3$ have been written [1–5]. In addition to being essential as dietary precursors for the *structurally* essential very long-chain $n-6$ PUFAs, e.g. arachidonic acid (20:4 $n-6$), adrenic acid (22:4 $n-6$) and one of the docosapentaenoic acids (22:5 $n-6$), 18:2 $n-6$ itself is also structurally essential in the skin [6, 7] without consideration of its metabolism to longer chain products. In contrast, 18:3 $n-3$ is usually present in low amounts in organ lipids throughout the animal kingdom (less than 1% of total fatty acids) and, itself, has no known structural function [8]. At present, it is believed that the essentiality of 18:3 $n-3$ consists of it being a precursor of the longer chain $n-3$ PUFAs, especially the metabolically important eicosapentaenoic acid (20:5 $n-3$) and the structurally important docosahexaenoic acid (22:6 $n-3$) [9–11].

Thus, the very-long-chain PUFAs are the 20 and 22 carbon fatty acids of both the $n-6$ and $n-3$ fatty acid series of which the most studied and best understood are 20:4 $n-6$, 20:5 $n-3$ and 22:6 $n-3$. To further establish the characteristics, functions and practical uses of very-long-chain PUFAs, this review will describe the role and importance of the 18 carbon PUFAs as well.

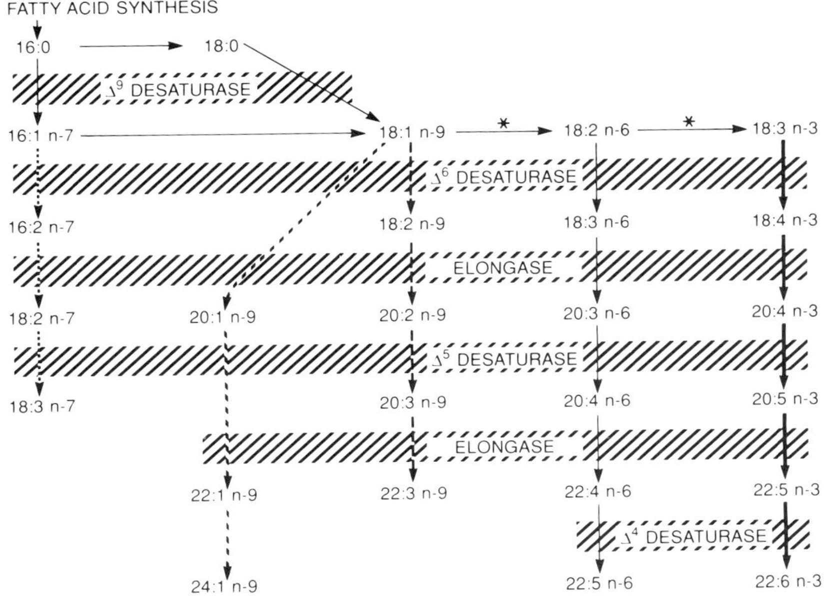

Figure 1. Pathways of long-chain fatty acid desaturation and elongation. Long-chain saturated fatty acids can be synthesized *de novo* and converted to monounsaturated acids (δ9 desaturase) in plants and animals. Oleic acid (18:1 $n-9$) can be further desaturated to linoleic acid (18:2 $n-6$ and α-linolenic acid (18:3 $n-3$) in plants only (*). Palmitoleic acid (16:1 $n-7$), 18:1 $n-9$, 18:2 $n-6$ and 18:3 $n-3$ can be further desaturated (δ6, δ5 and δ4 desaturases) and chain elongated (elongase) to longer chain more unsaturated fatty acids. Given equimolar concentrations of precursor, the more unsaturated the precursor, the more rapid the desaturation/elongation. The darker arrow therefore indicates proportionally more metabolism than the lighter and intermittent arrows; hence, longer chain, more unsaturated products of 16:1 $n-7$ and 18:1 $n-9$ are rarely found in mammals except under conditions of dietary deficiency of 18:2 $n-6$ and/or 18:3 $n-3$. Monounsaturated longer chain products of 18:1 $n-9$ are abundant in myelin and are synthesized by two-carbon elongation of 18:1 $n-9$ through repeated elongation steps without desaturation.

2. SOURCES AND CONCENTRATIONS

Marine phytoplankton and terrestrial plants have the ability to synthesize 18:2 $n-6$ and 18:3 $n-3$ from oleic acid (18:1 $n-9$); vertebrates, especially mammals, have either lost this ability by having ubiquitous dietary sources of 18:2 $n-6$ and 18:3 $n-3$ throughout evolution or never had it in the first place. In the marine food chain, long-chain derivatives of 18:3 $n-3$, e.g.

20:5 $n-3$ and 22:6 $n-3$, are the dominant PUFAs, and 18:3 $n-3$ is present in organ lipids of fish and marine mammals in only small amounts [12, 13]. In the terrestrial food chain, both 18:2 $n-6$ and 18:3 $n-3$ are produced by higher plants, the former being retained primarily in seeds and the latter mainly in the structure of photosynthetic tissues, but also in the seeds of some species [8] capable of producing commercial oils.

Although all the PUFAs can be chemically synthesized, for most purposes of isolation, purification or labelling, the best sources of individual PUFAs may be biological. Continuous effort is being expended on determining better ways of obtaining these fatty acids from existing sources or obtaining genetic variants better capable of producing individual PUFAs. At present, 18 carbon PUFAs are best isolated from plant seed oils, e.g. sunflower, corn and safflower for 18:2 $n-6$, evening primrose and borage for γ-linolenic acid (18:3 $n-6$) [14, 15], linseed, canola and soy for 18:3 $n-3$, and blackcurrant for stearidonic acid (18:4 $n-3$) [16]. The 20 and 22 carbon $n-6$ PUFAs are best isolated from terrestrial animal sources, e.g. 20:4 $n-6$ from liver or brain, 22: 4$n-6$ from brain or adrenals and 22:5 $n-6$ from rat testes. The 20 and 22 carbon $n-3$ PUFAs are best isolated from fats of marine fish or marine mammals, e.g. 20:5 $n-3$ and 22:6 $n-3$ from mackerel, salmon, tuna, sardine, menhaden or seal oil [13]. While these are the traditional biological sources, modern technology has brought other sources to potential commercialization, e.g. selected marine algae for 20:5 $n-3$ [17, 18] and some fungi and algae for 18:3 $n-6$ [19, 20].

2.1 Vegetable Oils

Society is accustomed to the triglyceride oils which not only supply energy, but also provide foods with palatability, flavour and a popular way of cooking in the form of deep fat frying. The natural seed oils (Table 1) include many traditional oils, each defining a modest range of fatty acids in proportions slightly dependent on climate, geographical area and cultivar. As a result of genetic manipulation we also now have high oleic (18:1 $n-9$) sunflower and safflower oil varieties, plus a recently announced high oleic 18:1 $n-9$ (85%) canola variety, the low-linolenic canola variety Stellar, and the low 18:3 $n-3$ flaxseed developed in Australia. It is to be hoped that the consumer will be accurately informed on the container label as to the actual oil supplied. A further issue has been the wide use of a proprietary 'specially processed' soybean oil, lightly hydrogenated and winterized. This reduces 18:3 $n-3$ acid to about 3%, but may give additional *trans* geometric isomers to add to those produced by conventional deodorization processes [21, 22]. The effect of these geometric isomers of the natural all-*cis* C_{18} polyethylenic fatty acids were not known a decade ago [23], but at least one

Table 1. Proportions (w/w%) of principle fatty acids in some common vegetable oils[a]

	Fatty acid								
Oil source	16:0	18:0	18:1	18:2n−6	18:3n−3	20:0	20:1	22:0	22:1
Soybean	11.3	4.6	25.0	51.4	7.2	0.2	–	0.2	–
Canola	3.1	1.5	60.0	20.2	12.0	0.4	1.3	0.2	1.0
Canola var. Stellar	4.2	2.1	59.5	26.3	2.4	0.6	1.3	0.6	0.4
Rapeseed	3.2	1.2	23.0	13.8	7.1	0.6	11.1	0.6	38.9
Mustard	3.3	0.8	9.1	12.7	13.1	0.5	11.1	1.2	44.6
Sunflower	5.5	3.4	48.4	42.3	–	0.2	–	0.2	–
High oleic sunflower	3.0	5.0	79.0	12.0	–	–	–	–	–
Safflower	6.5	2.3	12.0	77.6	0.5	0.4	0.3	0.3	0.1
High oleic safflower	5.0	TR	80.0	15.0	–	–	–	–	–
Corn (maize)	10.4	2.8	26.9	56.2	2.1	0.6	0.4	0.3	0.1
Peanut	10.9	2.3	48.1	30.8	0.8	1.2	1.4	2.9	0.1
Olive	9.5	3.3	79.6	5.8	0.6	0.3	0.3	0.1	–
Flaxseed	5.2	3.4	24.7	15.9	50.9	–	–	–	–
Flaxseed Australian	8.7	5.5	33.5	50.7	1.6	–	–	–	–

[a] Analyses courtesy of J. K. Daun except for 'high oleic' varieties; adapted from Ackman and Ratnayake [303]. Australian 'low-linolenic' flaxseed from Green [304].

such 18:3 $n-3$ isomer has more recently been found to be elongated *in vivo* to a 20:5 $n-3$ isomer [24, 25].

There is no apparent need for concentration of 18:2 $n-6$ or 18:3 $n-3$. They have always formed part of the natural diet of man and 18:2 $n-6$ is plentiful, perhaps even excessively, available [12]. Moreover these polyunsaturated vegetable fatty acids are readily digested, even to the point that 18:3 $n-3$ has been used as a marker to evaluate the co-absorption of other fatty acids [26].

This subject of digestion in mammals has recently been reviewed [27], but novel findings continue to be reported. An example is the recent observation that gastric lipase is very active against triglycerides in the neonatal piglet [28]. However, in man, this pre-intestinal process may be inactive after age 60 [29]. This topic of absorption will be discussed in greater depth after consideration of fish oil concentrates.

Table 2. Proportions (w/w%) of principle fatty acids in typical commercial fish and marine mammal oils, and in a retail fish oil concentration.

Oil Source	Menhaden[a] (U.S.A.)	Herring[a] (Norway)	Seal[b,c] (Canada)	Whale[b,d] (Canada)	Cod liver oil[e] (Icelandic)	Retail concentrate[e]
14:0	11.0	8.5	5.8	7.2	4.7	7.5
16:0	19.9	14.3	10.2	16.4	11.3	9.6
Σ *saturates*	*38.6*	*23.8*	*18.5*	*24.8*	*19.7*	*19.1*
16:1	13.7	6.1	21.4	0.	9.5	11.5
18:1	10.9	9.7	26.5	32.4	22.2	10.9
20:1	1.3	15.5	11.3	9.4	12.0	1.4
22:1	–	21.8	3.6	5.9	8.5	1.2
Σ *monoenoic*	*27.6*	*51.2*	*63.6*	*59.4*	*53.5*	*37.7*
18:2 $n-6$	1.2	1.4	0.9	1.6	1.2	1.2
18:3 $n-3$	1.1	1.5	0.2	0.7	0.6	1.1
20:4 $n-6$	0.6	0.4	0.3	0.2	0.4	0.8
20:5 $n-3$	14.6	5.1	6.4	2.2	8.3	27.8
22:5 $n-3$	1.5	0.8	2.6	1.3	1.1	2.0
22:6 $n-3$	7.5	14.6	5.1	3.5	9.2	11.4
Σ *polyenoic*	*33.8*	*25.0*	*17.9*	*15.8*	*26.8*	*43.2*

[a]Adapted from Ackman [31]. [b]Adapted from Ackman et al. [37]. [c]Commercial oil from harp seal pup blubber. [d]Commercial oil from North Atlantic finwhales. [e]Ackman, unpublished results.

2.2 Aquatic Sources of ω-3 Fatty Acids

2.2.1 The Temperate Zone

For many years it was assumed that the conventional C_{18}–C_{22} fatty acid pathways were operative in the marine world. Plants produced both classes ($n-6$, $n-3$) of C_{18} PUFAs and fish elongated these to the long-chain PUFA composition 'peculiar' to fish, seal and whale oils [30]. Table 2 shows that most common marine oils include roughly one-third each of saturated, monounsaturated and PUFA. More detailed studies have been published [31]. This focus on fish depot fats masked the basic facts of the types of fatty acids produced by the marine plants [32–34]. Algal lipids could include preformed long-chain PUFAs (mostly 20:5$n-3$, with some 22:6$n-3$), but inevitably had much more C_{18} PUFA of both the $n-6$ and $n-3$ families. The latter are not well represented in the fatty acids of fish oils (Table 2). As appears to be the case in man, where human depot fats have a relatively constant composition [35], surpluses of 18:2$n-6$ and 18:3$n-3$ are merely consumed for energy and not stored. The consequences of the multilayered and even circular food systems operative in the oceans are that in temperate latitudes the surplus C_{18} $n-6$ acids are usually consumed by invertebrate

Table 3. Principal and biochemically interesting fatty acids of muscle from a retail farmed Atlantic salmon *Salmo salar* and whole meats of laboratory-reared European oysters *Ostrea edulis*

Fatty acid	Salmon[a]		Oyster[b] total lipid
	Triglyceride	Phospholipid	
14:0	8.62	3.46	2.3
16:0	16.13	23.13	16.6
18:0	2.03	1.87	5.3
Σ saturates	*26.78*	*28.46*	*24.2*
16:1 $n-7$	8.10	2.39	5.3
18:1 $n-9$	11.80	5.23	3.8
18:1 $n-7$	2.56	1.80	3.3
Σ 20:1	11.14	2.11	5.6
Σ 22:1	10.07	1.05	–
Σ monoenoic	*43.67*	*12.58*	*18.0*
18:2 $n-6$	4.67	1.55	2.0
18:3 $n-3$	0.91	0.46	1.4
18:4 $n-3$	1.44	0.35	1.4
20:4 $n-6$	0.27	0.51	6.6
20:4 $n-3$	0.93	0.65	—
20:5 $n-3$	5.26	10.65	14.2
22:5 $n-6$	0.06	0.29	1.8
22:5 $n-3$	1.78	2.98	1.0
22:6 $n-3$	6.60	33.18	16.2
Σ polyenoic	*21.92*	*50.62*	*44.6*
Others	7.59	8.33	13.2

[a] S. M. Polvi and R. G. Ackman, unpublished results.
[b] From Napolitano et al. [305].

mulate as C_{20} and C_{22} fatty acids for membrane use. Most of these animals do not use depot fats for energy storage (e.g. the oyster, Table 3). If they do so, it may be as wax esters, a source of the long-chain (20:1, 22:1) monoethylenic fatty acids found in some marine fats [36]. In brief, 20:5 $n-3$ tends to accumulate in filter-feeding invertebrates such as oysters, but as animals become more organized for movement and also carnivorous the proportion of 22:6 $n-3$ increases, perhaps simply because it is preserved in triglycerides after assimilation from the neural systems of food animals. Finally, in fish muscle phospholipids (Table 3) 22:6 $n-3$ becomes dominant over 20:5 $n-3$. At the same time the 18:2 $n-6$ and 20:4 $n-6$ in total fatty acids diminish to about 1 and 0.5%, respectively, of fatty acids in cold water fish depot fats [31]. Farmed fish (Table 3) may, however, accumulate some 18:2 $n-6$ as a result of high proportions in their rations.

It will be recalled that seals, and probably whales, were originally terrestrial animals that returned to the sea. Their vital organ and membrane lipids need to be sharply differentiated from the depot fats, which are essentially based on the dietary marine fatty acids available from fish [37, 38]. The internal vital organ lipids show strong characteristics of retention of a metabolism based on $n-6$ PUFAs [38, 39]. The positions of the C_{20} and C_{22} $n-3$ fatty acids on the depot fat triglyceride differ in the marine mammals (positions 1 and 3) from that (position 2) characteristic of fish triglycerides, with potentially important consequences for the Inuit diet–good health hypotheses [40, 41].

2.2.2 The Tropics

A follow-up to the famous Kromhout *et al.* [42] long-term study of the diet–cardiovascular event risks in Zutphen, the Netherlands, was the publication of several letters in the 1985 September 26 issue of the *New England Journal of Medicine*. In reassessing independent studies, three groups supported the Kromhout *et al.* findings that a high intake of fish (presumably equal to or exceeding an average of 30 g day^{-1}) reduced cardiovascular mortality by up to 50%. The dissenting letter was from Hawaii, where no such effect could be observed.

It is therefore necessary to point out that in tropical waters there is an approximately 1:1 ratio of 20:4 $n-6$ to 20:5 $n-3$ and 22:6 $n-3$, and moreover the fish are usually very lean [43, 44]. Table 4 is a report of fatty acids in the muscle (edible part) of mahi-mahi, a fish popular in the Hawaiian diet. The phospholipids of the wild and captive fish (fed squid and herring in a 60:40 ratio) are similar and somewhat resemble the salmon phospholipids of Table 3. The triglycerides of mahi-mahi of both origins show very little long-chain $n-3$ fatty acids compared to the salmon triglyceride or the fish oils of Table 2. The striking difference is in the component labelled 20:3 $n-6$. There is no precedent for this fatty acid being important in fish lipids and it has sometimes been confused with either 22:1 or 20:4 $n-6$ [45]. In view of numerous reliable Australian fatty acid analyses reporting 20:4 $n-6$ [44] it is almost certain that the triglycerides of the wild mahi-mahi contain a very high content of 20:4 $n-6$. Weber [46] has proposed that there is a relationship between the proportions of families of circulating C_{20} fatty acids ($n-6$ or $n-3$) and coronary artery disease (Table 5). The health benefits of eating tropical fish, for example in Hawaii, might well be less than those from eating cold-water fish with their higher fat content. In any case the fats of the latter are low in 20:4 $n-6$ and high in 20:5 $n-3$ and 22:6 $n-3$ (Tables 2 and 3). Unlike food tropical fish, their 20:4 $n-6$ is not enough to be seriously competitive with 20:5 $n-3$. Otherwise 20:4 $n-6$ is not important in our diets, except from meats and eggs.

Table 4. Proportions of fatty acids in muscle lipids of female, captive and wild, mahi-mahi (*Coryphaena hippurus*) from Hawaii.[a,b]

Fatty acid	Phospholipid		Triglyceride	
	captive	wild	captive	wild
14:0	1.94	—	6.45	2.06
16:0	21.27	16.62	26.92	37.76
18:0	8.86	9.85	6.16	1.53
$18:1n-9$	21.27	14.89	31.07	29.32
$18:2n-6$	1.86	1.69	0.41	0.26
$18:3n-3$	—	—	0.13	—
$20:3n-6$	—	—	0.33	15.32[c]
$20:4n-6$	2.11	2.93	0.86	(15.32)[c]
$20:5n-3$	4.89	2.44	0.70	0.23
$22:5n-3$	1.06	2.02	0.38	0.72
$22:6n-3$	19.80	23.09	0.93	0.63
Σ	*83.06*	*73.53*	*74.34*	*87.83*

[a] Adapted from Ostrowski and Divakaran [306].
[b] Gram per cent mean of three replicate fish (original includes ± SD).
[c] Clearly the listed value of 15.32 should be for $20:4n-6$ and not $20:3n-6$ (cf. Ackman [44, 45]).

2.3 Dosage

The original views of Kromhout *et al.* [42] have been extended and somewhat modified [47] but the basic conclusion remains valid—consumption of 30 g day^{-1} of fish decreases mortality from coronary heart disease by about 50% compared to eating no fish at all. Various people have computed that in Zutphen the 30 g day^{-1} of fish corresponds to about 150 mg day^{-1} of $20:5n-3$. This is in fact the dose administered to elderly diabetics [48] with beneficial results. Croset *et al.* [49] were able to detect a slight but measurable effect of 100 mg day^{-1} $20:5n-3$ on platelet aggregation in elderly subjects. These studies seem to clinically establish a floor value for $20:5n-3$ but the

Table 5. Platelet phospholipids and frequency of death from coronary artery disease (CAD) in different populations.[a]

	Platelet phospholipid fatty acids[b]			
	$20:4n-6$	$20:5n-3$	Ratio $n-6:n-3$	CAD (% death)
Europe/U.S.A.	20–26	0.1–0.7	50	40
Japan	18–22	1.0–2.5	12	12
Greenland Eskimo	8.3–9.0	6.4–8.0	1.2	7

[a] Adapted from Weber [46]. [b] Approximate values.

effect of 22:6 $n-3$ is more difficult to estimate since there is retroconversion to 20:5 $n-3$ [50, 51]. Since 22:6 $n-3$ is inescapably included with 20:5 $n-3$ in fish as food in an approximately 1:1 ratio such figures match the basic 300 mg day^{-1} proposed by Simopoulos [52] for the sum of 20:5 $n-3$ plus 22:6 $n-3$ useful in the diet.

2.4 Encapsulated Products

Those with long memories will recall taking cod liver oil by the spoonful, or one or two emulsions with allegedly better flavours. These were frankly sources of vitamins A and D and were recognized as such in the official pharmacopoeia. As shown in Table 2, cod liver oil contains approximately 20% $n-3$ fatty acids, whereas regular menhaden oil has somewhat more. With 'winterization' [53], and other technology, oils such as menhaden, Japanese sardine or anchovy can be enriched to give, in practice, 180 mg g^{-1} of 20:5 $n-3$ and 120 mg g^{-1} of 22:6 $n-3$. These figures were on the label of a leading product, MaxEPA, licensed by the UK Government for treatment of hypertriglyceridaemia [54] and became the basic industry standard for 1 g capsules [53, 55, 56]. Thus a nominal 300 mg of 20:5 $n-3$ plus 22:6 $n-3$ has been available in the form of 1 g capsules of fish 'oil' for some years.

A practical limit of enrichment of natural triglycerides is set by the biosynthetic process for fish oils [41] whereby most of the PUFAs (essentially 20:5 $n-3$ and 22:6 $n-3$) is located in the 2-position [13, 40, 41]. Essentially this 1/3 total corresponds to the 300 mg g^{-1} already mentioned (or more if 22:5 $n-3$ and 18:4 $n-3$ are included). The further enrichment of triglycerides then depends on enzymatic processes which can introduce additional $n-3$ fatty acids from enriched individual fatty acid materials [57]. The retail ethyl and methyl ester products analyzed to date [53, 56] are nominally 50% 20:5 $n-3$ plus 22:6 $n-3$, but corresponding free acids are readily obtained from almost any fish oil by urea complexing at concentrations of up to 65% 20:5 $n-3$ plus 22:6 $n-3$ [55] and ethyl esters can be equally well concentrated [58] by this technique.

A number of considerations lead to the view that all four forms of presenting $n-3$ fatty acids (natural oils, synthetic oils, ethyl (or methyl) esters, free acids) are sufficiently well absorbed to be considered for therapeutic use [13, 41]. Absorption of ethyl esters may have an upper limit since 6 g day^{-1} is absorbed but not 18 g [59]. However, even if delayed, the ethyl esters contribute more 20:5 $n-3$ and 22:6 $n-3$ to circulating phospholipids [60] than to chylomicrons [26, 61, 62]. Thus, this is an option which may be useful for therapy directed to particular organs or problem sites. For example methyl esters ingested by patients at 7.5 g of $n-3$ fatty acids per day decreased plasma triglyceride and plasma cholesterol, as well as VLDL and HDL cholesterol [63]. The role of gastric mucosa [28, 64] needs to be considered.

There may be opportunities to directly initiate modification of prostaglandin (PG) synthesis [65] and improve ulcer conditions [66]. Although a few patients do not tolerate high doses of fish oils well, less is known about esters and free acids. These, at higher concentrations, offer the potential to reduce doses while maintaining an effective level of therapy [67].

It is important to note that in the past there have been problems with the 20:5 $n-3$ and 22:6 $n-3$ label claims made for some fish oil products. The only reliable way to present these fatty acids is as milligrams per gram of sample measured by capillary gas–liquid chromatography (GLC) with an internal standard technique [68, 69]. Diluents, polymers and additives are thus eliminated from consideration, and the number is the free acid so that different products can be compared for potential biochemical efficiency, subject to this efficiency being modified if the acids are absorbed by different routes. Capillary GLC also reveals thermal artifacts created by concentration steps [70].

2.5 Organ Distribution

The content of 18:2 $n-6$, 18:3 $n-3$ and the very-long-chain PUFAs in organ lipids of mammals is governed by both dietary intake and genetics. Thus, while the level of 18:2 $n-6$ and 18:3 $n-3$ in organ lipids is partially a function of their intake, it is also clear that genetics predetermines that some species will accumulate significant amounts of one series even against high dietary odds, e.g. the dolphin liver accumulates high amounts of 20:4 $n-6$ in viscera on a diet virtually devoid of 18:2 $n-6$, and the zebra liver accumulates relatively low amounts of 18:3 $n-3$ and very-long-chain $n-3$ fatty acids despite a leafy diet rich in 18:3 $n-3$ [71]. Furthermore, even consuming the same dietary proportions or ratio of 18:2 $n-6$ and 18:3 $n-3$, different species will accumulate different amounts of 20:4 $n-6$ and 22:6 $n-3$, especialy in liver [72–75]. Other species such as the cats (domestic and wild) are unable to metabolize sufficient 18:2 $n-6$ to meet their demands for 20:4 $n-6$; in this case 20:4 $n-6$ is both structurally and *dietarily* essential [76, 77]. However, the mammalian brain has evolved with a remarkably similar fatty acid composition regardless of species and dietary availability of longer or shorter chain PUFAs [78, 79]. This suggests that the lipid prerequisites for neuronal function are set within narrowly defined limits; intelligence seems to be more related to total neurons developed (brain size) rather than to a difference in neuronal lipid (PUFA) composition.

All cell types appear to have a structural requirement for several long-chain fatty acids for membrane integrity and function, including palmitic acid (16:0), 18:1 $n-9$, 18:2 $n-6$, 20:4 $n-6$ and 22:6 $n-3$. The proportion of these and other long-chain fatty acids differs widely in different phospholipids within a given membrane, within different membrane types of the same cell and within different organs of the body. Hence, even ignoring effects of differences in dietary fat, the occurrence of individual very-long-chain PUFAs is highly specific to the function of the membrane or lipid class in which they are located. Including the effect of dietary fat, some membranes have a programmed requirement for a specific fatty acid composition and are highly resistant to altering their fatty acid composition despite a radical change in availability (or dietary lack) of fatty acids that may be integral to the function of those membranes.

For instance, neuronal membrane phospholipids in adult mammals show remarkably little change in the composition of very-long-chain PUFAs, e.g. 22:6 $n-3$, even in the face of a significant increase in 20:5 $n-3$ or 22:6 $n-3$ in the diet [9, 80, 81]. However, platelet phospholipid fatty acid profiles are highly plastic and rapidly reflect an increase in the intake of 20:5 $n-3$, replacing 20:4 $n-6$ with 20:5 $n-3$ and 22:6 $n-3$ [82, 83].

3. BIOCHEMISTRY

The pathway of sequential desaturation followed by elongation is fundamental to the conversion of 18:2 $n-6$ and 18:3 $n-3$ to the very-long-chain $n-6$ and $n-3$ PUFAs, respectively (Figure 1). This pathway has been extensively described elsewhere [5, 84–87] but it is worth elaborating on several aspects.

3.1 Rate-limiting Steps

Both desaturation and elongation are enzyme-catalysed steps but only desaturation, which is a relatively slow step in relation to elongation, is considered rate limiting [84, 87]. One common desaturation/elongation pathway is generally thought to be responsible for synthesis of both $n-6$ and $n-3$ very-long-chain PUFAs in mammalian species. However, in culture, different cell lines may be able to metabolize the $n-3$ series but not the $n-6$ series, or *vice versa* [88], suggesting that variants of the desaturase enzyme may exist or that, *in vitro*, they may be susceptible to local membrane conditions which determine the degree of metabolism of the available substrate. Since desaturase proteins are not readily isolated and purified to homogeneity [89, 90], evidence for a single $\delta 6$ or $\delta 5$ desaturase protein metabolizing both $n-6$ and $n-3$ fatty acids is equivocal.

There are two desaturation/elongation steps to synthesize 20:4 $n-6$ from 18:2 $n-6$ and 20:5 $n-3$ from 18:3 $n-3$, and there are three desaturation/elongation steps to synthesize 22:6 $n-3$ from 18:3 $n-3$. Since substrate excess, disease processes and availability of cofactors all affect this pathway (see later), it has become clear that (1) this pathway does not always operate optimally and (2) one cannot necessarily equate adequate dietary availability of the substrate fatty acids with sufficient synthesis of the very-long-chain PUFAs. Thus, availability of the initial substrate fatty acids (18:2 $n-6$, 18:3 $n-3$) cannot be equated with production of the very-long-chain PUFAs, e.g. 20:4 $n-6$ or 22:6 $n-3$. In fact, *in vivo*, under optimal conditions, the conversion rate of 18:2 $n-6$ to 20:4 $n-6$ in rats is of the order of 30:1; e.g. it takes 30 mol of 18:2 $n-6$ to produce 1 mol of 20:4 $n-6$ [78]. Therefore, it is evident that other metabolic pathways, especially β-oxidation, compete with desaturation/elongation for the available 18:2 $n-6$ and 18:3 $n-3$ and consequently play a critical role in determining synthesis of very-long-chain PUFAs.

3.2 Differences Between Species, Organs and Cell Type

Although hepatic microsomal activity of the desaturases (especially $\delta 6$ and $\delta 5$) is demonstrable in virtually all species studied, it is clear that wide variation in relative desaturase activity exists between species. The most rapid (efficient) conversion of 18:2 $n-6$ to 20:4 $n-6$ is seen in the laboratory rat while the least efficient is seen in the cat family which is recognised as having negligible $\delta 6$ and $\delta 5$ desaturase activity [76]. Until the cat studies provided the sobering evidence of dramatic species differences, it was widely thought that desaturase activity was relatively efficient because most studies were conducted using the rat. It has become clear that, amongst mammals and based on a standard method of hepatic microsomal isolation and incubation, the rat and cat probably represent the two extremes of relative desaturation efficiency. In betwen lie most other species studied (mainly rodents) on a scale of the following order: rat > mouse > hamster > guinea pig > rabbit > cat. Due to their consumption of lean meat, cats have higher 20:4 $n-6$ in membrane lipids than rabbits, but rely on dietary sources for this 20:4 $n-6$; synthesis of 20:4 $n-6$ from 18:2 $n-6$ appears negligible in cats.

The question of where to place humans on the inter-species scale of 'relative $\delta 6$ desaturase activity' is intriguing. Although humans could be placed near the midpoint of the rodent ranking (see above), this would mainly be because they have higher 20:4 $n-6$ and 22:6 $n-3$ in phospholipids of serum red cells and liver than do rabbits and guinea pigs, but lower levels of these fatty acids than rats or mice. Since dietary intake of 18:2 $n-6$ itself, as well as of 20:4 $n-6$ and all $n-3$ fatty acids, influences desaturation/elongation of 18:2 $n-6$ to 20:4 $n-6$, placement of humans (who do not eat a standardized fat intake) in an inter-species ranking of relative desaturase capacity would be based on somewhat equivocal and arbitrary data.

In fact, experimental studies of $\delta 6$ and $\delta 5$ desaturase activity in isolated

Table 6. Desaturate activity in humans.

Site	Reference	Samples	Activity	Substrate
Liver	[92]	$n = 4$	< 1%	14-C 18:2 $n-6$
	[92]		5–60 pmol min^{-1} mg $protein^{-1}$	14-C 20:3 $n-6$
	[93]	$n = 5$	7–60 pmol min^{-1} mg $protein^{-1}$	14-C 18:2 $n-6$
	[93]		7–60 pmol min^{-1} mg $protein^{-1}$	14-C 18:3 $n-3$
Leucocytes	[91]	$n = 6$	1.8% (intact)	14-C 18:2 $n-6$
	[91]		2.7% (microsomes)	14-C 18:2 $n-6$
Platelets	[102]		ND[a]	14-C 20:3 $n-6$
Platelets	[72]	$n = 2$	ND[a]	14-C 20:3 $n-6$
Whole body	[94]	$n = 2$	< 0.0024% (serum)	2-H 18:2 $n-6$
Rat liver	[121]	$n = 6$	50 pmol min^{-1} mg $protein^{-1}$ (pregnant)	14-C 18:2 $n-6$
	[91]		11%	14-C 18:2 $n-6$
	[263]		32% (*in vivo*)	14-C 18:2 $n-6$
	[263]		92% (*in vivo*)	14-C 18:3 $n-3$
	[72]		50% (*in vivo*)	14-C 20:3 $n-6$

[a] Not detected.

human white blood cells [91], in human liver biopsies [92, 93], and in mass spectrometry studies of 13–C–18:2 $n-6$ metabolism in humans [94], have corroborated extensive dietary data [95] suggesting low to minimal $\delta 6$ desaturase activity in disease-free adult humans consuming 'Western' diets (Table 6). As a result, it has been commonly accepted that *adult* humans on a free-living Western-style diet have little or no ability to synthesize very-long-chain PUFAs from dietary 18:2 $n-6$ or 18:3 $n-3$. Nevertheless, human foetal liver has significant $\delta 6$ and $\delta 5$ desaturase activity [96] as have umbilical microvessels [97] and adult skin fibroblasts obtained from punch biopsies [98].

The implications of relatively low desaturase activity in humans may not be important in healthy individuals consuming a dietary source of 20:4 $n-6$ (meat) and 20:5 $n-3$ or 22:6 $n-3$ (fish). It is not clear whether strict vegetarians (consuming no meat, fish or dairy products) have desaturase activity as low as meat- and fish-eating humans or whether there are in fact some risks associated with this lifestyle. Alternatively, negligible dietary intake of 20 and 22 carbon PUFAs may actually increase desaturase activity, i.e. desaturase activity may be inversely proportional to dietary availability of long-chain PUFAs [99], or inversely proportional to the dietary level of competing substrates. In view of the practical clinical benefits of 20:5 $n-3$ and 22:6 $n-3$ and their presence in human milk, interest has been expressed in recommendations which would include a dietary source of the latter two fatty acids, especially in the neonatal period.

The liver is the most commonly studied organ in lipid/fatty acid metabolism studies and, in adult animals, it appears to have the most efficient desaturation/elongation. Other organs with active desaturation include testes, heart, kidney and gut. In comparison, adult rat brain has virtually no desaturase activity despite being rich in very-long-chain PUFAs [100]. Much of the very-long-chain PUFA in brain is acquired during early development when brain desaturation/elongation activity rivals that of liver [101]. In addition, in early life, the brain may obtain a significant proportion of its very-long-chain PUFAs pre-formed by the liver [78]. Human cell types with the appropriate cellular organelles (endoplasmic reticulum) but with little apparent desaturase activity include platelets [102] and peripheral blood leucocytes [91]. As in the cat, inactive desaturases in platelets and peripheral blood leucocytes would not be expected in view of the availability of long-chain PUFAs in serum lipids.

3.3 Substrate Competition

Substrate competition is one of the main features of the desaturation/elongation pathway and has a major influence on the synthesis of very-long-chain PUFAs [103, 104]. The 'non-essential' fatty acid substrates of $\delta 6$ desaturation, e.g. palmitoleic acid (16:1 $n-7$) and 18:1 $n-9$ inhibit desaturation of the 'essential' fatty acid substrates, i.e. 18:2 $n-6$ and 18:3 $n-3$. Since 16:1 $n-7$ and, especially, 18:1 $n-9$ are both present in the diet, and are readily synthesized by $\delta 9$ desaturation of 16:0 and 18:0, respectively, dietary excess of saturated and/or monounsaturated fatty acids impairs the synthesis of very-long-chain PUFAs [3]. The 18:2 $n-6$ and 18:3 $n-3$ are the preferred substrates for the $\delta 6$ desaturase so when present with equimolar amounts of 16:1 $n-7$ and 18:1 $n-9$, there is little interference with desaturation of 18:2 $n-6$ or 18:3 $n-3$ from the monoethylenic fatty acids [104]. In practice (especially in Western countries), the intake/synthesis of 16:0, 18:0, 18:1 $n-9$ and 18:2 $n-6$ is probably sufficient to impair 18:3 $n-3$ metabolism even in the presence of amounts of dietary 18:3 $n-3$ which would otherwise be sufficient to ensure adequate very-long-chain $n-3$ PUFA production [105].

Since the unsaturation of the substrate determines the substrate preference of the $\delta 6$ and $\delta 5$ desaturases, equimolar amounts of 18:3 $n-3$ are converted to longer chain $n-3$ PUFAs more readily than are 18:2 $n-6$ to longer chain $n-6$ PUFAs [6]. In humans, a dietary 18:2 $n-6$:18:3 $n-3$ ratio of about 5:1 appears to permit an appropriate balance of 18:2 $n-6$ and 18:3 $n-3$ metabolism for the synthesis of optimal amounts of both $n-3$ and $n-6$ very-long-chain PUFAs [78]. Rat studies suggest that prostacyclin production is optimal at a dietary ratio of 18:2 $n-6$:18:3 $n-3$ of 5:1 [106]. In fact, an intake of up to 2% of calories as dietary 18:3 $n-3$ is adequate to insure synthesis of long-chain $n-3$ PUFAs without significantly inhibiting production of long-chain $n-6$ PUFAs [107]. Thus, a dietary intake ratio of 5:1 would appear to be appropriate for production of adequate amounts of long-chain $n-6$ and $n-3$ PUFAs for both structural (membrane) and metabolic (PGs) purposes. In practice, the current consumption of 18:2$n-6$:18:3$n-3$ in Western countries is of the order of 10:1, resulting in a lower production of very-long-chain $n-3$ PUFAs. In the absence of adequate dietary intake of 20:5$n-3$ and 22:6$n-3$, the clinical implications of this distorted balance of very-long-chain PUFAs are becoming apparent [105, 108] (see Section 4).

3.4 Substrate and Product Availability: Effect on Desaturation

A high dietary intake of 18:2$n-6$ in relation to 18:3$n-3$ affects not only 18:3$n-3$ metabolism but may also affect its own metabolism. The effect of excess substrate causing substrate inhibition is accepted in both rodent studies and *in vitro* studies [87, 109] of desaturase activity and has recently been reported in human nutrition [110]. Thus, the lower levels of 20:4$n-6$ and other long-chain $n-6$ PUFAs which are reported in some diseases, e.g. atopic eczema [111] or hypertension [112], may be a function of the disease process inhibiting desaturation of 18:2$n-6$ but may also be partly a function of enhanced sensitivity to excess intake of 18:2$n-6$. This is easier to demonstrate in 18:3$n-3$ metabolism than in 18:2$n-6$ metabolism because, in humans, the usual dietary intake of 18:3$n-3$ is much lower than that of 18:2$n-6$ and the levels of $n-3$ fatty acids in human serum lipids are lower than those of $n-6$ fatty acids. Hence, increases of 3 to 5-fold in blood organ levels of 18:3$n-3$ can readily be seen after supplementation with 18:3$n-3$ whereas 18:2$n-6$ levels are unlikely to rise more than 10–25% after 18:2$n-6$ supplementation.

Virtually all studies of the metabolism of 18:3$n-3$ in healthy adult humans using dietary supplementation with 18:3$n-3$ (typically with linseed oil) have shown negligible changes in 20:5$n-3$ or 22:6$n-3$ in all serum lipid classes [112–116] (see below). In these studies, doses of linseed oil have ranged up to 60 ml day^{-1}, usually over a period of 2–4 weeks. Similar data indicating a relative inability to increase 20:4$n-6$ in serum phospholipids are available from studies of increased dietary intake of 18:2$n-6$ [95]. These data may indicate that desaturation/elongation is inefficient in humans or that daily consumption of 18:2$n-6$ or 18:3$n-3$ is sufficiently high to inhibit synthesis of long-chain $n-6$ or $n-3$ PUFAs.

Contrary to the classical concept of negligible capacity to synthesize 20:5$n-3$ from dietary 18:3$n-3$, recent human research suggests that synthesis of 20:5$n-3$, 22:5$n-3$ and 22:6$n-3$ *can* be increased in adult humans after consuming increased dietary 18:3$n-3$ and that long-chain $n-3$ PUFAs appear in serum under these conditions [116–118]. This effect has been seen in both normolipidemic and hyperlipidemic individuals and suggests that the form of the 18:3 $n-3$ supplement (linseed oil versus whole ground flaxseed), the feeding period or the amount of the supplement may affect that degree to which the very-long-chain $n-3$ PUFAs are synthesized and appear in serum lipids.

3.5 Cofactors

To confound further the delicate mechanism governing synthesis of very-long-chain PUFAs, it has become clear that several nutrients are essential cofactors in long-chain fatty acid metabolism. The only nutrient known to be an integral structural component of the desaturase protein is non-heme iron [89, 90]. Nevertheless, studies of experimental nutrient deficiencies, genetic abnormalities causing deficient absorption of specific nutrients, and enzyme studies have shown that deficient or excess intakes of metals (zinc, copper and magnesium) [119–122] and vitamins (pyridoxine and α-tocopherol) [123, 124] modulate synthesis of very-long-chain PUFAs in characteristic ways. Thus, deficiencies of zinc, magnesium and pyridoxine impair synthesis of 20:4 $n-6$ and 22:6 $n-3$ while deficiency of copper impairs 18:1 $n-9$ synthesis but increases levels of very-long-chain $n-6$ and $n-3$ PUFAs in both humans and experimental animals. The mechanism(s) through which these cofactors affect synthesis of very-long-chain PUFAs act are poorly understood but appears to involve modulation of electron transport from NADH to the terminal desaturase protein [125, 126]. It is becoming evident that the interaction of these cofactors in PUFA metabolism is of clinical and possibly diagnostic significance [120, 127, 128] and should not be overlooked in understanding abnormalities of PUFA metabolism in human diseases.

3.6 Disease

In humans, dietary deficiency of 18:2 $n-6$ and/or 18:3 $n-3$ causes a range of symptoms that affect especially the skin but include the immune response, the cardiovascular system, the eye and the peripheral nervous system [6, 129, 130]. In adult humans, symptoms of dietary PUFA deficiency generally only occur under one of three conditions: (1) when dietary PUFA availability is essentially zero, e.g. patients on total parenteral nutrition not containing short- or long-chain PUFA [131, 132]; (2) significant gut and digestive dysfunction, e.g. in cystic fibrosis [133, 134]; or (3) in the presence of an underlying disease affecting nutritional status in general, e.g. anorexia or alcoholism [135–137]. Hence the adult human is relatively resistant to dietary PUFA deficiency and this is probably achieved largely through the storage of 18:2 $n-6$ and 18:3 $n-3$ in adipose tissue.

Table 7. Polyunsaturated fatty acids in serum phospholipids in acrodermatitis enteropathica before and after zinc supplementation.

	18:2*n*−6			20:4*n*−6			22:6*n*−3		
Source	before	after	ref.[a]	before	after	ref.[a]	before	after	ref.[a]
Mack et al. [128]	15.3	23.7	(20.7)	4.8	13.1	(7.5)	0.4	1.0	(1.7)
Cunnane and Krieger [140]	27.4	37.3	(22.8)	5.9	12.2	(11.4)	1.0	2.9	(3.5)
Koletzko et al. [307]	26.7	22.4	(23.0)	9.3	9.2	(8.0)	4.0	7.2	(1.6)
Cash and Berger [308]	22.0	26.0[b]		2.0	4.5		NR	NR	

[a]Normal. [b]Serum total lipids. NR: not reported.

Since PUFA metabolism is dependent on nutritional cofactors which may themselves become limiting through disease or nutritional deficiency, it is now clear that an adequate dietary supply of 18:2*n*−6 and 18:3*n*−3 does not guarantee adequate synthesis of the longer chain PUFAs. A case in point is the rare, autosomal recessive disease—acrodermatitis enteropathica—which is characterized by a defect in zinc absorption [138, 139]. The symptoms only occur after weaning from breast milk (transition from high to low intake of long-chain PUFAs) and include severe skin lesions, gut pathology and growth retardation. The disease usually causes low serum zinc levels but this is not always observed. Regardless of serum zinc status, supplemental zinc is corrective. Also regardless of serum zinc status, long-chain PUFAs (especially 20:4*n*−6, 22:6*n*−3) in serum phospholipids are invariably 50–60% below normal and return to normal within 1–2 weeks of zinc supplementation (Table 7). The 18:2*n*−6 in serum phospholipids is usually above normal prior to zinc supplementation and often remains elevated after zinc supplementation [128, 140]. Despite the lower levels of 20:4*n*−6, eicosatrienoic acid (20:3*n*−9), which denotes dietary 18:2*n*−6 deficiency, remains undetected or at low levels (less than 0.2% of fatty acids in serum phospholipids). Hence, when desaturation is impaired, 20:3*n*−9 will not be synthesized and its levels do not adequately represent the deficiency of 20:4*n*−6. Thus, elevated levels of 20:3*n*−9 require both dietary deficiency of 18:2*n*−6 or 18:3*n*−3 *and* a functioning desaturase/elongase system.

In acrodermatitis enteropathica, the rapid response of serum long-chain PUFAs occurs in the absence of any change in diet except for zinc supplementation. This example serves to illustrate how a disease affecting absorption of a specific nutritional cofactor required in PUFA metabolism can impair the synthesis of 20:4*n*−6 and 22:6*n*−3 even in the presence of adequate dietary availability of 18:2*n*−6 and 18:3*n*−3. No other diseases specifically affecting the absorption or metabolism of nutrients required as cofactors in PUFA metabolism are as well characterized as acrodermatitis enteropathica. However, if specific deficiencies of pyridoxine, magnesium or vitamin E were to occur independently of a generalized state of malnutrition, experimental data suggests that a similar impairment of synthesis of 20:4 $n-6$ and 22:6 $n-3$ would be observed even in the presence of adequate dietary intake of 18:2 $n-6$ and 18:3 $n-3$. Conversely, in copper deficiency, the inhibitory effect is focused on desaturation of 18:1 $n-9$ and *increased* levels of long-chain PUFAs are observed in both experimental and clinical copper deficiency, e.g. Menke's disease [125, 141, 142].

Some diseases may be clinically responsive to supplementation with encapsulated oils containing high amounts of specific PUFAs even in the presence of ostensibly adequate intakes of 18:2 $n-6$ and 18:3 $n-3$. Diseases in the cardiovascular and immune/autoimmune spectrum have as yet no well-defined nutritional cofactor defect or other abnormality specifically related to PUFA metabolism so the benefit may be pseudo-pharmacological, e.g. by flooding enzymes in the PUFA or PG pathways. This effect of PUFA supplementation is discussed in Sections 4 and 5.

4. FUNCTION

4.1 Membranes

The location of very-long-chain PUFAs in membranes within the cell provides the strongest clue as to their function, that of modulation of membrane-related activity, e.g. ion gradients, enzyme activity, receptor configuration and activity [143–146]. Membranes with high activity tend to have the highest amount of very-long-chain PUFAs, e.g. photoreceptor and neuronal membranes which are especially enriched in 22:6 $n-3$ [9, 147, 148]. Although membrane attributes such as fluidity have been widely investigated in relation to the composition of very-long-chain PUFAs, it has become clear that the issue is very complex and that the presence of increased PUFA, for example, is not an overriding determinant of a change in fluidity. Thus, increasing PUFA concentration in a membrane does not necessarily mean that fluidity will increase because, at least in some cases, there are compensatory mechanisms which operate to adapt the membrane to increased availability of PUFAs, e.g. by increasing cholesterol content [149]. This would, in fact, be expected because the functions that membranes perform require a stable membrane environment and should therefore be resistant to wide variation in composition through dietary manipulation or these functions would be expected to fail. Thus, despite concerted effort and a wealth of experimental data in which enzyme modulation of $\pm$25–100% has been demonstrated [143, 147], it has been difficult to ascribe the presence of very-long-chain PUFAs in membranes to specific biologically significant functions. This has arisen in part through a dearth of information concerning the actual orientation of very-long-chain PUFAs in membranes and hence their physical relation to cholesterol and intrinsic proteins and enzymes in membranes.

For instance, 22:6 $n-3$ has received considerable recent attention as sophisticated new approaches are used to understand its function in photoreceptor membranes. Computer modelling and non-invasive studies of model membranes suggest that the high unsaturation and length of this fatty acid do not necessarily give it the anticipated increase in physical space which would in turn presumably increase the fluidity of membranes in which it is present [10, 150]. Rather, it appears that in the phosphatidylcholine of retinal photoreceptor membranes in which 22:6 $n-3$ may account for as much as 60% of the total fatty acids, and in which 22:6 $n-3$ is typically found paired with 16:0, the fluidity and melting point of this phosphatidylcholine species differs little from that of phosphatidylcholine in which 16:0 is paired with 16:1 $n-7$. Furthermore, 22:6 $n-3$ paired with 16:0 in phosphatidylcholine may not occupy much more space than 16:0 paired with 16:1 $n-7$ by virtue of the proposed helical configuration of 22:6 $n-3$ predicted by computer modelling [10, 150].

Hence, the very nature of membranes, e.g. a design and structure providing a constant cellular and intracellular environment in the face of natural or imposed variation in membrane substrate availability, has thwarted our in-depth understanding of the functions within membranes which are directly or even indirectly attributable to very-long-chain PUFAs.

4.2 Prostaglandins/Leukotrienes

Although 22:5 $n-6$, 22:5 $n-3$ and 22:6 $n-3$ appear at present to be restricted to only having structural functions, some 20 and 22 carbon PUFAs have mixed functions. Thus, 20:4 $n-6$ and 22:4 $n-6$ have both structural and metabolic functions as precursors/substrates of other compounds such as PGs and leukotrienes (LTs) [7, 151–153]. The 20:3 $n-6$ and 20:5 $n-3$ appear unlikely to have significant structural roles in membranes but both are important precursors for PGs and other oxygenated derivatives [50, 154]. Of the two essential dietary PUFAs (18:2 $n-6$ and 18:3 $n-3$), only 18:2 $n-6$ appears to have a significant structural role but both are recently recognized as being readily oxidized through β-oxidation [117] and may have biologically important effects in that metabolic route [135] as well as through their more familiar roles as precursors for longer chain PUFAs.

The ubiquitous 20:4 $n-6$ is the substrate for the '20:4 $n-6$ cascade', which has grown in the past two decades from a pathway involving endoperoxides as intermediates in the synthesis of half a dozen 'classical' PGs (e.g. E_2, $F_2\alpha$, D_2) to a series of pathways each leading to uniquely oxygenated products originating via lipoxygenases which are specific for each of the four double bonds in 20:4 $n-6$ (Figure 2). As a result, in addition to the classical PGs,

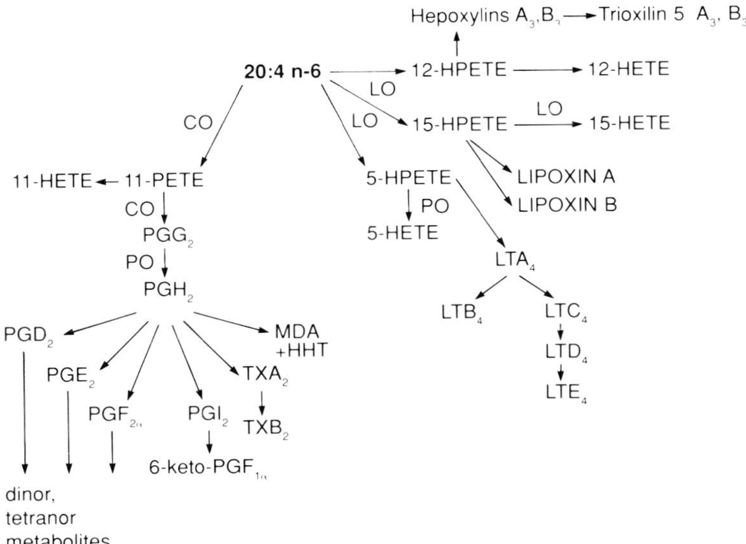

Figure 2. The arachidonic acid cascade. After appropriate experimental or pathological stimulation, free arachidonic acid (20:4 $n-6$) is released from membrane phospholipids or other cellular lipids and metabolized via the cyclooxygenase (CO) or one of several lipoxygenases (LO) pathways to a variety of oxygenated derivatives which differ according to the cell type in question. In each pathway, a short-lived peroxy intermediate is the first product, e.g. 11-PETE (11-peroxyeicosatetraenoic acid) or 5-HPETE, 12-HPETE or 15-HPETE (hydroperoxyeicosatetraenoic acid). 11-PETE has two further products, 11-HETE (11-hydroxyeicosatetraenoic acid) or the endoperoxide (PGG_2) which is converted to a second endoperoxide (PGH_2) via peroxidase (PO). Prostanoids derived from PGH_2 include the classical 2-series prostaglandins (PGD_2, PGE_2, $PGF_{2\alpha}$), as well as prostacyclin (PGI_2) and thromboxane (TXA_2). Dinor- and tetranor-PG derivatives are the products of the classical PGs excreted in the urine. Non-prostanoid products include malondialdehyde (MDA) and hydroxyheptadecatrienoic acid (HHT). Leukotrienes (LT) and the more stable 5-HETE are derived from 5-HPETE. Hepoxylins, trioxylins and lipoxins are derived from 12-HPETE and 15-HPETE, respectively, via correspondingly numbered lipoxygenases.

pathways for the synthesis of prostacyclin, thromboxanes, LTs, hepoxilins and lipoxins have themselves each literally developed a 'cascade' resulting in at least 30 chemically defined products [7, 153]. Some are produced through biological or pathological stimuli and have significant biological roles, e.g. prostacyclin, thromboxane and some LTs.

Like 20:4 $n-6$, 20:3 $n-6$, 20:5 $n-3$ and 22:4 $n-6$ are also substrates for the cyclooxygenase resulting in production of the 1-series, 3-series and

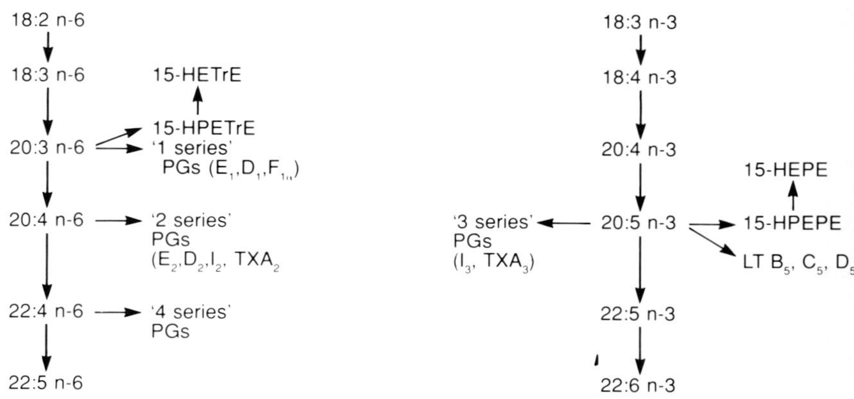

Figure 3. Oxygenated products of 20 and 22 carbon PUFAs. In addition to the extensively studied arachidonic acid cascade (see Figure 2) dihomo-γ-linolenic acid (20:3 $n-6$), adrenic acid (22:4 $n-6$) and eicosapentaenoic acid (20:5 $n-3$) are known to produce oxygenated products. Products from 20:3 $n-6$ include the classical 1-series PGs, 15-hydroperoxyeicosatrienoic acid (15-HPETrE) and 15-hydroxyeicosatrienoic acid (15-HETrE). Products derived from 20:5 $n-3$ include the 3-series PGs, prostacyclin (PGI_3), thromboxane (TXA_3), leukotrienes (LTs) derived via the 5-lipoxygenase and 15-hydroperoxyeicosapentaenoic acid (15-HEPE) and 15-hydroxyeicosapentaenoic acid (15-HEPE). Since the cyclooxygenase and lipoxygenase enzymes are thought to be used in common by both the $n-6$ and $n-3$ fatty acids, synthesis of PGs, LTs or related products from $n-6$ fatty acids is usually competitively inhibited by increased availability of $n-3$ fatty acids and *vice versa*. Much of the clinical/nutritional basis for dietary intervention with $n-3$ or $n-6$ fatty acids is related to manipulation of synthesis of the 1-, 2- and/or 3-series PGs and related products.

4-series PGs (Figure 2). PGE_1 derived from 20:3 $n-6$ was the first PG to be isolated and chemically characterized and was the first to be commercialized for a specific clinical use (closing the ductus arteriosus of newborn infants and, more recently, to prevent gastric mucosal irritation/bleeding caused by chronic use of non-steroidal anti-inflammatory drugs). Like prostacyclin, PGE_1 has vasodilatory and anti-aggregatory properties, and is also an important cellular immune stimulator [4]. Tissue 20:3 $n-6$ can be raised by various dietary sources of 18:3 $n-6$ which have differential effects on release of 1-series and 2-series PGs [155]. Also, 20:3 $n-6$ can be metabolized by the 15-lipoxygenase to 15-hydroxyeicosatrienoic acid which is a potent inhibitor of 5-lipoxygenation of 20:4 $n-6$, an intermediate in the synthesis of LTs from 20:4 $n-6$ [156].

Release of 3-series PGs (TXA_3 , PGI_3) and 5-lipoxygenase products is increased when dietary intake of 20:5 $n-3$ is raised [154, 157, 158]. The biological significance of these PGs is not fully understood. What is probably a biologically more important function of 20:5 $n-3$ than producing 3-series PGs is the competitive inhibition of the 20:4 $n-6$ 'cascade' [83]. Thus, along with aspirin and other non-steroidal anti-inflammatory drugs, 20:5 $n-3$ is considered to be one of the most effective inhibitors of 20:4 $n-6$ metabolism. Indeed, like all $n-3$ PUFA, increased dietary availability or synthesis of 20:5 $n-3$ inhibits desaturation of $n-6$ fatty acids. Hence, 20:5 $n-3$ inhibits both synthesis of 20:4 $n-6$ and metabolism of 20:4 $n-6$ through the 20:4 $n-6$ 'cascade'. Furthermore, in combination with a dietary source of 18:3 $n-6$, 20:5 $n-3$ may favourably enhance synthesis of 1-series PGs while contributing to the inhibition of 20:4 $n-6$ metabolism to 2-series PGs [159].

4.3 Oxidation/Energy Metabolism

Studies in the rat [160] and in humans [117, 161–163] indicate that 18:2 $n-6$ and 18:3 $n-3$ are more readily oxidized to carbon dioxide than are long-chain saturated fatty acids. In fact, *in vivo*, only 18:1 $n-9$ appears to be more easily oxidized than either 18:2 $n-6$ or 18:3 $n-3$. On the face of it, this would appear to be a wasteful use of fatty acids which mammals cannot synthesize, and whose longer chain products are structurally and metabolically essential. All of 16:0, 16:1 $n-7$, 18:0 and 18:1 $n-9$ can be readily synthesized and, being either fully saturated or only monounsaturated, should theoretically be easier to oxidize than PUFAs such as 18:2 $n-6$.

While an explanation for either the biological importance of oxidizing 18:2 $n-6$ or 18:3 $n-3$ or the cellular mechanism/organelles (peroxisomes) through which this is achieved is not as yet apparent, two important points in support of the ready oxidation of 18:2 $n-6$ and 18:3 $n-3$ are evident. (1) Consumption of a Western-type diet (high fat, high linoleic acid) provides at least 10% of calories as 18:2 $n-6$ when only 1–3% is required for structural/metabolic roles [164]. In individuals maintaining body weight, this leaves 70–90% of dietary 18:2 $n-6$ to be stored or oxidized. If body fat content is maintained constant and its fatty acid composition is stable (15–20% 18:2 $n-6$) [165], most dietary 18:2 $n-6$ consumed by adults must be oxidized. (2) Regardless of dietary intake, the structural/metabolic setpoint for 18:3 $n-3$ is lower than for 18:2 $n-6$ leading to lower levels of 18:3 $n-3$ in all organs of most species (18:3 $n-3$ is rarely above 1% except in adipose tissue whereas 18:2 $n-6$ is rarely below 8–10% of total fatty acids in any given organ except brain). One plausible explanation for the generally low organ levels of 18:3 $n-3$ is that this is achieved through oxidation. Even when 18:3 $n-3$ is supplemented at 10–20 times the average Western intake (less than 1 g day^{-1}), its amounts in serum and red cells lipids plateau within weeks at less than 3–4% of total fatty acids. If 18:3 $n-3$ does not accumulate beyond this point, the reason must involve substantial oxidation.

It is presently unknown whether oxidation of 18:2$n-6$ and 18:3$n-3$ is a route of their metabolism 'essential' to the organism, e.g. because it is a more efficient source of carbon for other metabolic processes than saturated fatty acids (for instance), or whether this route is important because, at the level consumed, 18:2$n-6$ (and to a lesser extent 18:3$n-3$) are xenobiotics. In addition to 18:1$n-9$, both 18:2$n-6$ and 18:3$n-3$ are preferentially released into the blood when there is a demand for free fatty acids, e.g. during fasting [117], suggesting that the role of 18:2$n-6$ and 18:3$n-3$ as oxidizable substrates may be physiologically important.

5. NUTRITIONAL/CLINICAL APPLICATIONS

The practical application of PUFAs is primarily in the field of clinical and nutritional therapeutics. In this context, the PUFA most frequently studied in concentrated or semi-purified form are 18:2$n-6$, 18:3$n-6$, and the very-long-chain PUFA, 20:5$n-3$ and 22:6$n-3$. Interesting and valuable studies have also been done with pure 20:3$n-6$ [72, 166] and 18:3$n-3$ [105, 129]. In situations requiring enteral (naso-gastric) or total parenteral nutrition, the inclusion of 18:2$n-6$ at 1–3% of total calories and 18:3$n-3$ at 0.5% of calories [108, 129] is well recognized as essential for the prevention of symptoms of 18:2$n-6$ and/or 18:3$n-3$ deficiency, e.g. hair loss and dry, scaly dermatitis [131, 132]. Interestingly, recent human studies of 18:3$n-3$ supplementation in the presence of adequate 18:2$n-6$ but low 18:3$n-3$ intake suggest that 18:3$n-3$ deficiency may in some way divert 18:2$n-6$ metabolism so that 18:2$n-6$ is utilized differently resulting in symptoms of 18:2$n-6$ deficiency, e.g. dermatitis; when 18:3$n-3$ is then supplemented, symptoms of 18:2$n-6$ deficiency are then reduced and conversion of 18:2$n-6$ to metabolites such as 20:4$n-6$ and 2-series PGs may actually *increase* transiently [105, 129].

Symptoms of 18:2$n-6$ deficiency may occur despite its presence in adipose tissue of infants, children and adults with clinical symptoms of deficiency of either 18:2$n-6$ or 18:3$n-3$ [167]. Thus, the adipose tissue pools of 18:2$n-6$ and 18:3$n-3$ appear to be unavailable or inadequate to meet the systemic needs during dietary deficiency of these fatty acids. The 20 and 22 carbon PUFAs derived from 18:2$n-6$ and 18:3$n-3$ are not known to have been included in enteral or parenteral feeding regimens to date so it is presently unclear whether they may be more effective in preventing the symptoms of $n-3$ or $n-6$ fatty acid deficiency than the parent fatty acids.

5.1 Neonatal Period

Human breast milk from mothers with term infants is a uniquely complete source of shorter and long-chain PUFAs from 18:2$n-6$ and 18:3$n-3$

through to 22:5 $n-6$ and 22:6 $n-3$ [168]. Although preterm milk has been reported to have higher total fat than term milk and to have all the longer chain $n-6$ and $n-3$ PUFAs in equal proportions to term milk [169], prematurity is associated with low 20:5 $n-3$ and 22:6 $n-3$ in serum [170]. Infant milk formulae containing 20:5 $n-3$ and 22:6 $n-3$ have recently been shown to increase the proportion of these two fatty acids in the serum and red blood cells of neonates not given a dietary source of these fatty acids, e.g. not breast-fed [171, 172]. In infants, 20:5 $n-3$ and 22:6 $n-3$ are lower in serum and red blood cells in those infants not receiving a dietary source of very-long-chain PUFAs, e.g. in infants given conventional formula containing PUFAs as only 18:2 $n-6$ and 18:3 $n-3$ [173, 174] or in infants born to vegans [175]. However, the significance of lower levels of 20:5 $n-3$ and 22:6 $n-3$ in neonatal blood is still unclear. While there is no question of the essentiality of 22:6 $n-3$ and 20:4 $n-6$ for normal brain and behavioural development [11, 73, 78, 100, 148, 176–178], the level in blood at which these fatty acids can be said to be limiting for normal growth, development and behaviour in humans, e.g. the margin of safety, is still unknown.

Also unknown is the extent to which 18:2 $n-6$ and 18:3 $n-3$ in human milk are used for structural versus metabolic purposes in the newborn, i.e. is 18:2 $n-6$ in milk mainly incorporated unchanged into newly synthesized organ lipids, is it metabolized to 20:4 $n-6$ and other long-chain $n-6$ PUFAs, or is it oxidized? Oxidation of 18:2 $n-6$ and 18:3 $n-3$ in neonates may not be important or it may be more important than it is in adults, e.g. for thermogenesis. These questions need to be addressed before the essentiality of 20 and 22 carbon PUFAs in human milk can be fully evaluated.

Since breast fed infants do obtain very-long-chain PUFAs in breast milk (Table 8) and, throughout the world, human breast milk is relatively uniform in its content of 20 and 22 carbon PUFAs [179], the intake and serum levels of PUFAs in breast fed infants should be considered the normal and desirable level. If, due to practical difficulties of manufacturing an infant formula containing very-long-chain PUFAs, 18:2 $n-6$ and 18:3 $n-3$ are to be the only PUFAs given in infant formulae, their concentrations should reflect the requirement for these fatty acids and the risks associated with high intakes of 18:2 $n-6$, i.e. inhibition of $n-3$ fatty acid metabolism and substrate inhibition of $n-6$ fatty acid metabolism.

5.2 Dietary Management of Disease

In addition to meeting the essential fatty acid requirements of bottle-fed infants and enterally or parenterally fed patients, PUFAs have been studied for their possible practical benefits in treating or preventing several classes of disorders seen in clinical medicine. If the effective dose is above the normal requirement this may reflect a tendency for long-term deficiency when reverting

Table 8. Polyunsaturated fatty acid composition of human milk.

	Australia[a]					U.K.[d]	
	C[e]	M[f]	Thailand[b]	Hungary[b]	U.S.A.[c]	vegans	omnivores
18:2 $n-6$	7.8	10.8	6.6	9.3	16.0	31.7	6.9
20:3 $n-6$	0.5	0.3	1.5[g]	1.4[g]	0.4	0.3	0.2
20:4 $n-6$	0.7	0.4	NR	NR	0.6	0.7	0.5
22:4 $n-6$	0.4	0.1	NR	NR	0.2	0.1	0.1
22:5 $n-6$	NR	NR	NR	NR	0.2	0.1	0.2
18:3 $n-3$	0.4	0.6	0.4	0.4	0.6	1.5	0.8
20:5 $n-3$	0.4	0.2	0.7[h]	0.3[h]	< 0.1	0.2	< 0.1
22:5 $n-3$	0.4	0.2	NR	NR	0.1	0.3	0.5
22:6 $n-3$	0.6	0.3	NR	NR	0.2	0.2	0.6

[a]Gibson and Kneebone [168]. [b]Crawford [78]. [c]Carlson et al. [170]. [d]Sanders et al. [309].
[e]Colostrum. [f]Mature milk. [g]Total $n-6$ metabolites. [h]Total $n-3$ metabolites.
NR: not reported.

to normal PUFA intake, e.g. high saturated fat intake, or the extent to which a disease has progressed and has distorted the utilization of or requirement for those PUFA present in the diet, e.g. hyperlipidaemia which traps relatively high amounts of 18:2$n-6$ esterified in serum lipids [180].

Therapeutic effects of PUFA are thought to occur through two major and possibly related mechanisms: (1) correction of abnormal function or structure of membranes and (2) modulation of PG and/or LT synthesis. Little progress has been made in identifying specific structural or functional aspects of membranes that are related to PUFA metabolism. However, as a prelude to establishing such a relationship, considerable effort has been expended in determining the nature of abnormal serum fatty acid profiles in some diseases and whether they are correctable with PUFA supplementation.

The relation of disease processes to PG synthesis, metabolism and excretion has also received much attention. It is now becoming clear that altering the intake of specific PUFAs and, especially the ratio of $n-3$ to $n-6$ PUFAs, will modulate the synthesis of the PGs derived from 20:3$n-6$, 20:4$n-6$ and 20:5$n-3$. For instance, the production of 2-series PGs and related active metabolites of 20:4$n-6$ can be inhibited by increasing the membrane content of 20:5$n-3$. This is achieved most effectively by increasing dietary intake of 20:5$n-3$ but may also be achieved by increasing dietary 18:3$n-3$ [181, 182]. This will also result in an increase in the ratio of 3-series/2 series PGs [157]. In the presence of a dietary source of 18:3$n-6$ or 20:3$n-6$, synthesis of 1-series PGs (primarily PEG_1) is increased and the ratio of PGE_1/PGE_2 can be significantly increased [166]. Further dietary manipulation of the ratio of 1-series/2-series PGs can be achieved by increasing both dietary 18:3$n-6$ and

20:5 $n-3$ which tends to decrease production of 2-series PGs while increasing 20:3 $n-6$ and 1-series PGs due to decreased synthesis of 20:4 $n-6$ [159].

The significance of these manipulations lies in the relation of altered PG synthesis/excretion to disease. PGE_1 is vasodilatory, antiaggregatory and anti-inflamatory while, with the exception of PGI_2 (prostacyclin), 2-series PGs have the opposite effects. Hence, most therapeutic strategies involving attempts to reduce the risk/incidence of vasoconstriction, platelet aggregation or inflammation involve a reduction of 2-series PGs whether through diet or drugs. The main undesirable cardiovascular side effect of reducing synthesis of 2-series PGs with dietary intervention is that vascular endothelial synthesis of prostacyclin (which is vasodilatory and anti-aggregatory) is usually inhibited along with inhibition of thromboxane and LT synthesis [4].

5.2.1 Cardiovascular Disease

Coronary vascular disease ultimately kills about 35% of the population in Western/developed countries. Its origins are diverse and involve interactions of lifestyle, diet and genetics. The dietary aspects involve long-chain PUFA synthesis and metabolism to a considerable extent. Two of the major risk factors are cholesterol and saturated fatty acids (12:0, 14:0, 16:0); these impair synthesis of long-chain PUFAs [6, 183], and both contribute to hyperlipidaemia and platelet hyperaggregability.

Early studies of the effects of $n-3$ fatty acids on platelet function showed that, in comparison with Japanese farmers and Danes, Japanese fishermen and Inuit consuming *traditional* diets had lower indices of coronary vascular disease [184, 185]. These studies initiated the interest in marine fish/fish oils in relation to cardiovascular health, especially in relation to platelet function and lipemia. While reference to fish/fish oils in this section will be exclusively to those of *marine* origin, it should be noted that *freshwater* fish (with low $n-3$ fatty acid content) also appear to have an important therapeutic value in relation to cardiovascular health [186].

Tables 1–3 show that vegetable oils contain lower levels of saturated fatty acids than do fish oils, or fish fats themselves. Some concentrates of fish oils contain very low levels of saturated fatty acids, but this is an inadvertent result of concentration processes and not a deliberate policy. Despite the presence of saturated fatty acids in fish oils and fish lipids there is good evidence for beneficial effects from the $n-3$ fatty acids in the same oil or lipid, probably by direct action on the epithelial cells of the vascular system. The heart itself may or may not benefit from reduced arrhythmia, but in a recent rat study with vegetable oils [187] the authors concluded from functional and morphological studies:

> In summary, the functional properties of the hearts of rats fed various high-fat diets were not significantly altered, despite recognizable morphological changes. The results of this

study indicate that diets low in saturated fats may have potential long-term adverse effects on the myocardium. Although we recognize that there are considerable species differences between rats and humans, the findings raise questions regarding the present recommendations for increased consumption of unsaturated fats and reduced consumption of saturated fats as a means of reducing the incidence of cardiovascular heart disease. Although avoidance of saturated fats may reduce cholesterol levels in humans, the benefits of such dietary changes have to be established more definitively with respect to myocardial lesions when recommendations are made for entire populations.

Intervention studies based on serum cholesterol reduction by dietary means have recently been severely criticized [188]. However, a weakness in the epidemiological approach to the cardiovascular benefits of $\omega 3$ fatty acids has been the difficulty in precisely defining, in clinically measurable terms, the factors responsible for the known benefits of a steady moderate fish consumption [189, 190]. The effect is clearcut from 0 to 30 g day^{-1} of fish but non-linear in the range of 30–150 g day^{-1} [47].

The literature on $n-3$ fatty acids is now so vast that this section will have to omit such promising research areas as reduced arrhythmias in rats [191]. As is typical of many fish oil studies the rat may not be a good model for man, where no similar beneficial effect was observed when cod liver oil was administered [192].

Human intervention studies with either fish or fish oils do not usually last long enough to modify arterial disease death rates, although they may cast light on blood chemistry changes [189]. However, a recent study conducted with men recovering from myocardial infarctions [193] showed that subjects advised to eat fatty fish (or to take fish oil capsules) had a 29% reduction in 2-year all-cause mortality compared with those not so advised. Moreoever this trend began to appear after approximately 100 days into the trial. The inevitable series of letters of comment [194] elicited a neutral reply but are themselves indicative of the wide interest in the topic.

Certain very recent papers and reports [195–200] strongly support the concept that one of the risk reduction factors associated with $n-3$ fatty acids is the production of prostaglandin PGI_3 directly in the endothelial cells of the vascular wall. Thus in a decade the wheel has come full circle and one of the original hypotheses [201] on the reason for the health benefits of $n-3$ fatty acids, production of PGI_3, is restored to prominence. This effect is difficult to test in man but cultured porcine endothelial cells respond differently even to $18{:}2n-6$ and $18{:}3n-3$ [202].

The pig could be a better model for studying atherogenic processes than the rat or rabbit, although in the latter animal fish oil decreased atherosclerosis [203]. The spectacular results achieved by Weiner *et al.* [204], with pigs fed an atherogenic diet with and without cod liver oil, were partially explained by additional pig work [205, 206]. A more recent sudy with normolipidemic pigs suggests that mackerel oil in the diet may reduce or inhibit platelet aggregation and hence provides cellular proliferation at a flow restriction site [207]. Fish oil was shown to retard the progression of, and cause regression of, coronary atherosclerosis in pigs [208]. An extensive swine study by Kim *et al.* [209] showed reduction in atherosclerotic lesion development compared to a butter-based diet, by 71–94% on all sites. One of the observations in this study that is sure to generate further controversy was that overall LDL cholesterol levels increased and some HDL levels decreased, concurrent with marked retardation of atherogenesis. Foxall and Shwaery [210] have also viewed fish oils negatively as a result of cholesterol profiles in pigs.

It is typical of such novel treatment with a 'natural' therapeutic agent, in lieu of a synthetic drug, that restenosis following coronary angioplasty has also become a controversial area, with favourable [211] and unfavourable [212] results. As this is a fairly common operation statistics should be readily available, but the habitual use of a variety of drugs post-operatively causes unusual complications in comparing such studies. Follow up letters [213, 214] did not clarify this situation.

Hyperlipidaemia. Isocaloric substituion of dietary saturated fatty acids by 18:2 $n-6$ (corn or sunflower oil) is hypocholesterolemic in hyperlipidemic and normolipidemic adult humans [162, 215]. Some evidence indicates that the hypocholesterolemic effect of 18:3 $n-6$ and 20:4 $n-6$ is greater than that of 18:2 $n-6$ [216, 217] suggesting that to be hypocholesterolemic, 18:2 $n-6$ must be converted to 18:3 $n-6$ and presumably to longer chain $n-6$ PUFAs, particularly 20:4 $n-6$, but possibly 22:4 $n-6$ or 22:5 $n-6$ as well. However, 20:3 $n-6$ (an obligatory intermediate between 18:3 $n-6$ and 20:4 $n-6$) has been reported to not affect serum cholesterol in normolipidemics or hyperlipidemic insulin-dependent diabetics [218]. Furthermore, serum levels of 20:4 $n-6$ are almost impossible to alter by dietary supplementation with either 18:2 $n-6$ alone or with 18:2 $n-6$ and 18:3 $n-6$ combined as evening primrose oil [95]. Whether dietary 18:2 $n-6$ needs to be metabolized to longer chain $n-6$ PUFAs to exert a hypocholesterolemic effect therefore remains to be established.

Linseed oil (50–60%, 18:3 $n-3$; Table 1) is hypolipidemic in type IIa and IIb hyperlipidaemia [113, 219]. As in the case of dietary supplementation with 18:2 $n-6$, this effect appears to occur in the absence of changes in longer chain $n-3$ PUFAs, even at a dose of 60 ml day^{-1} of linseed oil for 2 weeks providing a minimum 25 g day^{-1} 18:3 $n-3$. Since 18:2 $n-6$ is oxidized more slowly in hyperlipidemics [162], the relatively high rate of oxidation of 18:2 $n-6$ and 18:3 $n-3$ (compared to saturated fatty acids and longer chain PUFAs) may be an important factor in their hypocholesterolemic effects. This would account for the lack of change in serum lipid levels of the longer chain PUFAs derived from both these fatty acids and the relative ineffectiveness of 20:3 $n-6$ as a hypocholesterolemic agent. It might also account for the approximately equipotent hypocholesterolemic effects of dietary 18:2 $n-6$ and 18:1 $n-9$ since both have similar rates of whole body oxidation [160].

As noted earlier [54], one fish oil product is approved in the United Kingdom for treatment of hypertriglyceridemia under the auspices of the National Health Service. The long-term fish oil therapy results of Schectman *et al.* [220], discussed under diabetes, are therefore difficult to evaluate as an 11% reduction in triglyceride would basically seem to be important enough to warrant government approval and expenditures. As noted by Havel [221], triglycerides are not universally accepted as a risk factor in atherosclerosis. Harris [222], in his review of human studies and fish oils, goes into more detail to link triglycerides to VLDL lipoproteins. Type V hyperlipidemic patients are subject to a high risk of acute pancreatitis and their immediate benefit from the effect of fish oils, even at high doses, should outweigh any risks from the fish oil. Nephrotic syndrome patients may also benefit from reduction in triglycerides [223].

That fish oil $\omega 3$ fatty acids act in the liver to suppress intrahepatic assembly of VLDL triglycerides and cholesterol esters is well accepted [224, 225], and this has been linked to an acyl-coenzymeA:1,2-diacylglycerol acyltransferase [226]. Fish oils are not known for their reduction in cholesterol levels in normals, any such effect being nominal and often contradictory [222]. Overall this survey showed no effect on LDL cholestrol in normals and a slight effect (+3.4%) on HDL cholesterol. The effects in type IIb patients were marginal increases in LDL and HDL cholesterol, but in types IV and V the LDL went up 29.9% as triglyceride fell by 52.2%. At this time [221] the complexity of the interrelations is daunting and it is not unusual to find reports adverse in one or more details for specific clinical hyperlipidemias on administration of fish oils [227]. Those authors, however, concluded with the statement 'The results of this study, therefore, do not exclude the possibility of an *overall* beneficial cardiovascular effect of fish oil, due to these or other mechanisms of action'. This seems to summarize the current situation for hyperlipidaemia. The opportunities for drawing sweeping generalizations are now less obvious than formerly. In addition to gender differences in response of hypercholesterolaemic subjects [228] the genetic variations between individuals means that results of fish oil treatment can be 'variable and unpredictable' [229].

Hypertension. Adipose tissue content of 18:3 $n-3$ has been shown to be significantly inversely associated with blood pressure. Thus, each 1% increase in 18:3 $n-3$ in adipose tissue was associated with a decrease of 5 mm Hg in systolic, diastolic and composite mean arterial blood pressure [230]. Dietary intake of 18:3 $n-3$ has also been independently reported to be inversely related to blood pressure [231].

The extensive fish-feeding studies of Singer *et al.* [232] with consumption of mackerel itself reducing high blood pressure have created a favourable impression for fish oils generally. The $n-6$ blood fatty acid changes have just been published and discussed [219]. The most specific study on hypertension published recently, with either 10 or 50 ml of fish oil (MaxEPA) fed to patients with essential hypertension, is that of Knapp and Fitzgerald [233]. The high dose group showed a reduction in both systolic and diastolic pressures. Neither the low dose nor the control groups showed such an effect. An alteration in the synthesis of vasodilator PGs was found to *not* be a direct mechanism accounting for this effect. The PGI_3 metabolite excretion did increase, on a relative basis, in both dose levels. There are clearly many facets to this medical problem and $(n-3)$ fatty acid concentrates may reduce the volume of 'fish oil' to more manageable levels. The curious differences between mackerel and herring reported by Singer [219] may be $n-3$ dose related, but in any event the mackerel lipid $n-3$ fatty acids available per day were clearly less than in the study of Knapp and Fitzgerald [233], yet had a measurable effect on hypertension.

The recent work of Bonaa *et al.* [234] showed realistic reduction in both systolic and diastolic blood pressures with administration of 6 g day^{-1} of 85% (EPA plus DPA plus DHA) in ethyl ester form given to hypertensive subjects. Those who ate fish three or more times a week as part of their usual diet did not change mean blood pressure. Apparently this related to their plasma phospholipid $n-3$ fatty acids, already raised by diet, since decreases in blood pressure were larger as concentrations of plasma phospholipid $n-3$ fatty acid increased.

Platelet function. Sustained platelet activation and aggregation are important in the development of thrombi and atherosclerotic lesions [184]. Platelet activation stimulates the aggregation and binding of further platelets and also induces vascular smooth muscle cell proliferation leading to thickening of the vessel wall and narrowing of the lumen. The long-chain PUFAs of both the $n-6$ and $n-3$ series are implicitly involved in the regulation of platelet aggregation tendency.

Platelet number is widely reported as being decreased by up to 10% by fish oil consumption but platelet size is not decreased. *In vitro* data from several well-controlled studies on aggregation of platelet suspensions of platelet-rich plasma induced by ADP, collagen, adrenaline and thrombin to individuals consuming fish oils are clear regarding a platelet suppression effect of fish oils [50, 235–237]. Studies employing the traditional Inuit diet, or dietary supplementation with marine fish, fish oil or fish oil concentrates usually suggest that platelet clotting *in vivo* is impaired by enhanced availability of long-chain $n-3$ PUFAs. Nevertheless, some data (especially *in vivo*) does not support a beneficial effect of fish oils in suppressing platelet aggregation [238] and it is possible that better controls in some studies, e.g. of previous $20{:}5\,n-3$ intake, would clarify the matter [184]. Recent research suggests that habitual high fish intake (33 g day^{-1}) compared to low fish intake (2 g day^{-1}) does not necessarily result in longer cutaneous bleeding time, lower platelet number or reduction of platelet aggregation induced *in vitro* [190]. Also, restenosis time in patients undergoing coronary angioplasty does not appear to be reduced by fish oils [212], nor is bleeding time in stroke patients reduced by 1.8 g day^{-1} $20{:}5\,n-3$ for 6 weeks [239]. Other indices of platelet aggregation/activation tendency such as lower subendothelial platelet deposition in pigs, circulating levels of platelet aggregates, or lower levels of platelet-specific plasma proteins, including fibrinogen, have been observed with consumption of fish oils (up to 50 ml day^{-1} for 4 weeks in humans) [184, 240].

The reasons for some of the clinically-relevant *in vivo* indices of platelet aggregation tendency appearing to be equivocal may relate to differences in experimental design and control over factors such as type and quality of fish oil or fish used, e.g. $20{:}5\,n-3$ content, the presence of fish protein, time period and dose of feeding, etc. In view of the evidence for an inverse relation between: (1) blood and body fat content of $n-6$ PUFAs ($18{:}2\,n-6$ and $20{:}3\,n-6$) and newly diagnosed coronary heart disease [241, 242], (2) fish consumption and 20 year mortality from coronary heart disease [42], (3) between platelet and adipoise tissue levels of $20{:}5\,n-3$ and risk of angina pectoris or a first acute myocardial infarction [243], it would seem prudent to encourage fish consumption while further deliberating on the possible merits of encapsulated fish oils in the modulation of platelet function.

While effects on platelet function *in vivo* may still be unclear, it is evident that addition of long-chain $n-3$ PUFAs to the diet increases their levels in both plasma and platelets and reduces the level of $n-6$ PUFAs, especially $20{:}4\,n-6$ [82, 184, 244]. While effects of dietary supplements of $18{:}3\,n-3$ on platelet fatty acid composition and aggregation have not been as frequently studied, $18{:}3\,n-3$ does inhibit platelet aggregation [182, 245] and recent evidence suggests that flax as a dietary source of $18{:}3\,n-3$ does raise long chain PUFAs in platelets [246]. Fish oils typically contain approximately equivalent amounts of $22{:}6\,n-3$ and $20{:}5\,n-3$. While thrombin-stimulated platelets from fish-oil fed subjects release $20{:}5\,n-3$ and release in addition $20{:}4\,n-6$ [247], $22{:}6\,n-3$ appears to be equipotent to $20{:}5\,n-3$ in inhibiting $20{:}4\,n-6$ utilization by the cyclo- and lipoxygenases without itself being released from platelets [248].

Platelets are a uniquely rich source of thromboxane A_2 and the activation/aggregation of platelets can be induced by TXA_2. The majority of human studies addressing the *in vitro* release of TXA_2 (measured as TXB_2) during clotting indicate that a fish oil supplement inhibits TXA_2 synthesis by, and release from, platelets [184]. Nevertheless, *in vitro*, collagen-induced release of TXA_2 is not necessarily inhibited by fish oil supplementation. Fish oils also tend to reduce the endothelial synthesis of prostacyclin (PGI_2) in humans. *In vivo*, fish oils inhibit systemic release/excretion of 2-series PGs and their metabolites (including the metabolites of TXB_2) but this is not always observed [50, 238, 249].

Platelets are capable of producing 3-series PGs from 20:5 $n-3$ but this pathway has low activity [157]. Systemic production of PGI_3 (antithrombotic and vasodilatory) [250] and urinary excretion of the metabolite of PGI_3 is measurable and reproducibly increased after fish oil consumption [251] as was the metabolite of TXA_3 [236].

Platelet PG production and aggregability are also modified by 20:3 $n-6$ [154, 166]. In humans, orally administered 20:3 $n-6$ is accumulated by platelet lipids but is not significantly metabolized to 20:4 $n-6$ in platelets [72]. Nevertheless, in addition to increased production of PGE_1, PGE_2 was increased after maximal aggregatory stimulation with thrombin. This suggests that the amount of 20:3 $n-6$ conversion to 20:4 $n-6$ in platelets may be below the limits of detection of the fatty acids or enzyme assay methods and yet still result in increased production of 2-series PGs detectable by radioimmunoassay.

Diabetes. The controversy affecting the health benefits attributed to fish oils (essentially to the long-chain $\omega 3$ fatty acids) extends to diabetes. In this increasingly important disease the susceptible population falls into two groups, insulin-dependent diabetes mellitus (Type 1) and non-insulin-dependent diabetes mellitus (Type 2, usually referred to as NIDDM). A recent study [252] pointed out that the rat may not be a good model for NIDDM in humans and by comparing both a fish oil concentrate and safflower oil concluded that the *added* fat (10 g of each fat), supplemental to the standard diabetic diet, was responsible for the negative effects on glucose metabolism in NIDDM cases. There are many experimental designs and some give favourable results [253]. However, there have also been rather strong adverse reports for fish oils [254], based on increasing fasting blood glucose and impaired insulin secretion. Recently a comparison of 31 healthy subjects and 22 NIDDM patients for arterial compliance measured by Doppler ultrasonography came out in favour of fish eaters, compared to non-fish eaters, in both groups [197]. The usual caveat that caution should be observed promptly appeared [198], but it would seem reasonable to assume that at least in late stages of diabetes improved circulation would be a very positive benefit [48, 255] achievable with very small doses ($150\,mg\,day^{-1}$ of EPA). In the study of Borkman *et al.* [252] the 20:5 $n-3$ administered was presumably 10 times as much. In Type 1 (insulin dependent cases) 4 g of 20:5 $n-3$ per day was administered in one of the first such studies [256]. Although short term (8 weeks), the results were favourable including lowered blood viscosity thought to be achieved through red blood cell deformability. With 9 g of fish oil concentrate providing $1.8\,g\,day^{-1}$ of $20{:}5\,n-3$ and $0.9\,g\,day^{-1}$ of $22{:}6\,n-3$, Rillaerts *et al.* [257] produced results for Type 1 patients showing that the reduction in blood viscosity at low shear rates was due to plasma and not erythrocytes *per se*. In this study isocaloric exchange with $n-3$ replacing $n-6$ fatty acids improved some lipid and hemorheological parameters over 10 weeks.

Considerable interest was generated by Schectman *et al.* [220] in a longer term study which included diabetics. The $n-3$ fatty acids administered (Omega-500, Omegacaps Ltd, St Louis, MO.) provided 300 mg of $20{:}5\,n-3$ and 200 mg of $22{:}6\,n-3$ per gram capsule, presumably in methyl or ethyl ester form [*cf.* 55, 56, 258, 259]. This is one confusing element in their study, since differences in absorption routes hinder comparisons with triglyceride products [41]. The other confusing element was a change in dose level from 15 g of 'fish oil' to 6 g after 3 months of study, just as plasma triglyceride studies were apparently stabilizing. The subsequent increase in plasma triglyceride over 3 months could reflect the decreased dose. The tendency has been however to oversimplify the results to a cessation of the hypotriglyceridemic effect. Moreover as noted above, absorption of ethyl esters, for example, may be complete at $6\,g\,day^{-1}$ but not at $18\,g\,day^{-1}$ [59]. Thus new variables continue to be introduced to confound comparisons just as possible definitive studies with humans are expected. Recommendations to eat more fish in lieu of taking fish oil capsules [260] may give people the illusion of a major health benefit and have essentially nil risk [222]. Defined *clinical* benefits seem to be impossible to determine in the population at large for genetic reasons [261]. Moreover, dietary limitations for diabetics may make it impossible to comply by eating fatty fish [193]. Appropriate fish oil supplements should contribute negligible amount of cholesterol and saturated fat to the diet, allowing the $n-3$ fatty acids to be considered the sole active agents [258].

Despite the apparent lack of success in treating non-insulin dependent diabetes with very-long-chain $n-3$ PUFAs, it may be worth considering the use of $18{:}3\,n-3$ (in linseed oil capsules or dietary flax supplements). Studies in which $18{:}3\,n-3$ intake has been increased using linseed oil clearly show that platelet reactivity is diminished compared to controls and that elevated blood lipids have been reduced, both effects reportedly occurring in the absence of a change in long-chain $n-3$ PUFAs [113, 114, 181]. These effects would be desirable in diabetics in whom atherosclerosis, peripheral vascular disease, retinopathy and nephropathy are the most debilitating aspects of the disease.

Effects of pure $20{:}3\,n-6$ have also been studied in relation to platelet PGE_1 and platelet aggregability in normals and in insulin-dependent diabetics [218]. After $20{:}3\,n-6$ supplementation, healthy subjects had increased $20{:}3\,n-6$ and $20{:}4\,n-6$ in red blood cells, increased PGE_1 but not PGE_2 or TXA_2 in clotted blood, and significantly reduced ADP-induced platelet aggregation. Diabetics also had increased 20:3 *n* − 6 in red cells but no change in red cell 20:4 *n* − 6, no change in 1- or 2-series PGs, and no change in platelet aggregation after oral 20:3 *n* − 6. Serum lipids, glucose and haemoglobin A1 remained unchanged in both groups compared to pretreatment values. These data suggest that diabetics do not synthesize or incorporate 20:4 *n* − 6 into their red blood cells to the same extent as normals. However, this does not appear to affect production of 2-series PGs in clotting blood [218].

It has recently become evident that a dietary supplement of flax (without purification of the linseed oil) is an effective means of increasing very-long-chain *n* − 3 PUFAs (20:5 *n* − 3, 22:5 *n* − 3 and 22:6 *n* − 3) in serum, red cells and platelets in both normolipidemics and hyperlipidemics [117, 118, 246]. Furthermore, unlike fish oils given to non-insulin-dependent diabetics, short-term intake of flax by non-insulin-dependent diabetics improves rather than worsens glucose tolerance (Jenkins and Cunnane, unpublished). If 18:3 *n* − 3 in flax or linseed oil may be of benefit in raising 20:5 *n* − 3 and 22:6 *n* − 3 in diabetes, the presence of 18:4 *n* − 3 (stearidonic acid) in blackcurrant seed oil may be an even more effective way to increase, through endogenous synthesis, the lipid incorporation of long-chain *n* − 3 PUFAs without consuming 20:5 *n* − 3 or 22:6 *n* − 3 directly in the diet. Restoring endogenous synthesis of 20:5 *n* − 3 with the consequence of providing appropriate amounts of the longer chain *n* − 3 PUFAs in the lipid classes may be as important, and possibly more effective and less toxic, than providing the end-product *n* − 3 PUFAs pre-formed as dietary encapsulated fish oils. Present data suggests that the mechanism of apprehended fish oil toxicity in non-insulin-dependent diabetes is related to the high levels of 20:5 *n* − 3 in the liver needing to be oxidized (along with the other fatty acids in fish oils, e.g. 16:0 and 18:1 *n* − 9), resulting in impaired glucose oxidation and/or increased hepatic synthesis and output of glucose [262].

The issue of disease-related inhibition of the PUFA desaturases must also be considered in evaluating the potential therapeutic efficacy of the precursor fatty acids (18:2 *n* − 6 or 18:3 *n* − 3) versus the longer chain PUFAs (20:3 *n* − 6, 20:4 *n* − 6, 20:5 *n* − 3 or 22:6 *n* − 3). Experimental studies in animal models of insulin-dependent diabetes [84, 263] indicate that alloxan- or streptozotocin-induced diabetes causes inhibition of desaturation. The metabolism of 20:3 *n* − 6 to 20:4 *n* − 6 has been shown to be impaired in non-insulin-dependent diabetes [218] and the metabolism of 18:3 *n* − 3 to 20:5 *n* − 3 has been reported to be impaired in insulin-independent diabetes [264]. As a result it has been suggested that treatment of the cardiovascular complications of diabetes using PUFAs should involve the longer chain derivatives, e.g. 20:3 *n* − 6 or 20:5 *n* − 3 [4].

However, human diabetes does not appear to necessarily result in lower levels of very-long-chain PUFAs in serum [218, 265], red blood cells [266] or retinal vascular lipids [267]. How the serum levels of longer chain PUFAs can be sustained at normal levels in diabetes when synthesis appears to be impaired is not understood.

Two long-term trials involving modification of dietary PUFA intake in diabetes (maturity-onset, non-insulin-dependent) have been conducted. One involved $18{:}2\,n-6$ which was given at 5 (control) or $20\,\mathrm{g\,day^{-1}}$ (experimental) in an otherwise similar diet for 6 years [268, 269]. The incidence and/or severity of several clinical indices including retinopathy, cardiac ischemia, high blood glucose and glucose intolerance, and hyperlipidaemia, was markedly less in the high compared to the low $18{:}2\,n-6$ group. The other was a 7-year trial in which dietary PUFA modification was done on the basis of advice to one group to maintain total fat intake at less than 30% of total calories while increasing the PUFA:saturated fat ratio from an average of 0.3 to 0.9 [270]. In this study, progression of retinopathy over 7 years was also reduced but not as effectively as in the trial reported by Houtsmuller [269]. The less significant effect of increased PUFA intake relative to other fat in the latter trial was probably due to less controlled dietary modification and less significant changes in $18{:}2\,n-6$ of serum lipids. Neither study reported a significant change in $20{:}4\,n-6$ after long-term supplementation with $18{:}2\,n-6$, but this might not be required to change the metabolically active pool of $20{:}3\,n-6$ or $20{:}4\,n-6$, e.g. the pool supplying substrate for synthesis of PGE_1 or prostacyclin.

5.2.2 *Inflammatory Diseases*

The traditional treatment for a wide range of inflammatory diseases has been to reduce PG synthesis with drugs such as aspirin and thereby reduce the chemotactic effects of PGs. Although this approach is often particularly effective for pain relief, stomach ulcers and gastrointestinal intolerance to non-steroidal anti-inflammatory drugs are a frequent side-effect of inhibiting PG synthesis since PGs are required to maintain the integrity of the mucosal surface of the gut.

As an alternative to drugs, dietary intervention with encapsulated PUFA supplements containing $18{:}3\,n-6$ and/or $20{:}5\,n-3$ in inflammatory disease has been vigorously pursued in the past decade because of the potential benefits of: (1) raised PGE_1 and (2) effective inhibition of synthesis of 2-series PGs by $20{:}5\,n-3$ and, possibly, $22{:}6\,n-3$. Furthermore, fish oils markedly inhibit human neutrophil chemotaxis and inhibit neutrophil production of tumour necrosis factor and interleukin-1 [271]. Despite effectiveness in animal models [272] and anti-PG/LT effects in double-blind placebo-controlled trials, in general, fish oils have not provided sustained relief in rheumatoid arthritis [273–275].

Nevertheless, beneficial trends in lowering of joint tenderness and fatigue have been noted with a weak correlation to lower neutrophil LTB_4 production. Similarly, supplementation with 18:2 $n-6$ and 18:3 $n-6$ (evening primrose oil), which increases PGE_1 production in animal models [155], did not alter clinical or biochemical parameters in rheumatoid arthritis [151].

5.2.3 Cancer

Epidemiological studies associate the higher cancer risk in the Western world with higher total dietary fat intake. This association has been strongest for colon and breast cancer [276, 277]. The role of total fat intake has been clearly demonstrated by the effectiveness in reducing breast cancer risk of reduction of dietary fat from 40 to 20% of calories [278]. The high fat intake of Westerners now includes much more 18:2 $n-6$, assessed as up to 10 times the normal requirement. This level of dietary 18:2 $n-6$ in animals given tumorigenic chemicals frequently leads to a higher incidence of tumours and the implication that PUFAs (especially 18:2 $n-6$) promote an increased risk of cancer [279]. Tumour models in the rat secrete high amounts of 2-series PGs and, in the rat, dietary 18:2 $n-6$ is readily converted to 20:4 $n-6$, so that a positive relation between dietary 18:2 $n-6$ and tumour growth in the rat seems evident. It is also clear that a proportionately high intake of 18:2 $n-6$ in a low fat diet is not necessarily tumour-promoting [280], so reduced total calorie (fat) intake in which 18:2 $n-6$ is an adequate to high proportion must still be considered a logical and beneficial therapeutic strategy in the prevention or treatment of both breast and colon cancer. Tumour incidence is lower in different target organs in several experimental models when the animals are fed fish oils at as much as 20% of calories [281]. The mechanism of this effect is thought to involve a lowering of 2-series PGs [282].

Increased intake of 18:3 $n-3$ has also been shown to be beneficial in reducing tumour growth in the rat [283]. Whether this occurs through inhibition of PG synthesis in the tumour is not known. Another aspect of the anti-tumour effectiveness of 18:3 $n-3$ which has gained prominence recently is the relation between fatty acid unsaturation and tumour toxicity. PUFAs with 3–5 double bonds have a marked cytotoxicity to a range of tumour cells grown *in vitro*; this effect is specific for tumour cell lines as demonstrated by the virtual absence of cytotoxicity to normal cell lines grown in the presence of equal starting numbers of tumour cells. The mechanism appears not to involve inhibition of PG synthesis but may include manipulation of peroxide products of PUFA which appear cytotoxic to tumour cells [284, 285, 286].

5.2.4 Dermatology

Two aspects of PUFA metabolism are of particular relevance to the skin: (1) 18:2 $n-6$ is a unique component of the waxy lipids of the epidermis which constitute the barrier to body water loss and infection across the skin [3, 6, 7], and (2) adverse stimuli affecting the skin induce metabolic transformations of 20:4 $n-6$ via the cycloxygenase and lipoxygenase pathways, resulting in erythema, edema, pain and neutrophil infiltration of affected sites [287, 288]. Severe skin diseases such as psoriasis and epidermolysis bullosa frequently involve increased synthesis of 12-hydroxyeicosatetraenoic acid, a lipoxygenase product of 20:4 $n-6$, and/or 2-series PGs [287, 289]. In the case of psoriasis, a moderate, dose-dependent overall benefit of fish oils has been reported, with significant improvement in some patients [290].

Atopic eczema has long been associated with dietary deficiency of 18:2 $n-6$ because of the similarity of its dermal symptoms to those of 18:2 $n-6$ deficiency [288]. In fact, recent evidence suggests that there is no dietary deficiency of 18:2 $n-6$ in atopic eczema but, rather, a possible impairment of conversion of 18:2 $n-6$ or 18:3 $n-3$ to longer chain PUFAs [111]. Oral 18:2 $n-6$ and 18:3 $n-6$ (evening primrose oil) or 20:5 $n-3$ produce clinical improvement, an effect correlated with increased PUFA ($n-6$ or $n-3$, depending on the trial) in serum lipids [288, 291, 292].

6. CONCLUSIONS

The past decade has seen important advances in our understanding of the biochemistry, metabolism and clinical/nutritional significance of long-chain PUFAs. In this period, new plant and marine sources of PUFAs have been commercialized for specific nutritional/clinical applications, e.g. evening primrose, borage, blackcurrant and microalgae. These applications should be seen as an addition to the more conventional PUFA intake derived from widely used edible oils e.g. soya, Canola, sunflower, etc. The significance of factors modulating desaturase activity in humans, e.g. cofactor nutrients, disease, substrate competition and substrate-induced inhibition, have also been more fully recognized and documented. These advances have led to an unprecedented expansion in the use of encapsulated PUFAs (both $n-6$ and $n-3$) in the treatment and prevention of disease.

While some success in this direction is recently evident, and has been documented here and elsewhere [293–298], it is clear that we still need much more basic information concerning the absorption, distribution and metabolism of PUFAs in humans before the potential of this form of disease management is optimized. In particular, it is evident that stable isotopes of PUFAs labelled at sites other than the carboxyl terminal need to be developed, as well as systems to monitor PUFA metabolism in humans, e.g. *in vivo* NMR spectroscopy [299]. Data derived from the few studies to date involving stable isotopes have suggested that shorter chain PUFAs, e.g. 18:2 $n-6$ and 18:3 $n-3$, may have important roles through oxidation; these data need further documentation.

Currently, the main nutritional/clinical applications for long-chain PUFAs are in the cardiovascular, neonatal and anti-inflammatory fields; whether encapsulated PUFA-rich oils remain commercialized will depend to a large extent on advances in our understanding of PUFA metabolism in these areas. The focus on fats and cardiovascular problems [300] has led to an overwhelming but possibly misguided attempt to intervene in public health almost solely through serum cholesterol control [189, 301]. This bias has masked other health problems related to our current dietary intakes of the two series of polyunsaturated fatty acids [12, 302]. The II International Conference on the Health Effects of ω3 Polyunsaturated Acids in Seafoods, held in Washington in March 1990, issued a final statement reading in part 'The researchers urged that appropriate government agencies officially recognize the vitally important differences between ω3 and ω6 polyunsaturated fatty acids'. This consensus is an indication of the need for clearer recognition of the health benefits of polyunsaturated fatty acids at a time when the public attitude towards dietary fats is one of apprehension bordering on fear.

ACKNOWLEDGMENTS

R.G.A. and S.C.C. acknowledge support from the Natural Sciences Engineering and Research Council of Canada. Janet O'Dor, Elizabeth Pollock and Adina Petroff assisted with the preparation of the manuscript.

REFERENCES

[1] Aaes – Jorgensen, E. Essential fatty acids. *Physiol. Rev.*, 41 (1961) 1–51.

[2] Alfin-Slater, R. B. and Aftergood, L. Essential fatty acids reinvestigated. *Biol. Rev.*, 48 (1968) 758–784.

[3] Holman, R. T. Nutritional and metabolic interrelationships between fatty acids. *Fedn. Proc.*, 23 (1964) 1062–1067..

[4] Horrobin, D. F. The regulation of prostaglandin biosynthesis by the manipulation of essential fatty acid metabolism. *Rev. Pure Appl. Pharm. Sci.*, 4 (1983) 339–383.

[5] Jeffcoat, R. and James, A. T. The regulation of desaturation and elongation of fatty acids in mammals. In Numa, S. (ed.), *Fatty Acid Metabolism and Its Regulation*. Elsevier, Amsterdam, 1984, pp. 85–112.

[6] Holman, R. T. Essential fatty acid deficiency. In Holman, R. T. (ed.), *Progress in the Chemistry of Fats and Other Lipids*. Pergamon Press, Oxford, 1971, Vol. 9, pp. 275–348.

[7] Hansen, H. S. Dietary essential fatty acids and *in vivo* prostaglandin production in mammals, *Wld. Rev. Nutr. Diet.*, 42 (1983) 102–134.

[8] Tinoco, J. Dietary requirements and functions of alpha-linolenic acid in animals. *Prog. Lipid Res.*, 21 (1982) 1–45.

[9] Salem, N., Kim, H. Y. and Yergey, J. A. Docosahexaenoic acid: membrane function and metabolism. In Simopoulos, A., Kifer, R. R. and Martin, R. E. (eds), *Health Effects of Polyunsaturated Fatty Acids in Seafoods*. Academic Press, Orlando, 1986, pp. 263–317.

[10] Dratz, E. A. and Deese, A. J. The role of docosahexaenoic acid (22:6ω3) in biological

membranes: Examples from photoreceptors and model membrane bilayers. In Simopoulos, A. *et al.* (ed.), *Health Effects of Polyunsaturated Fatty Acids in Seafoods*. Academic Press, Orlando, 1986, pp. 319–351.

[11] Neuringer, M., Connor, W. E., Lin, D. S., Barstad, L. and Luck, S. Biochemical and functional effects of prenatal and postnatal omega-3 fatty acid deficiency on retina and brain in rhesus monkeys. *Proc. Natl. Acad. Sci. USA*, 83 (1986) 4021–4025.

[12] Lands, W. E. M. *Fish and Human Health*, Academic Press, Orlando, 1986.

[13] Ackman, R. G. and Ratnayake, W. M. N. Fish oils, seal oils, esters and acids—are all forms of omega-3 intake equal? In Chandra, R. K. (ed.), *Health Effects of Fish and Fish Oils*. ARTS Biomedical Publishers, St. John's, Newfoundland, 1989, pp. 373–393.

[14] Miller, R. W., Earle, F. R. and Wolff, I. A. Search for new seed oils. XV. Oils of Boraginacae. Lipids, 3 (1968) 43–45.

[15] Hudson, B. J. F. Evening primrose (*Oenothera* spp.) oil and seed. *J. Am. Oil Chem. Soc.*, 61 (1984) 540–543.

[16] Traitler, H., Winter, H., Richli, U. and Ingenbleek, Y. Characterization of gamma-linolenic acid in Ribes seed. *Lipids*, 19 (1984) 923–928.

[17] Heoksema, S. D., Behrens, P. W., Gladue, R., Arnėtt, K. L., Cole, M. S., Rutten, J. M. and Kyle, D. J. An EPA-containing oil from microalgae in culture. In Chandra, R. K. (ed.), *Health Effects of Fish and Fish Oils*. ARTS Biomedical publishers, St. John's, Newfoundland, 1989, pp. 337–347.

[18] Behrens, P. W., Heoksema, S. D., Arnett, K. L., Cole, M. S., Heubner, T. A., Rutten, J. M. and Kyle, D. J. Eicosapentaenoic acid from microalgae. In Somkuti, G. A. (ed.), *Biotechnology of Microbiol Products*, in press.

[19] Shaw, R. The occurrence of gamma-linolenic acid in fungi. *Biochim. Biophys. Acta.*, 98 (1965) 230–237.

[20] Hudson, B. J. F. and Karis, I. G. Effect of crop maturity on leaf lipids. *J. Sci. Food Agric.*, 25 (1974) 1491–1502.

[21] Ackman, R. G., Hooper, S. N. and Hooper, D. L. Linolenic acid artifacts from the deodorization of oils. *J. Am. Oil Chem. Soc.*, 51 (1974) 42–49.

[22] Grandgirard, A., Sebedio, J.-L. and Fleury, J. Geometrical isomerization of linolenic acid during heat treatment of vegetable oils. *J. Am. Oil Chem. Soc.*, 61 (1984) 1563–1568.

[23] Applewhite, T. H. Nutritional effects of hydrogenated soya oil. *J. Am. Oil Chem. Soc.*, 58 (1981) 260–269.

[24] Piconneaux, A., Grandgirard, A. and Sebedio, J.-L. Identification of an unusual polyunsaturated fatty acid in liver of rats fed with heated linseed oil. *C. R. Acad. Sci.*, Ser III, No. 8, 300 (1985) 353–358.

[25] Grandgirard, A., Piconneaux, A., Sebedio, J.-L., O'Keefe, S. F., Semon, E. and Le Quere, J.-L. Occurrence of geometrical isomers of eicosapentaenoic and docosahexaenoic acids in liver lipids of rats fed heated linseed oil. *Lipids*, 24 (1989) 799–804.

[26] Lawson, L. D. and Hughes, B. G. Human absorption of fish oil fatty acids as triacylglycerols, free acids, or ethyl esters. *Biochem. Biophys. Res. Comm.*, 152 (1988) 328–335.

[27] Nelson, G. J. and Ackman, R. G. Absorption and transport of fat in mammals with emphasis on $n-3$ polyunsaturated fatty acids. *Lipids*, 23 (1988) 1005–1014.

[28] Chiang, S.-H., Pettigrew, J. E., Clarke, S. D. and Cornelius, S. G. Digestion and absorption of fish oil by neonatal piglets. *J. Nutr.*, 119 (1989) 1741–1743.

[29] Moreau, H., Gargouri, Y., Bernadal, A., Pieroni, G., Verger, R., Laugier, R., Sauniere, J.-F., Sarles, H., Lecat, D. and Junien, J.-L. Etude biochimique et physiologique des lipases preduodenales d'origines animale et humane: Revue. *Rev. Franc. Corps Gras.*, 35 (1988) 169–176.

[30] Hilditch, T. P. and Williams, P. N. *The Chemical Constitution of Natural Fats*, 4th edn. Chapman and Hall, London, 1964.

[31] Ackman, R. G. Fatty acid composition of fish oils. In Barlow, S. M. and Stansby, M. E. (eds), *Nutritional Evaluation of Long-Chain Fatty Acids in Fish Oils*. Academic Press, London, 1982, pp. 25–88.

[32] Ackman, R. G., Fatty acid metabolism of bivalves. In Pruder, G. D., Langdon, C. J. and Conklin, D. E. (eds), *Proceedings of the Second International Conference on Aquaculture Nutrition: Biochemical and Physiological Approaches to Shellfish Nutrition*. Louisiana State University, Special Publ. No. 2, Baton Rouge, LA, 1982, pp. 358–376.

[33] Ackman, R. G. Algae as sources of edible lipids. In Pryde, E. H., Princen, L. H. and Mukherjee, K. D. (eds), *New Sources of Fats and Oils*. American Oil Chemists' Society, Champaign, 1981, pp. 189–220.

[34] Ackman, R. G., Tocher, C. S. and McLachlan, J. Marine phytoplankter fatty acids. *J. Fish. Res. Bd. Canada*, 25 (1968) 1603–1620.

[35] Hayes, K. C. Dietary saturated fatty acids and low-density or high-density lipoprotein cholesterol. *New Eng. J. Med.*, 322 (1990) 402–403.

[36] Ackman, R. G., Sebedio, J.-L. and Kovacs, M. I. P. Role of eicosenoic and docosenoic fatty acids in freshwater and marine lipids. *Mar. Chem.*, 9 (1980) 157–164.

[37] Ackman, R. G., Epstein, S. and Eaton, C. A. Differences in the fatty acid compositions of blubber fats from northwestern Atlantic finwhales (*Balaenoptera physalus*) and harp seals (*Pagophilus groenlandica*). *Comp. Biochem. Physiol.*, 40B (1971) 683–697.

[38] Ackman, R. G. and Hooper, S. N. Long-chain monoethylenic and other fatty acids in heart, liver and blubber lipids of two harbor seals (*Phoca vitulina*) and one grey seal (*Halichoerus grypus*) *J. Fish. Res. Bd. Canada*, 31 (1974) 333–341.

[39] Ackman, R. G. and Lamothe, F. Marine mammals. In Ackman, R. G. (ed.), *Marine Biogenic Lipids*. CRC Press, Boca Raton, FL, 1989, Vol. II, pp. 179–381.

[40] Ackman, R. G. Some possible effects on lipid biochemistry of differences in the distribution on glycerol of long-chain $n-3$ fatty acids in the fats of marine fish and marine mammals. *Atherosclerosis*, 70 (1988) 171–173.

[41] Ackman, R. G. Problems in fish oils and concentrates. In Cambie, R. C. (ed.), *Fats for the Future*. Ellis Horwood, Chichester, 1989, pp. 189–204.

[42] Kromhout, D., Bosschieter, E. B. and de Lezenne Coulander, C. The inverse relation between fish consumption and 20-year mortality from coronary heart disease. *New Eng. J. Med.*, 312 (1985) 1205–1209.

[43] Brown, A. J., Roberts, D. C. K. and Truswell, A. S. Fatty acid composition of Australian marine finfish: A review. *Food Australia*, March (1989) 655–666.

[44] Ackman, R. G. Nutritional composition of fats in seafoods. *Prog. Food Nutr. Sci.*, 13 (1989) 161–241.

[45] Ackman, R. G. Simplification of analysis of fatty acids in fish lipids and related lipid samples. *Acta. Med. Scand.*, 222 (1987) 99–103.

[46] Weber, P. C. Clinical studies on the effects of $n-3$ Fatty acids on cells and eicosanoids in the cardiovascular system. *J. Int. Med.*, 225 Suppl. 1 (1989) 61–68.

[47] Kromhout, D. $n-3$ Fatty acids and coronary heart disease: Epidemiology from Eskimos to Western populations. *J. Int. Med.*, 225 Suppl. 1 (1989) 47–51.

[48] Driss, F., Vericel, E., Lagarde, M., Dechavanne, M. and Darcet, Ph. Inhibition of platelet aggregation and thromboxane synthesis after intake of small amounts of icosapentaenoic acid. *Thromb. Res.*, 36 (1984) 389–396.

[49] Croset, M., Véricel, E., Rigaud, M., Hanss, M., Courpron, Ph., Dechavanne, M. and Lagarde, M. Functions and tocopherol content of blood platelets from elderly people after low intake of purified eicosapentaenoic acid. *Thromb. Res.*, 57 (1990) 1–12.

[50] von Schacky, C. and Weber, P. C. Metablism and effects on platelet function of the purified eicosapentaenoic and docosahexaenoic acid in humans. *J. Clin. Invest.*, 76 (1985) 2446–2450.

[51] Fischer, S., Vischer, A., Preac-Mursic, V. and Weber, P. C. Dietary docosahexaenoic acid is retroconverted in man to eicosapentaenoic acid, which can be quickly transformed to prostaglandin I_3. *Prostaglandins*, 34 (1987) 367–375.

[52] Simopoulos, A. P. Summary of the NATO advanced research workshop on dietary $\omega 3$ and $\omega 6$ fatty acids: biological effects and nutritional essentiality. *J. Nutr.*, 119 (1989) 521–528.

[53] Ackman, R. G. The year of the fish oils. *Chem. Ind.*, March 17 (1988) 139–145.

[54] Anonymous, Maxepa gets a product licence for hyperlipidaemia. *Pharmaceut. J.*, May 9 (1987) 587.

[55] Ratnayake, W. M. N., Olsson, B., Matthews, D. and Ackman, R. G. Preparation of omega-3 PUFA concentrates from fish oils via urea complexation. *Fat Sci. Technol.*, 90 (1988) 381–386.

[56] Ackman, R. G., Ratnayake, W. M. N. and Macpherson, E. J. EPA and DHA contents of encapsulated fish oil products. *J. Am. Oil. Chem. Soc.*, 66 (1989) 1162–1164.

[57] Haraldsson, G. G., Hoskuldsson, P.-A., Sigurdsson, S.Th., Thorsteinsson, F. and Gudbjarnason, S. The preparation of triglycerides highly enriched with ω-3 polyunsaturated fatty acids via lipase catalyzed interesterification. *Tetrahedron Lett.* 30 (1989) 1671–1674.

[58] Nilsson, W. B., Gauglitz Jr, E. J., Hudson, J. K., Stout, V. F. and Spinelli, J. Fractionation of menhaden oil ethyl esters using supercritical fluid CO_2. *J. Am. Oil Chem. Soc.*, 65 (1988) 109—117.

[59] Hawthorne, A. B., Daneshmend, T. K. and Hawkey, C. J. Letter, under 'Fish Oil Revisited', Lancet, September 30 (1989) 811.

[60] Hamazaki, T., Urakaze, M., Makuta, M., Ozawa, A., Soda, Y., Tatsumi, H., Yano, S., Kumagai, A. Intake of different eicosapentaenoic acid-containing lipids and fatty acid pattern of plasma lipids in the rats. *Lipids*, 22 (1987) 994–998.

[61] El-Boustani, S., Colette, C., Monnier, L., Descomps, B., Crastes de Paulet, A. and Mendy, F. Enteral absorption in man of eicosapentaenoic acid in different chemical forms. *Lipids*, 22 (1987) 711–714.

[62] Monnier, L., El Boustani, S., Crastes de Paulet, A., Descomps, B. and Mendy, L. Aspects du métabolisme des acides gras polyinsaturés chez des sujets témoins et diabétiques. Implications nutritionnelles. *Rev. Franc. Corps Gras.*, 36 (1989) 3–10.

[63] Subbaiah, P. V., Davidson, M. H., Ritter, M. C., Buchanan, W. and Bagdade, J. D. 'Effects of dietary supplementation with marine lipid concentrate on the plasma lipoprotein composition of hypercholesterolemic patients. *Atherosclerosis*, 79 (1989) 157–166.

[64] Vamecq, J., Grataroli, R., Draye, J.-P., Lafont, H. and Nalbone, G. 'Effects of dietary corn oil and salmon oil on the oxidation of fatty acids and prostaglandin E_2 in rat gastric mucosa. *Prostaglandins*, 37 (1989) 335–344.

[65] de la Hunt, M. N., Hillier, K. and Jewel, R. Modification of upper gastrointestinal prostaglandin synthesis by dietary fatty acids. *Prostaglandins*, 35 (1988) 597–608.

[66] Szabo, S. and Rogers, C. Diet, ulcer disease, and fish oil. *Lancet*, January (1988) 119.

[67] Smith, P., Arnesen, H., Opstad, T., Dahl, K. H. and Eritsland, J. Influence of highly concentrated $n-3$ fatty acids on serum lipids and hemostatic variables in survivors of myocardial infarcation receiving either oral anticoagulants or matching placebo. *Thromb. Res.*, 53 (1989) 467–474.

[68] Einig, R. G. and Ackman, R. G. Omega-3 PUFA in marine oil products. *J. Am. Oil Chem. Soc.*, 64 (1987) 499–502.

[69] AOCS. Official method Ce 1b–89: Fatty acid composition by GLC. In *Official Methods and Recommended Practices of the American Oil Chemists' Society*, 4th edn. American Oil Chemists' Society, Champaign, 1989.

[70] Wijesundera, R. C., Ratnayake, W. M. M. and Ackman, R. G. Eicosapentaenoic acid geometrical isomer artifacts in heated fish oil esters. *J. Am. Oil Chem. Soc.*, 66 (189) 1822–1830.

[71] Williams, G. and Crawford, M. A. Comparison of the fatty acid component in structural lipids from dolphins, zebra and giraffe: possible evolutionary implications. *J. Zool.*, 213 (1987) 673–684.

[72] Stone, K. J., Willis, A. L., Hart, M., Kirtland, S. J., Kernoff, P. B. A. and McNicol, G. P. The metabolism of dihomo-gamma-linolenic acid in man. *Lipids*, 14 (1979) 174–180.

[73] Crawford, M. A., Hassam, A. G., Williams, G. and Whitehouse, W. L. Essential fatty acids and fetal brain growth. *Lancet*, February 28 (1976) 452–453.

[74] Horrobin, D. F., Huang, Y. S., Cunnane, S. C. and Manku, M. S. Essential fatty acids in plasma, red blood cells and liver phospholipids in common laboratory animals as compared to humans. *Lipids*, 19 (1984) 806–811.

[75] Cunnane, S. C., Huang, Y. S. and Manku, M. S. Species comparison of the triacylglycerol content of arachidonic acid: Inverse correlation with total triacylglycerol in liver and plasma. *Biochim. Biophys. Acta*, 876 (1986) 183–186.

[76] Rivers, J. P. W., Sinclair, A. J. and Crawford, M. A. Inability of the cat to desaturate essential fatty acids. *Nature*, 258 (1975) 171–173.

[77] Hassam, A. G., Rivers, J. P. W. and Crawford, M. A. The failure of the cat to desaturate linoleic acid; Its nutritional implications. *Nutr. Metab.*, 21 (1977) 321–328.

[78] Crawford, M. A., Doyle, W., Williams, G. and Drury, P. J. The role of fats and EFAs for energy and cell structures in the growth of fetus and neonate. In Vergroesen, A. J. and Crawford, M. (eds), *The Role of Fats in Human Nutrition*, 2nd edn. Academic Press, London, 1989, pp. 81–115.

[79] Crawford, M. A., Costeloe, K., Doyle, W., Ferguson, A., Hatton, E. P., Leaf, A., Lennon, E. A. and Meadows, N. $N-6$ and $N-3$ fatty acids in the food chain. In Chandra, R. K. (ed.), *Health Effects of Fish and Fish Oils*, ARTS Biomedical Publishers, St. John's, Newfoundland, 1989, pp. 219–238.

[80] Galli, C. and Socini, A. Dietary lipids in pre- and post-natal development. In Perkins, E. G. and Visek, W. J. (eds), *Dietary Fats and Health*, Proceedings of the American Oil Chemists' Society Conference, Chicago, 1983, Vol. 16, pp. 278–301.

[81] Greenwod, C. E., McGee, C. D. and Dyer, J. R. Influence of dietary fat on brain membrane phospholipid fatty acid composition and neuronal function in mature rats. *Nutrition*, 5 (1989) 278–281.

[82] Holub, B. J., Celi, B. and Skeaff, C. M. The alkenylacyl class of ethanolamine phospholipid represents a major form of eicosapentaenoic acid (EPA)-containing phospholipid in the platelets of human subjects consuming a fish oil concentrate. *Thromb. Res.*, 50 (1988) 135–143.

[83] Weaver, B. J. and Holub, B. J. Health effects and metabolism of dietary eicosapentaenoic acid. *Progr. Food Nutr. Sci.*, 12 (1988) 111–150.

[84] Brenner, R. R. The oxidative desaturation of unsaturated fatty acids in animals. *Mol. Cell Biochem.*, 3 (1974) 41–52.

[85] Brenner, R. R. Factors influencing fatty acid chain elongation and desaturation. In Vergroesen, A. J. and Crawford, M. (eds), *The Role of Fats in Human Nutrition*, 2nd edn. Academic Press, London, 1989, pp. 45–79.

[86] Sprecher, H. and James, A. T. Biosynthesis of long chain fatty acids in mammalian systems. In Emken, E. A. and Dutton, H. J. (eds), *Geometrical and Positional Fatty Acid Isomers*. American Oil Chemists' Society, Champaign, 1979, pp. 303–338.

[87] Voss, A. C. and Sprecher, H. Regulation of the metabolism of linoleic acid to arachidonic acid in rat hepatocytes. *Lipids*, 23 (1988) 660–665.

[88] Maeda, M., Doi, O. and Akamatsu, Y. Metabolic conversion of polyunsaturated fatty acids in mammalian cultured cells. *Biochim. Biophys. Acta*, 530 (1978) 153–164.

[89] Strittmatter, P., Spatz, L., Corcoran, D., Rogers, M. J., Setlow, B. and Redline, R. Purification and properties of rat liver microsomal stearoyl coenzyme A desaturase. *Proc. Natl. Acad. Sci. USA*, 71 (1974) 4565–4569.

[90] Okayasu, T., Nagao, M., Ishibashi, T., and Imai, Y. Purification and partial characterization of linoleoyl-CoA desaturase from rat liver microsomes. *Arch. Biochem. Biophys.*, 206 (1981) 21–28.

[91] Cunnane, S. C., Keeling, P. W. N., Thompson, R. P. H. and Crawford, M. A. Linoleic and arachidonic acid metabolism in human peripheral blood leucocytes: Comparison with the rat. *Br. J. Nutr.*, 51 (1984) 209–217.

[92] Blond, J.-P., Lemarchal, P. and Spielmann, D. Comparison of *in vitro* desaturation of linoleic and dihomogammalinolenic acids by homogenates of human liver. *C.R. Acad. Sci. Paris*, 292 (III) (1981) 911–914.

[93] De Gomez Dumm, I. N. T. and Brenner, R. R. Oxidative desaturation of alpha-linolenic, linoleic, and stearic acids by human liver microsomes. *Lipids*, 10 (1975) 315–317.

[94] Emken, E. A., Rohwedder, W. K., Adolf, R. O., Rakoff, H. and Gulley, R. M. Metabolism in *cis*-12, *trans*-15-octadecadienoic acid relative to palmitic, stearic, oleic, and linoleic acids. *Lipids*, 22 91987) 495–504.

[95] Manku, M. S., Morse-Fisher, N. and Horrobin, D. F. Changes in human plasma essential fatty acid levels as a result of administration of linoleic acid and gamma-linolenic acid. *Eur. J. Clin. Nutr.*, 42 (1988) 55–60.

[96] Chambaz, J., Ravel, D., Manier, M. C., Pepin, D., Mulliez, N., and Bereziat, G. Essential fatty acids interconversion in the human fetal liver. *Biol. Neonate*, 47 (1985) 136–140.

[97] Moore, S. A., Yoder, E. and Spector, A. A. Human microvascular endothelium produces arachidonate (20:4$n-6$), eicosapentaenoate (20:5$n-3$), and docosahexaenoate (22:6$n-3$) from essential fatty acid precursors. *FASEB J.*, 3 (1989) A705.

[98] Aeberhard, E. E., Corbo, L. and Menkes, J. H. Polyenoic acid metabolism in cultured human skin fibroblasts. *Lipids*, 13 (1978) 758–767.

[99] Siguel, E. N. and Maclure, M. Relative activity of unsaturated fatty acid metabolic pathways in humans. *Metabolism*, 36 (1987) 664–669.

[100] Naughton, J. M. Supply of polyenoic fatty acids to the mammalian brain. *Int. J. Biochem.*, 13 (1981) 21–32.

[101] Cook, H. W. *In vitro* formation of polyunsaturated fatty acids by desaturation in rat brain: some properties of the enzymes in developing brain and comparisons with liver. *J. Neurochem.*, 30 (1978) 1327–1334.

[102] Bills, T. K., Smith, B. and Silver, M. J. Selective release of arachidonic acid from the phospholipids of human platelets in response to thrombin. *J. Clin. Invest.*, 60 (1977) 1–6.

[103] Mohrhauer, H. and Holman, R. T. Effect of linolenic acid upon the metabolism of linoleic acid. *J. Nutr.*, 81 (1963) 67–74.

[104] Sprecher, H. Biosynthesis of polyunsaturated fatty acids and its regulation, In Kunau, W. H. and Holman, R. T. (eds), *Polyunsaturated Fatty Acids*. American Oil Chemists' Society, Champaign, 1977, pp. 1–18.

[105] Bjerve, K. S., Mostad, I. L. and Thorensen, L. Alpha-linolenic acid deficiency in patients on long-term gastric-tube feeding: Estimation of linolenic acid and long-chain unsaturated $n-3$ fatty acid requirement in man. *Am. J. Clin. Nutr.*, 45 (1987) 66–77.

[106] Lee, J. H., Fukumoto, M., Nishida, H., Ikeda, I. and Sugano, M. The interrelated effects of $n-6/n-3$ and polyunsaturated/saturated ratios of dietary fats on the regulation of metabolism in rats. *J. Nutr.*, 119 (1989) 1893–1899.

[107] Nouvelot, A., Delbart, C. and Bourre, J. M. Hepatic metabolism of dietary alpha-linolenic acid in suckling rats, and its possible importance in polyunsaturated fatty acid uptake by the brain. *Ann. Nutr. Metab.*, 30 (1986) 316–323.

[108] Holman, R. T., Johnson, S. B. and Hatch, T. F. A case of human linolenic acid deficiency involving neurological abnormalities. *Am. J. Clin. Nutr.*, 35 (1982) 617–623.

[109] Marcel, Y. L., Christiansen, K. and Holman, R. T. The preferred metabolic pathway from linoleic acid to arachidonic acid *in vitro*. *Biochim. Biophys. Acta*, 164 (1968) 25–34.

[110] Lasserre, M., Mendy, F., Spielmann, D. and Jacotot, B. Effects of different dietary intake of essential fatty acids on C20:3$n-6$ and C20:4$n-6$ serum levels in human adults. *Lipids*, 20 (1985) 227–233.

[111] Manku, M. S., Horrobin, D. F., Morse, N. L., Wright, S. and Burton, J. L. Essential fatty acids in the plasma phospholipids of patients with atopic eczema. *Br. J. Dermatol.*, 110 (1984) 643–648.

[112] Singer, P., Jaeger, W., Voigt, S. and Thiel, H., Defective desaturation and elongation of $n-6$ and $n-3$ fatty acids in hypertensive patients. *Prostaglandins Leuk. Med.*, 15 (1984) 159–165.

[113] Singer, P., Berger, I., Wirth, M., Godicke, M., Jaeger, W. and Voigt, S. Slow desaturation and elongation of linoleic and alpha-linolenic acids as a rationale of eicosapentaenoic acid-rich diet to lower blood pressure and serum lipids in normal, hypertensive and hyperlipidemic subjects. *Prostaglandins Leuk. Med.*, 24 (1986) 173–193.

[114] Sanders, T. A. B. and Younger, K. M. The effect of dietary supplements of omega-3 polyunsaturated fatty acids on the fatty acid composition of platelets and plasma choline phosphoglycerides. *Br. J. Nutr.*, 45 (1981) 613–616.

[115] Sanders, T. A. B. and Roshanani, F. The influence of different types of omega-3 polyunsaturated fatty acids on blood lipids and platelet function in healthy volunteers. *Clin. Sci.*, 64 (1983) 91–99.

[116] Adam, O., Wolfram, G. and Zollner, N. Effect of alpha-linolenic acid in the human diet on linoleic acid metabolism and prostaglandin biosynthesis. *J. Lipid Res.*, 27 (1986) 421–426.

[117] Cunnane, S. C., Wolever, T. M. S., Armstrong, J. K. and Jenkins, D. J. A. Differential mobilization of essential fatty acids into the serum free fatty acid pool in response to glucose ingestion. In Galli, C. and Simopoulos, A. P. (eds), *Dietary Omega-3 and Omega-6 Fatty Acids* (series A: Life Sciences, Vol. 171). Plenum Press, New York, 1989, pp. 385–387.

[118] Ganguli, S., Cunnane, S. C., Armstrong, J. K., Wolever, T. M. S. and Jenkins, D. J. A. Effect of flax on serum cholesterol and plasma and red cell essential fatty acids in humans. In *12th International Conference on Social and Preventive Medicine*, Montreal, August, 1989.

[119] Cunnane, S. C. Evidence that the adverse effects of zinc deficiency on EFA composition in rats are independent of food intake. *Br. J. Nutr.*, 59 (1988) 273–278.

[120] Cunnane, S. C. Role of zinc in lipid and long chain fatty acid metabolism and in membranes. *Progr. Food Nutr. Sci.*, 12 (1988) 151–188.

[121] Cunnane, S. C. and Wahle, K. W. J. Zinc deficiency increases the rate of delta-6 desaturation of linoleic acid in rat mammary tissue. *Lipids*, 16 (1981) 771–774.

[122] Mahfouz, M. M. and Kummerow, F. A. Effect of magnesium deficiency on delta-6 desaturase activity and fatty acid composition of rat liver microsomes. *Lipids*, 24 (1989) 727–732.

[123] Cunnane, S. C., Horrobin, D. F. and Manku, M. S. Accumulation of linoleic and gamma-linolenic acids in tissue lipids of pyroxidine deficient rats. *J. Nutr.*, 114 (1984) 1754–1761.

[124] Cunnane, S. C. Vitamin E intake affects serum thromboxane and tissue essential fatty acid composition in the rat. *Ann. Nutr. Metab.*, 32 (1988) 90–96.

[125] Wahle, K. W. J. and Davies, N. T. Effects of dietary copper deficiency in the rat on fatty acid composition of adipose tissue and desaturase activity of liver microsomes. *Br. J. Nutr.*, 34 (1975) 105–112.

[126] Hammermueller, J. D., Bray, T. M. and Bettger, W. J. Effect of zinc and copper deficiency on microsomal NADPH-dependent active oxygen generation in rat lung and liver. *J. Nutr.*, 117 (1987) 894–901.

[127] Krieger, I., Alpern, B. E. and Cunane, S. C. Transient neonatal zinc deficiency. *Am. J. Clin. Nutr.*, 43 (1986) 955–958.

[128] Mack, D., Koletzko, B., Cunnane, S., Cutz, E. and Griffiths, A. Acrodermatitis enteropathica with normal serum zinc levels: diagnostic value of small bowel biopsy and essential fatty acid determination. *Gut*, 30 (1989) 1426–1429.

[129] Bjerve, K. S., Fischer, S. and Alme, K. Alpha-linolenic acid deficiency in man: Effect of ethyl linolenate on plasma and erythrocyte fatty acid composition and biosynthesis of prostanoids. *Am. J. Clin. Nutr.*, 46 (1987) 570–576.

[130] Bjerve, K. S., Fischer, S., Wammer, F. and Egeland, T. Alpha-linolenic acid and long-chain omega-3 fatty acid deficiency: Effect on lymphocyte function, plasma and red cell lipids, and prostanoid formation. *Am. J. Clin. Nutr.*, 49 (1989) 290–300.

[131] Miller, D. G., Williams, S. K., Palombo, J. D., Griffin, R. E., Bistrian, B. R. and Blackburn, G. L. Cutaneous application of safflower oil in preventing essential fatty acid deficiency in patients on home parenteral nutrition. *Am. J. Clin. Nutr.*, 46 (1987) 419–423.

[132] Freund, H., Floman, N., Schwartz, B. and Fischer, J. E. Essential fatty acid deficiency in total parenteral nutrition. *Ann. Surg.*, 190 (1979) 139–143.

[133] Rivers, J. P. W. and Hassam, A. G. Defective essential-fatty-acid metabolism in cystic fibrosis. Lancet, October (1975) 642–645.

[134] Lloyd-Still, J. D., Johnson, S. B. and Holman, R. T. Essential fatty acids in cystic fibrosis and the effects of safflower oil supplementation. *Am. J. Clin. Nutr.*, 34 (1981) 1–7.

[135] Holman, R. T. Essential fatty acids and nutritional disorders. Nestle Nutr. Workshop Ser., *Lipids Mod. Nutr.*, 13 (1987) 157–171.

[136] Cabre, E., Peiago, J. L., Abad-Lacruz, A., Gil, A., Gonzalez-Huiz, F., Sanchez de Medina, F. and Gassul, M. A. Polyunsaturated fatty acid deficiency in liver cirrhosis: its relation to associated protein-energy malnutrition (preliminary report). *Am. J. Gastroenterol.*, 83 (1988) 712–717.

[137] Johnson, S. B., Gordon, E., McClain, C., Low, G. and Holman, R. T. Abnormal polyunsaturated fatty acid patterns of serum lipids in alcoholism and cirrhosis: Arachidonic acid deficiency in cirrhosis. *Proc. Natl. Acad. Sci. USA*, 82 (1985) 1815–1818.

[138] Moynahan, E. J. Acrodermatitis enteropathica: A lethal inherited zinc deficiency disorder. *Lancet*, August (1974) 399.

[139] Cunnane, S. C. *Zinc: Clinical and Biochemical Significance*, CRC Press, Boca Raton, 1988, p. 100.

[140] Cunnane, S. C. and Krieger, I. Fatty acids in serum phospholipids in Acrodermatitis Enteropathica before and after zinc treatment. *J. Am. Coll. Nutr.*, 7 (1988) 249–250.

[141] Hara, A. and Taketomi, T. Cerebral lipid and protein abnormalities in Menkes' steely-hair disease. *Jpn. J. Exp. Med.*, 56 (1986) 277–284.

[142] Cunnane, S. C. The profile of long chain fatty acids in serum phospholipids: A possible indicator of copper status in humans. *Am. J. Clin. Nutr.*, 48 (1988) 1475–1478.

[143] Stubbs, C. D. and Smith, A. D. The modification of mammalian membrane polyunsaturated fatty acid composition in relation to membrane fluidity and function. *Biochim. Biophys. Acta*, 779 (1984) 89–137.

[144] Shinitzky, M., Membrane fluidity and cellular functions, *Physiol. Membrane Fluidity*, 1 (1984) 1–51.

[145] Brenner, R. R. Effect of unsaturated acids on membrane structure and enzyme kinetics. *Progr. Lipid Res.*, 23 (1984) 69–96.

[146] Spector, A. A. and Yorek, M. A. Membrane lipid composition and cellular function. *J. Lipid Res.*, 26 (1985) 1015–1035.

[147] Youyou, A., Durand, G., Pascal, G., Piciotti, M., Dumont, O. and Bourre, J. M. Recovery of altered fatty acid composition induced by a diet devoid of $n-3$ fatty acids in myelin, synaptosomes, mitochondria, and microsomes of developing rat brain. *J. Neurochem.*, 46 (1986) 224–228.

[148] Bourre, J. M., Francois, M., Youyou, A., Dumont, O., Piciotti, M., Pascal, G. and Durand, G. The effects of dietary alpha-linolenic acid on the composition of nerve membranes, enzymatic activity, amplitude of electrophysiological parameters, resistance to poisons and performance of learning tasks in rats. *J. Nutr.*, 119 (1989) 1880–1882.

[149] Cooper, R. A. Abnormalities of cell-membrane fluidity in the pathogenesis of disease. *New Eng. J. Med.*, 297 (1977) 371–377.

[150] Applegate, K. R. and Glomset, J. A. Computer-based modelling of the conformation and packing properties of docosahexaenoic acid. *J. Lipid Res.*, 27 (1986) 658–680.

[151] Samuelsson, B. Leukotrienes: Mediators of immediate hypersensitivity reactions and inflammation. *Science*, 220 (1983) 568–575.

[152] Hansen, T. M., Lerche, A., Kassis, V., Lorenzen, I. and Sondergaard, J. Treatment of rheumatoid arthritis with prostaglandin E_1 precursors *cis*-linoleic acid and gamma-linolenic acid. *Scand. J. Rheumatol.*, 12 (1983) 85–88.

[153] Samuelsson, B., Dahlen, S. E., Lindgren, J., Rouzer, C. A. and Serhan, C. N. Leukotrienes and lipoxins: Structures, biosynthesis, and biological effects. *Science*, 237 (1987) 1171–1176.

[154] Smith, D. L., Willis, A. L., Nguyen, N., Conner, D., Zahedi, S. and Fulks, J. Eskimo plasma constituents, dihomo-gamma-linolenic acid, eicosapentaenoic acid and docosahexaenoic acid inhibit the release of atherogenic mitogens. *Lipids*, 24 (1989) 70–75.

[155] Jenkins, D. K., Mitchell, J. C., Manku, M. S. and Horrobin, D. F. Effects of different sources of gamma-linolenic acid on the formation of essential fatty acid and prostanoid metabolites. *Med. Sci. Res.*, 16 (1988) 525–526.

[156] Miller, C. C., Ziboh, V. A., Wong, T. and Fletcher, M. P. Dietary supplementation with oils rich in $(n-3)$ and $(n-6)$ fatty acids influences *in vivo* levels of epidermal lipoxygenase products in guinea pigs. *J. Nutr.*, 120 (1990) 36–44.

[157] Fischer, S. and Weber, P. C. Thromboxane A_3 (TXA_3) is formed in human platelets after dietary eicosapentaenoic acid (C20:5-omega-3). *Biochem. Biophys. Res. Comm.*, 116 (1983) 1091–1099.

[158] Hwang, D. H., Boudreau, M. and Chanmugam, P. Dietary linolenic acid and longer-chain $n-3$ fatty acids: Comparison of effects on arachidonic acid metabolism in rats, *J. Nutr.*, 118 (1988) 427–437.

[159] Nassar, B. A., Manku, M. S., Huang, Y. S., Jenkins, D. K. and Horrobin, D. F. The influence of dietary marine oil (Polepa) and evening primrose oil (Efamol) on prostaglandin production by the rat mesenteric vasculature. *Prostaglandins Leukotrienes Med.*, 26 (1987) 253–263.

[160] Leyton, J., Drury, P. J. and Crawford, M. A. Differential oxidation of essential and non-essential fatty acids *in vivo* in the rat. *Br. J. Nutr.*, 57 (1987) 383–393.

[161] Jones, P. J. H., Pencharz, P. B. and Clandinin, M. T. Absorption of ^{13}C-labelled stearic, oleic, and linoleic acids in humans: Application to breath tests. *J. Lab. Clin. Med.*, 105 (1985) 647–652.

[162] Nichaman, M. Z., Sweeley, C. C. and Olson, R. E. Plasma fatty acids in normolipemic and hyperlipemic subjects during fasting and after linoleate feeding. *Am. J. Clin. Nutr.* 20 (1967) 1057–1069.

[163] Nestel, P. J. and Barter, P. Metabolism of palmitic and linoleic acids in man: differences in turnover and conversion to glycerides. *Clin. Sci.*, 40 (1971) 345—350.

[164] Carroll, K. K. Essential fatty acids: What level in the diet is most desirable. In Bazan, N. G., Brenner, R. R. and Giusto, N. M. (eds), *Function and Biosynthesis of Lipids*, Plenum, New York, 1977, pp. 535–546.

[165] van Staveren, W. A., Deurenberg, P., Katan, M. B., Burema, L., De Groot, L. C. P. G. M. and Hoffmans, M. D. A. F. Validity of the fatty acid composition of subcutaneous fat tissue microbiopsies as an estimate of the long-term average fatty acid composition of the diet of separate individuals. *Am. J. Epidemiol.*, 123 (1986) 455–463.

[166] Willis, A. L., Comai, K., Kuhn, D. L. and Paulsrud, J. Dihomo-gamma-linolenate suppresses platelet aggregation when administered *in vitro* or *in vivo*. *Prostaglandins*, 8 (1974) 509.

[167] Hansen, A. E., Haggard, M. E., Boelsche, A. N., Adam, D. J. D. and Wiese, H. F. Essential fatty acids in infant nutrition III. Clinical manifestations of linoleic acid deficiency. *J. Nutr.*, 66 (1958) 565–576.

[168] Gibson, R. A. and Kneebone, G. M. Fatty acid composition of human colostrum and mature breast milk. *Am. J. Clin. Nutr.*, 34 (1981) 252–257.

[169] Lepage, G., Collet, S., Bougle, D., Kien, L. C., Lepage, D., Dallaire, L., Darling, P. and Roy, C. C. The composition of preterm milk in relation to the degree of prematurity. *Am. J. Clin. Nutr.*, 40 (1984) 1042–1049.

[170] Carlson, S. E., Rhodes, P. G. and Ferguson, M. G. Docosahexaenoic acid status of preterm infants at birth and following feeding with human milk or formula. *Am. J. Clin. Nutr.*, 44 (1986) 798–804.

[171] Putnam, J. C., Carlson, S. E., DeVoe, P. W. and Barness, L. A., The effect of variations in dietary fatty acids on the fatty acid composition of erythrocyte phosphatidylcholine and phosphatidylethanolamine in human infants. *Am. J. Clin. Nutr.*, 36 (1982) 106–114.

[172] Carlson, S. E., Carver, J. D. and House, S. G. High fat diets varying in ratios of polyunsaturated to saturated fatty acids and linoleic to linolenic acid: A comparison of rat neural and red cell membrane phospholipids. *J. Nutr.*, 116 (1986) 718–725.

[173] Sanders, T. A. B. and Naismith, D. J. A comparison of the influence of breast-feeding and bottle-feeding on the fatty acid composition of the erythrocytes. *Br. J. Nutr.*, 41 (1979) 619–623.

[174] Jackson, K. A. and Gibson, R. A. Weaning foods cannot replace breast milk as sources of long-chain polyunsaturated fatty acids. *Am. J. Clin. Nutr.*, 50 (1989) 980–982.

[175] Sanders, T. A. B., Ellis, F. R. and Dickerson, J. W. T. Polyunsaturated fatty acids and the brain. *Lancet*, April 2 (1977) 751–752.

[176] Clandinin, M. T., Chappell, J. E., Leong, S., Heim, T., Swyer, P. R. and Chance, G. W. Intra-uterine fatty acid accretion rates in human brain: Implications for fatty acid requirements. *Early Hum. Dev.*, 4 (1980) 121–129.

[177] Yamamoto, N., Saitoh, M., Moriuchi, A., Nomura, M. and Okuyama, H. Effect of dietary alpha-linolenate/linoleate balance on brain lipid compositions and learning ability of rats. *J. Lipid Res.*, 28 (1987) 144–151.

[178] Lamptey, M. S. and Walker, B. L. A possible essential role for dietary linolenic acid in the development of the young rat. *J. Nutr.*, 106 (1976) 86–93.

[179] Drury, P. J., Crawford, M. A. and Laurance, B. M. Comparison of the fatty acids in human milk from Hungary and Thailand. *Progr. Lipid Res.*, 25 (1986) 235–238.

[180] Nichaman, M. Z., Olson, R. E. and Sweeley, C. C. Metabolism of linoleic acid-1-^{14}C in normolipemic and hyperlipemic humans fed linoleate diets. *Am. J. Clin. Nutr.*, 20 (1967) 1070–1083.

[181] Budowski, P., Trostler, N., Lupo, M., Vaisman, N. and Eldor, A. Effect of linseed oil ingestion on plasma lipid fatty acid composition and platelet aggregability in healthy volunteers. *Nutr. Res.*, 4 (1984) 343–346.

[182] Budowski, P. and Crawford, M. A. Alpha-linolenic acid as a regulator of the metabolism of arachidonic acid: Dietary implications of the ratio, $n-6{:}n-3$ fatty acids. *Proc. Nutr. Soc.* 44 (1985) 221–229.

[183] Leikin, A. I. and Brenner, R. R. Cholesterol-induced microsomal changes modulate desaturase activities. *Biochim. Biophys. Acta.*, 992 (1987) 294–303.

[184] Hornstra, G. The significance of fish and fish-oil enriched food for prevention and therapy of ischaemic cardiovascular disease. In Vergroesen, A. J. and Crawford,

M. (eds), *The Role of Fats in Human Nutrition*. Academic Press, London, 1989, pp. 151–235.

[185] Sinclair, H. M. Advantages and disadvantages of an Eskimo diet. In Fumagalli, R., Kritchevski, D. and Paoletti, R. (eds), *Drugs Affecting Lipid Metabolism*. Elsevier/North Holland Biomedical Press, Amsterdam, 1980, pp. 363–370.

[186] Agren, J. J., Hanninen, O., Laitinen, M., Seppanen, K., Bernhardt, I., Fogelholm, L., Herranen, J. and Penttila, I. Boreal freshwater fish diet modifies the plasma lipids and prostanoids and membrane fatty acids in man. *Lipids*, 23 (1988) 924–929.

[187] Magnuson, B. A., Schiefer, H. B., Crichlow, E. C., Bell, J. M. and Olson, J. P. Effects of various high-fat diets on myocardial contractility and morphology in rats. *Drug-Nutrient Interact.*, 5 (1988) 213—226.

[188] Moore, T. G. *Heart Failure*. Random House, New York, 1989.

[189] Muller, A. D., van Houwelingen, A. C., van Dam-Mieras, M. C. E., Bas, B. M. and Hornstra, G. Effect of a moderate fish intake on haemostatic parameters in healthy males. *Thromb. Haemost.*, 61 (1989) 468–473.

[190] van Houwelingen, A. C., Hornstra, G., Kromhout, D. and de Lezenne Coulander, C. Habitual fish consumption, fatty acids of serum phospholipids and platelet function. *Atherosclerosis*, 75 (1989) 157–165.

[191] McLennan, P. L., Abeywardena, M. Y. and Charnock, J. S. Reversal of the arrythmogenic effects of long-term saturated fatty acid intake by dietary $n-3$ and $n-6$ polyunsaturated fatty acids. *Am. J. Clin. Nutr.*, 51 (1990) 53–58.

[192] Hardarson, T., Kristinsson, A., Skuladottir, G., Asvaldsdottir, H. and Snorrasson, S. P. Cod liver oil does not reduce ventricular extrasystoles after myocardial infarction. *J. Int. Med.*, 226 (1989) 33–37.

[193] Burr, M. L., Gilbert, J. F., Holliday, R. M., Elwood, P. C., Fehily, A. M., Rogers, S., Sweetnam, P. M. and Deadman, N.M. Effects of changes in fat, fish, and fibre intakes on death and myocardial reinfarction: diet and reinfarction trial (DART). *Lancet*, September 30 (1989) 757–761.

[194] *Lancet*. Fish and the heart (letters), December 16 (1989) 1450–1452.

[195] Abeywardena, M. Y., Fischer, S., Schweer, H. and Charnock, J. S. *In vivo* formation of metabolites of prostaglandins I_2 and I_3 in the marmoset monkey (*Callithrix jacchus*) following dietary supplementation with tuna fish oil. *Biochim. Biophys. Acta*, 1003 (1989) 161–166.

[196] Goldsmith, M. F. Endothelial dysfunction plays dynamic role in coronary artery disease. *J. Am. Med. Assoc.*, 263 (190) 789–790.

[197] Wahlqvist, M. L., Lo, C. S. and Myers, K. A. Fish intake and arterial wall characteristics in healthy people and diabetic patients. *Lancet*, October 21 (1989) 944–946.

[198] Bertherat, J., Belot-Veyssier, C., Baudin, E., Casanova, S. and Polak, M. Fish, diabetes, and the arterial wall. *Lancet*, December 2 (1989) 1332.

[199] Yerram, N. R. and Spector, A. A. Effects of omega-3 fatty acids on vascular smooth muscle cells: Reduction in arachidonic acid incorporation into inositol phospholipids. *Lipids*, 24 (1989) 594–602.

[200] Yerram, N. R., Moore, S. A. and Spector, A. A. Eicosapentaenoic acid metabolism in brain microvessel endothelium: effect on prostaglandin formation. *J. Lipid Res.* 30 (1989) 1747–1757.

[201] Dyerberg, J. and Jørgensen, K. A. Marine oils and thrombogenesis. *Progr. Lip. Res.* 21 (1982) 255–269.

[202] Hennig, B. and Watkins, B. A. Linoleic acid and linolenic acid: effect on permeability properties of cultured endothelial cell monolayers. *Am. J. Clin. Nutr.* 49 (1989) 301–305.

[203] Zhu, B.-Q., Sievers, R. E., Isenberg, W. M., Smith, D. L. and Parmley, W. W. Regression of atherosclerosis in cholesterol-fed rabbits: effects of fish oil and verapamil. *J. Am. Coll. Cardiol.*, 15 (1990) 231–237.

[204] Weiner, B. H., Ockene, I. S., Levine, P. H., Cuénoud, H. F., Fisher, M., Johnson, B. F., Daoud, A. S., Jarmolych, J., Hosmer, D., Johnson, M. H., Natale, A., Vaudreuil, C., Hoogasian, J. J. Inhibition of atherosclerosis by cod-liver oil in a hyperlipidemic swine model. *New Eng. J. Med.*, 315 (1986) 841–846.

[205] Hartog, J. M., Verdouw, P. D., Klompe, M. and Lamers, J. M. J. Dietary mackerel in pigs: effect on plasma lipids, cardiac sarcolemmal phospholipids and cardiovascular parameters. *J. Nutr.*, 117 (1987) 1371–1378.

[206] Hartog, J. M., Lamers, J. M. J., Montfoort, A., Becker, A. E., Klompe, M., Morse, H., ten Cate, F. J., van der Werf, L., Hulsmann, W. C., Hugenholtz, P. G. and Verdouw, P. D. Comparison of mackerel-oil and lard-fat enriched diets on plasma lipids, cardiac membrane phospholipids, cardiovascular performance, and morphology in young pigs. *Am. J. Clin. Nutr.*, 46 (1987) 258–266.

[207] Hartog, J. M., Lamers, J. M. J., Essed, C. E., Schalkwijk, W. P. and Verdouw, P. D. Does platelet aggregation play a role in localized intimal proliferation in normolipidemic pigs with fixed coronary artery stenosis fed dietary fish oil? *Atherosclerosis*, 76 (1989) 79–88.

[208] Sassen, L. M. A., Hartog, J. M., Lamers, J. M. J., Klompe, M., Van Woerkens, L. J. and Verdouw, P. D. Mackerel oil and atherosclerosis in pigs. *Eur. Heart J.*, 10 (1989) 838–846.

[209] Kim, D. N., Ho, H.-T., Lawrence, D. A., Schmee, J. and Thomas, W. A. Modification of lipoprotein patterns and retardation of atherogenesis by a fish oil supplement to a hyperlipidemic diet for swine. *Atherosclerosis*, 76 (1989) 35–54.

[210] Foxall, T. L. and Shwaery, G. T. Effects of dietary fish oil and butterfat on serum lipids and monocyte and platelet interactions with aortic endothelial cells. *Atherosclerosis*, 80 (1990) 171–179.

[211] Dehmer, G. J., Popma, J. J., van den Berg, E. K., Eichhorn, E. J., Prewitt, J. B., Campbell, W. B., Jennings, L., Willerson, J. T. and Schmitz, J. M. Reduction in the rate of early restenosis after coronary angioplasty by a diet supplemented with $n-3$ fatty acids. *New Eng. J. Med.*, 319 (1988) 733–740.

[212] Reis, G. J., Boucher, T. M., Sipperly, M. E., Silverman, D. I., McCabe, C. H., Baim, D. S., Sacks, F. M., Grossman, W. and Pasternak, R. C. Randomised trial of fish oil for prevention of restenosis after coronary angioplasty. *Lancet*, July 15 (1989) 177–181.

[213] *Lancet*. Letters, August 19 (1989) 443.

[214] *Lancet*. Letters, September 16 (1989) 693–694.

[215] Grundy, S. M., Bilheimer, D., Blackburn, H., Brown, V., Kwiterovich, P. O., Mattson, F., Schonfeld, G. and Weidman, W. H. Rationale of the statement of the American Heart Association. *Circulation*, 65 (1982) 839A–854A.

[216] Horrobin, D. F. and Manku, M. S. How do polyunsaturated fatty acids lower plasma cholesterol levels? *Lipids*, 18 (1983) 558–562.

[217] Kingsbury, K. J., Morgan, D. M., Aylott, C. and Emmerson, R. Effects of ethyl arachidonate, cod-liver oil, and corn oil on the plasma-cholesterol level. *Lancet*, April 8 (1961) 739–741.

[218] Mikhailidis, D. P., Kirtland, S. J., Barradas, M. A., Mahadeviah, S. and Dandona, P. The effect of dihomogammalinolenic acid on platelet aggregation and prostaglandin release, erythrocyte membrane fatty acids and serum lipids: Evidence for defects in PGE_1 synthesis and delta 5-desaturase activity in insulin-dependent diabetics. *Diabetes Res.*, 3 (1986) 7–12.

[219] Singer, P. Decrease of linoleic acid in serum lipids and increase of arachidonic acid in serum triglycerides after diets supplemented with $n-3$ fatty acids. *Akt. Ernahr.*, 14 (1989) 264–268.

[220] Schectman, G., Kaul, S., Cherayil, G. D., Lees, M. and Kissebah, A. Can the hypotriglyceridemic effect of fish oil concentrate be sustained? *Ann. Int. Med.*, 110 (1989) 346–352.

[221] Havel, R. J. Role of triglyceride-rich lipoproteins in progression of atherosclerosis. *Circulation*, 81 (1990) 694–696.

[222] Harris, W. S. Fish oils and plasma lipid and lipoprotein metabolsim in humans: A critical review. *J. Lipid Res.*, 30 (1989) 785–807.

[223] Bakker, D. J., Haberstroh, B. N., Philbrick, D. J. and Holub, B. J. Triglyceride lowering in nephrotic syndrome patients consuming a fish oil concentrate. *Nutr. Res.*, 9 (1989) 27–34.

[224] Anonymous. Fish oils reduce postprandial lipemia. *Nutr. Revs.*, 47 (1989) 211–213.

[225] Rustan, A. C. and Drevon, C. A. Eicosapentaenoic acid inhibits hepatic production of very low density lipoprotein. *J. Inter. Med.*, 225 Suppl. 1 (1989) 31–38.

[226] Rustan, A. C., Nossen, J. O., Christiansen, E. N. and Drevon, C. A. Eicosapentaenoic acid reduces hepatic synthesis and secretion of triacylglycerol by decreasing the activity of acyl-coenzyme A:1,2-diacylglycerol acyltransferase. *J. Lipid Res.*, 29 (1988) 1417–1426.

[227] Zucker, M. L., Bilyeu, D. S., Helmkamp, G. M., Harris, W. S. and Dujovne, C. A. Effects of dietary fish oil on platelet function and plasma lipids in hyperlipoproteinemic and normal subjects. *Atherosclerosis*, 73 (1988) 13–22.

[228] Dart, A. M., Riemersma, R. A. and Oliver, M. F. Effects of MaxEPA on serum lipids in hypercholesterolaemic subjects. *Atherosclerosis*, 80 (1989) 119–124.

[229] Molgaard, J., von Schenck, H., Lassvik, C., Kuusi, T. and Olsson, A. G. Effect of fish oil treatment on plasma lipoproteins in type III hyperlipoproteinaemia. *Atherosclerosis*, 81 (1990) 1–9.

[230] Berry, E. M. and Hirsch, J. Does dietary linolenic acid influence blood pressure? *Am. J. Clin. Nutr.*, 44 (1986) 336–340.

[231] Salonen, J. T., Salonen, R., Ihanainen, M., Parviainen, M., Seppanen, R., Kantola, M., Seppanen, K. and Rauramaa, R. Blood pressure, dietary fats, and antioxidants. *Am. J. Clin. Nutr.*, 48 (1988) 1226–1232.

[232] Singer, P., Wirth, M., Voigt, S., Richter-Heinrick, E., Godicke, W., Berger, I., Naumaun, E., Listing, J., Hartrodt, W. and Taube C. Blood pressure and lipid-lowering effect of mackerel and herring diet in patients with mild essential hypertension. *Atherosclerosis*, 56 (1985) 223–235.

[233] Knapp, H. R. and Fitzgerald, G. A. The antihypertensive effects of fish oil: A controlled study of polyunsaturated fatty acid supplements in essential hypertension. *New Eng. J. Med.*, 320 (1989) 1037–1043.

[234] Bonaa, K. H., Bjerve, K. S., Straume, B., Gram, I. T. and Thelle, D. Effect of eicosapentaenoic and docosahexaenoic acids on blood pressure in hypertension. *New Eng. J. Med*, 322 (1990) 795–801.

[235] Goodnight, S. H., Harris, W. S. and Connor, W. E. The effects of dietary omega-3 fatty acids on platelet composition and function in man: A prospective, controlled trial. *Blood*, 58 (1981) 880–885.

[236] Knapp, H. R., Reilly, I. A. G., Alessandrini, P. and Fitzgerald, G. A. *In vivo* indexes of platelet and vascular function during fish-oil administration in patients with atherosclerosis. *J. Am. Coll. Cardiol.*, 12 (1986) 1073–1078.

[237] Skeaff, C. M. and Holub, B. J. The effect of fish oil consumption on platelet aggregation responses in washed human platelet suspensions. *Thromb. Res.*, 51 (1988) 105–115.

[238] Herold, M. and Kinsella, J. E. Fish oil consumption and decreased risk of cardiovascular and human feeding trials. *Am. J. Clin. Nutr.*, 43 (1986) 566–598.

[239] Green, D., Barreres, L., Borensztajn, J., Kaplan, P., Reddy, M. N., Rovner, R. and Simon, H. A double-blind, placebo-controlled trial of fish oil concentrate (MaxEpa) in stroke patients. *Stroke*, 16 (1985) 706–709.

[240] Hostmark, A. T., Bjerkedal, T., Kierulf, P., Flaten, H. and Ulshagen, K. Fish oil and plasma fibrinogen. *Br. Med. J.*, 297 (1988) 180–181.

[241] Wood, D. A., Butler, S., Riemersma, R. A., Thomson, M., Oliver, M. F., Fulton, M., Birtwhistle, A. and Elton, R. Adipose tissue and platelet fatty acids and coronary heart disease in Scottish men. *Lancet*, July 21 (1984) 117–121.

[242] Simpson, H. R. C., Barker, K., Carter, R. D., Cassels, E. and Mann, J. I. Low dietary intake of linoleic acid predisposes to myocardial infarction. *Br. Med. J.*, 285 (1982) 683–684.

[243] Wood, D. A., Riemersma, R. A., Butler, S., Thomson, M., Macintyre, C., Elton, R. A. and Oliver, M. F. Linoleic and eicosapentaenoic acids in adipose tissue and platelets and risk of coronary heart disease. *Lancet*, January 24 (1987) 177–183.

[244] Aukema, H. M. and Holub, B. J., Effect of dietary supplementation with a fish oil concentrate on the alkenylacyl class of ethanolamine phospholipid in human platelets. *J. Lipid Res.*, 30 (1989) 59–64.

[245] Owren, P. A., Hellem, A. J. and Odegaard, A. Linolenic acid for the prevention of thrombosis and myocardial infarction. *Lancet*, November 7 (1964) 975–979.

[246] Holub, B. J. Effect of flax and dietary alpha-linolenic acid on blood platelets in human subjects. *Proceedings of the 53rd Flax Institute*, Fargo, ND, January 1990.

[247] Mahadevappa, V. G. and Holub, B. J. Quantitative loss of individual eicosapentaenoyl-relative to arachidonoyl-containing phospholipids in thrombin-stimulated human platelets. *J. Lipid Res.*, 28 (1987) 1275–1280.

[248] Croset, M., Guichardant, M. and Lagarde, M. Different metabolic behaviour of long-chain $n-3$ polyunsaturated fatty acids in human platelets. *Biochim. Biophys. Acta*, 961 (1988) 262–269.

[249] Hornstra, G. Regulation of eicosanoid production by fatty acid availability. *Adv. Inflamm. Res.*, 10 (1985) 1–6.

[250] Needleman, P., Raz, A., Minkes, M. S., Ferrendelli, J. A. and Sprecher, H. Triene prostaglandins; Prostacyclin and thromboxane biosynthesis and unique biological properties. *Proc. Natl. Acad. Sci. USA*, 76 (1979) 944–948.

[251] Fischer, S. and Weber, P. C. Prostaglandin I_3 is formed *in vivo* in man after dietary eicosapentaenoic acid. *Nature*, 307 (1984) 165–168.

[252] Borkman, M., Chisholm, D. J., Furler, S. M., Storlien, L. H., Kraegen, E. W., Simons, L. A. *et al.* Effects of fish oil supplementation on glucose and lipid metabolism in NIDDM. *Diabetes*, 38 (1989) 1314–1319.

[253] Popp-Snijders, C., Schouten, J. A., Heine, R. J., van der Meer, J. and van der Veen, E. A. Dietary supplementation of omega-3-polyunsaturated fatty acids improves insulin sensitivity in non-insulin-depedent diabetics. *Diabetes Res.*, 4 (1987) 141–147.

[254] Glauber, H., Wallace, P., Grive, K. and Brechtec, G., Adverse metabolic effect of omega-3 fatty acids in non-insulin-dependent diabetes mellitus. *Ann. Intern. Med.*, 108 (1988) 663–668.

[255] Driss, F. and Darcet, Ph., Effects of fish oils rich in $n-3$ fatty acids on cardiovascular risk factors. *Rev. Franc. Corps Gras*, 35 (1988) 7–11.

[256] Miller, M. E., Anagatgou, A. A., Ley, B., Marshall, P. and Steiner, M. Effect of fish oil concentrates on hemorheological and hemostatic aspects of diabetes mellitus: A preliminary study. *Thromb. Res.*, 47 (1987) 201—214.

[257] Rillaerts, E. G., Engelmann, G. J., van Camp, K. M. and de Leeuw, I. Effect of omega-3 fatty acids in diet of type 1 diabetic subjects on lipid values and hemorheological parameters. *Diabetes*, 38 (1989) 1412–1416.

[258] Harris, W. S., Dujorne, C. A., Zucker, M. and Johnson, B. Effects of a low saturated fat, low cholesterol fish oil supplement in hypertriglyceridemic patients. *Ann. Int. Med.*, 109 (1988) 465–470.

[259] Harris, W. S., Zucker, M. L., Dujorne, C. A. ω-3 fatty acids in hypertriglyceridemic patients: Triglycerides vs methyl esters. *Am. J. Clin. Nutr.*, 48 (1088) 992–997.

[260] Wolfram, G. ω-3- und ω-6-Fettsauren—biochemische besonderheiten und biologische wirkungen. *Fat Sci. Technol.*, 91 (1989) 459–468.

[261] McCormick, J. and Skrabanek, P. Coronary heart disease is not preventable by population interventions. *Lancet*, October 8, (1988) 839–841.

[262] Stacpoole, P. W., Alig, J., Kilgore, L. L., Ayala, C. M., Herbert, P. N., Zech, L. A. and Fisher, W. R. Lipodystrophic diabetes mellitus: Investigations of lipoprotein metabolism and the effects of omega-3 fatty acid administration in two patients. *Metabolism*. 37 (1988) 944–951.

[263] Poisson, J. P. Comparative *in vivo* and *in vitro* study of the influence of experimental diabetes on rat liver linoleic acid delta-6 and delta-5 desaturation. *Enzyme*, 34 (1985) 1–14.

[264] Beitz, J., Schimke, E., Liebaug, U., Block, H. U., Beitz, A., Honigmann, G., Sziegoleit, W., Muller, G., and Mest, H. J. Influence of a cod liver oil diet in healthy and insulin-dependent diabetic volunteers on fatty acid pattern inhibition of prostacyclin formation by low density lipoprotein (LDL) and platelet thromboxane. *Klin. Wocheschr.*, 64 (1986) 793–799.

[265] Faas, F. H., Dang, A. Q., Kemp, K., Norman, J. and Carter, W. J. Red blood cell plasma fatty acid composition in diabetes mellitus. *Metabolism*. 37 (1988) 711–713.

[266] Freyburger, G., Gin, H., Heape, A., Juguelin, H., Boisseau, M. R. and Cassagne, C. Phospholipid and fatty acid composition of erythrocytes in type I and type II diabetes. *Metabolism*, 38 (1989) 673–678.

[267] Futterman, S. and Kupfer, C. The fatty acid composition of the retinal vasculature of normal diabetic human eyes. *Invest. Opthalmol.*, 7 (1968) 105–108.

[268] Houtsmuller, A. J., van Hal-Ferwerda, J., Zahn, K. J. and Henkes, H. E. Favourable influences of linoleic acid on the progression of diabetic micro- and macroangiopathy in adult onset diabetes mellitus. *Progr. Lipid Res.*, 20 (1981) 377–386.

[269] Houtsmuller, A. J. Significance on linoleic acid in the metabolism and therapy of diabetes mellitus. *Wld. Rev. Nutr. Diet.*, 39 (1982) 85–123.

[270] Howard-Williams, J., Patel, P., Jelfs, R., Carter, R. D., Awdry, P., Bron, A., Mann, J. I. and Hockaday, T. D. R. Polyunsaturated fatty acids and diabetic retinopathy. *Br. J. Opthalmol.*, 69 (1985) 15–18.

[271] Endres, S., Ghorbani, R., Kelley, V. E., Georgilis, K., Lonnemann, G., Van der Meer, J. W. M., Cannon, J. G., Rogers, T. S., Klempner, M. S., Weber, P. C., Schaefer, E. J., Wolff, S. M. and Dinarello, C. A. The effect of dietary supplementation with $n-3$ polyunsaturated fatty acids on the synthesis of interleukin-1 and tumour necrosis factor by mononuclear cells. *New Eng. J. Med.*, 320 (1989) 265–271.

[272] Fernandes, G. Effect of dietary fish oil supplement on autoimmune disease: changes in lymphoid cell subsets, oncogene mRNA expression and neuroendocrine hormones. In Chandra, R. K. (ed.), *Health Effects of Fish and Fish Oils*. ARTS Biomedical Publishing, St. John's, Newfoundland, 1989, pp. 409–433.

[273] Kremer, J. M. Different doses of fish oil fatty acid in rheumatoid arthritis: effects on clinical and immunological parameters. In Chandra, R. K. *Health Effects of Fish and Fish Oils*. ARTS Biomedical Publishers, St. John's, Newfoundland, 1989, pp. 435–446.

[274] Kremer, J. M., Julsis, W., Michalek, A. V., *et al.* Fish oil fatty acid supplementation in active rheumatoid arthritis. *Ann. Inter. Med.*, 106 (1987) 497–503.

[275] Kremer, J. M., Bigauoette, J., Michalek, A. V., Timchalk, M. A., Lininger, L., Rynes, R. I., Huyck, C., Zieminski, J. and Bartholomew, L. E. Effects of manipulation of dietary fatty acids on clinical manifestations of rheumatoid arthritis. *Lancet*, January 26 (1985) 184–187.

[276] Jensen, O. M. Epidemiological evidence associating lipids with cancer causation. In Perkins, E. G. and Visek, W. J. (eds), *Dietary Fats and Health*. American Oil Chemists' Society, Champaign, 1983, pp. 698–709.

[277] Visek, W. J. and Clinton, S. K. Dietary fat and breast cancer. In Perkins, E. G. and Visek, W. J. (eds), *Dietary Fats and Health*. American Oil Chemists' Society, Champaign, 1983, pp. 721–740.

[278] Boyd, N. F., Shannon, P., Kriukov, V., Fish, E., Lockwood, G., McGuire, V., Cousins, M., Mahoney, L., Lickley, L. and Tritchler, D. Effect of a low-fat high-carbohydrate diet on symptoms of cyclical mastopathy. *Lancet*, July 16 (1988) 128–132.

[279] Carroll, K. K. The role of dietary fat in carcinogenesis, In Perkins, E. G. and Visek, W. J. (eds), *Dietary Fats and Health*. American Oil Chemists' Society, Champaign, 1983, pp. 710–720.

[280] Birt, D. F. The influence of dietary fat on carcinogenesis: lessons from experimental models. *Nutr. Rev.*, 48 (1990) 1–5.

[281] Carroll, K. K. Fish oils and cancer. In Chandra, R. K. (ed.), *Health Effects of Fish and Fish Oils*. ARTS Biomedical Publishers, St. John's, Newfoundland, 1989, pp. 395–408.

[282] Karmali, R. A. Fatty acids: Inhibition. *Am. J. Clin. Nutr.*, 45 (1987) 4714–4719.

[283] Kearney, R. Promotion and prevention of tumour growth effects of endotoxin, inflamation and dietary lipids. *Int. Clin. Nutr. Rev.*, 7 (1987) 157–168.

[284] Begin, M. E., Ells, G. and Das, U. N., *et al.* Differential killing of human carcinoma cells supplemented with $n-3$ and $n-6$ polyunsaturated fatty acids. *J. Natl. Cancer Inst.*, 77 (1986) 1053.

[285] Begin, M. E. Tumour cytotoxicity of essential fatty acids. *Nutrition*, 5 (1989) 258–260.

[286] Karmali, R. A., Marsh, J. and Fuchs, C. Effect of omega-3 fatty acids on growth of a rat mammary tumour. *J. Natl. Cancer Inst.*, 73 (1984) 457.

[287] Ziboh, V. A. Modulation of cutaneous inflammatory disorders by dietary fish oil omega-3 fatty acids. In Chandra, R. K. (ed.), *Health Effects of Fish and Fish Oils*. ARTS Biomedical Publishing, St. John's, Newfoundland, 1989, pp. 597–614.

[288] Burton, J. L. Dietary fatty acids and inflammatory skin disease. *Lancet*, January 7 (1989) 27–31.

[289] Cunnane, S. C., Kent, E. I., McAdoo, K. R., Caldwell, D., Lin, A. N. and Carter, D. M. Abnormalities of plasma and erythrocyte essential fatty acid composition in epidermolysis bullosa: Influence of treatment with diphenylhydantoin. *J. Invest. Dermatol.*, 89 (1987) 395–399.

[290] Ziboh, V. A., Cohen, K. A., Ellis, C. N., Miller, C., Hamilton, T. A., Kragballe, K., Hydrick, C. R. and Voorhees, J. J. Effects of dietary supplementation of fish oil on neutrophil and epidermal fatty acids. *Arch. Dermatol.*, 122 (1986) 1277–1282.

[291] Wright, S. and Burton, J. L. Oral evening-primrose-seed oil improves atopic eczema. *Lancet*, November 21 (1982) 1120–1122.

[292] Wright, S. Atopic dermatitis and essential fatty acids: A biochemical basis for atopy? *Acta Derm. Venereol.*, 114 (1985) 143–145.

[293] Bottiger, L. E., Dyerberg, J. and Nordoy, A. (eds), $n-3$ Fish oils in clinical medicine. (Acta Med. Scand. Supp. Ser. No. 5), *J. Int. Med.*, Symp. 731, 225 (1989) 1–238.

[294] Chandra, R. K. (ed.), *Health Effects of Fish and Fish Oils*. ARTS Biomedical Publishing, St., John's, Newfoundland, 1989.

[295] Cunnane, S. (ed.), *Symposium Proceedings: Essential Fatty Acids in Human Nutrition, Symposium Proceedings*, Toronto, May 9–11. *Nutrition*, 5 (1989) 246–281.

[296] Galli, C. and Simopoulos, A. P. (eds). *Dietary Omega-3 and Omega-6 Fatty Acids* (Series A: Life Sciences, Vol. 171). Plenum Press, New York, 1989.

[297] Lees, R. S. and Karel, M. (eds), *Omega-3 Fatty Acids in Health and Disease*. Marcel Dekker, New York, 1990.

[298] Vergroesen, A. J. and Crawford, M. (eds), *The Role of Fats in Human Nutrition*. Academic Press, London, 1989.

[299] Cunnane, S. C. Applications of NMR spectroscopy to the study of lipid composition and metabolism. *Nutrition*, 5 (1989) 253–255.

[300] Leaf, A. and Weber, P. C. Cardiovascular effects of $n-3$ fatty acids. *New Eng. J. Med.*, 318 (1988) 549–557.

[301] Leaf, A. Management of hypercholesterolemia. *New Eng. J. Med.* 321 (1989) 680–684.

[302] Renaud, S. Linoleic acid, platelet aggregation and myocardial infarction. Atherosclerosis, 80 (1990) 255–256.

[303] Ackman, R. G. and Ratnayake, W. M. N. Lipid Analyses: Part 1. Properties of fats, oils and lipids: Recovery and basic compositional studies with gas–liquid chromatography and thin-layer chromatography. In Vergroesen, A. J. and Crawford M. (eds), *The Role of Fats in Human Nutrition*, 2nd edn. Academic Press, London, 1989, p. 454–455.

[304] Green, A. G. A mutant genotype of flax (*Linum usitatissimum L.*) containing very low levels of linolenic acid in its seed oil. *Can. J. Plant Sci.*, 66 (1986) 499–503.

[305] Napolitano, G. E., Ratnayake, W. M. N. and Ackman, R. G. Fatty acid components of larval *Ostrea edulis* (L.): Importance of triacylglycerols as a fatty acid reserve. *Comp. Biochem. Physiol.*, 90B (1988) 875–883.

[306] Ostrowski, A. C. and Divakaran, S. The amino acid and fatty acid compositions of selected tissues of the dolphin fish (*Coryphaena hippurus*) and their nutritional implications. *Aquaculture*, 80 (1989) 285–299.

[307] Koletzko, B., Bretschneider, A. and Bremer, H. J. Fatty acid composition of plasma lipids in acrodermatitis enteropathica before and after zinc supplementation. *Eur. J. Pediatr.*, 143 (1985) 310–314.

[308] Cash, R. and Berger, C. K. Acrodermatitis enteropathica: Defective metabolism of unsaturated fatty acids. *J. Pediatr.*, 74 (1969) 717–729.

[309] Sanders, T. A. B., Ellis, F. R. and Dickerson, W. J. T. Studies of vegans: The fatty acid composition of plasma choline phosphoglycerides, erythrocytes, adipose tissue, and breast milk, and some indicators of susceptibility to ischemic heart disease in vegans and omnivore controls. *Am. J. Clin. Nutr.*, 31 (1978) 805–813.

SUPERCRITICAL FLUID EXTRACTION AND CHROMATOGRAPHY OF LIPIDS AND RELATED COMPOUNDS

Keith D. Bartle and Tony A. Clifford

OUTLINE

Advances in Applied Lipid Research, Volume 1, pages 217–264

ISBN: 1-55938-317-8

1. ADVANTAGES OF USING A SUPERCRITICAL FLUID FOR EXTRACTION AND CHROMATOGRAPHY

A supercritical fluid is a substance above its critical temperature and pressure. Above its critical temperature, it does not condense or evaporate to form a liquid or a gas, but is a fluid, with properties changing continuously from gas-like to liquid-like as the pressure increases. Table 1 shows the critical parameters of some compounds useful as supercritical fluids. One compound, CO_2, has so far been the most widely used, because of its convenient critical temperature, cheapness, non-explosive character and non-toxicity (of great importance in the food industry). Because the molecule is non-polar it is classified as a non-polar solvent, although it has some limited affinity with polar solutes because of its large molecular quadrupole. This is not too much of a problem in the lipid field, as many substances of interest are fairly soluble in CO_2, including fatty acids (which probably exist as dimers in the fluid) mono, di- and triglycerides, steroids, pesticides and PCBs, and much of the

Table 1. Critical parameters of selected substances useful as supercritical fluids

	T_c (°C)	p_c (atm)	ρ_c (10^3 kg m^{-3})
CO_2	31.3	72.9	0.47
N_2O	36.5	72.5	0.45
SF_6	45.5	37.1	0.74
NH_3	132.5	112.5	0.24
H_2O	374	227	0.34
n-C_2H_{10}	152	37.5	0.23
n-C_5H_{12}	197	53.3	0.23
Xe	16.6	53.4	1.10
CCl_2F_2	112	40.7	0.56
CHF_3	25.9	46.9	0.52

Table 2. Typical physical property values for gases, supercritical fluids and liquids

	Density (10^3 kg m^{-3})	Viscosity (mPa s)	Self-diffusion coefficient (10^4 m^2 s^{-1})
Gas			
30°C, 1 atm	$0.6–2 \times 10^{-3}$	$1–3 \times 10^{-2}$	0.1–0.4
Supercritical fluid			
near T_c, p_c	0.2–0.6	$1–3 \times 10^{-2}$	0.7×10^{-3}
near T_c, $4p_c$	0.4–0.9	$3–9 \times 10^{-2}$	0.2×10^{-3}
Liquid			
30°C, 1 atm	0.6–1.6	0.2–3	$0.2–2 \times 10^{-5}$

time work can be carried out with pure CO_2. For the extraction and chromatography of more polar molecues, it is common to add modifiers or entrainers, such as the lower alcohols, to CO_2, usually in small quantities. In such cases, it is important to be aware of the modifier–CO_2 phase diagram to ensure that the solvent is in one phase. For example, for methanol–CO_2 at 50°C there is only one phase above 95 bar whatever the composition, but below this pressure, certain compositions will be liquid, some gaseous (supercritical) and some separate into both of these phases [1]. It should be mentioned that, for both pure fluids and mixtures, many of the advantages of a supercritical fluid are possessed by liquids which are just sub-critical, and these are used in industrial processes, e.g. in the extraction of hops. The term near-critical is used to describe both situations and is preferred by some people.

Supercritical fluid extraction (SFE) and supercritical fluid chromatography (SFC) take advantage of the fact that a supercritical fluid can have properties intermediate between those of a liquid and gas, and that these properties can be controlled by pressure. Table 2 shows some rather approximate typical values of important properties: density (this is related to solvating power), viscosity (related to flow rates) and diffusion coefficients (related to mass transfer within the fluid). The principal advantage for SFE is that solubilities and, in particular, the relative solubilities of two compounds can be controlled via both pressure and temperature, making extraction selective to some extent. Other advantages are the relatively easy removal of the solvent and the non-toxicity and cheapness of CO_2. The advantage for SFC is that the higher diffusion coefficient gives narrower peaks and better separation for reasonable analysis times, while at the same time the solvating effect of the fluid permits chromatography of thermo-lable and higher molar-mass compounds impossible to analyse by gas chromatography. The disadvantage of using a supercritical fluid is that high-pressure technology is involved. Although SFE and SFC are the two areas where supercritical fluids have been widely exploited, research into the use of these fluids in other areas,

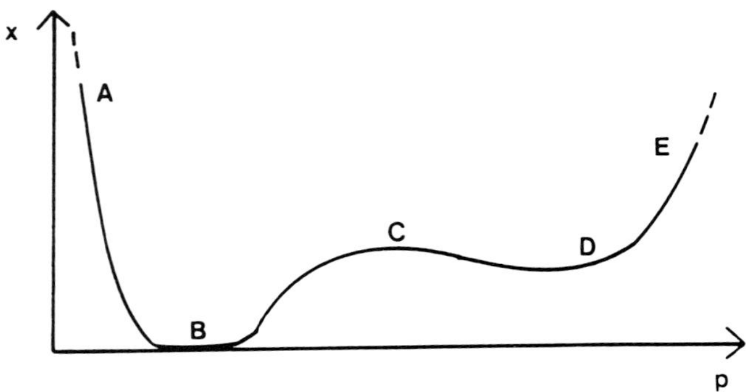

Figure 1. Features which can occur in solubility isotherms.

such as preparative SFC, chemical reactions, recrystallisation and electrochemistry is proceeding. It must be finally said that SFE and SFC are applicable to particular problems, where existing techniques are experiencing problems, and excessive euphoria as to their usefulness in any particular field must be avoided.

2. BASIC PRINCIPLES OF SUPERCRITICAL FLUID EXTRACTION

Extraction by a supercritical (or any) fluid is never complete in finite time, but may be considered to be successful under some conditions on the basis of either accuracy, in the case of analytical extraction, or economics, in the case of process extractions. For extraction to be successful two main criteria must be satisfied. The solute must, firstly, be sufficiently soluble in the supercritical fluid and, secondly, the solute must be transported sufficiently rapidly, by diffusion or otherwise, from the interior of the matrix in which it is contained.

Solubility is the subject of Sections 4, 5 and 6, and only a few major points will be made here. Firstly, the solubility can be described either as the mole fraction of solute in the solution, x, or by the amount (as moles or mass) of solute per unit volume, i.e. concentration, S, at saturation. The relationships between these quantities are trivial and involve molar masses and the density of the solution, which, as the solutions are dilute, is often taken to be the density of the pure fluid.

Solubility of a substance in a supercritical fluid is contributed to by two factors: the volatility of the substance and the solvating effect of the supercritical fluid. The latter factor, as we shall see more quantitively in the next section, is primarily a function of the density of the fluid. The solubility of a substance at constant temperature as a function of pressure has the schematic form of Figure 1. In section A–B, which is at very low pressures, which

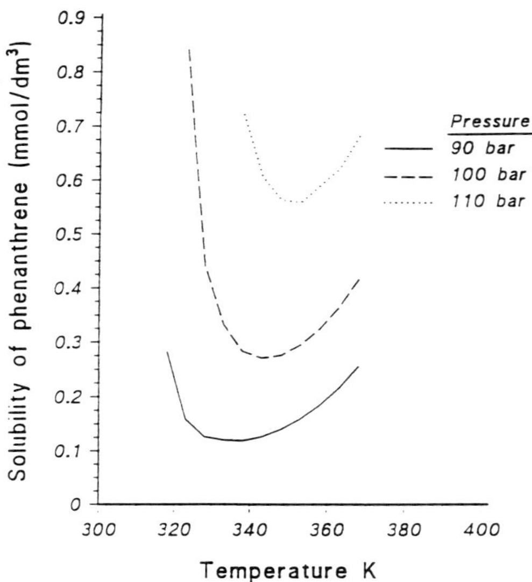

Figure 2. Predicted values of the solubility of phenanthrene.

are not of concern in most extraction processes, x falls as the solute is diluted by the fluid. In section B–C there is a rapid rise in x at a so-called threshold pressure characteristic of the solute–fluid system, which is a pressure somewhat above the critical pressure of the fluid. This occurs because of the rapid rise in the density, and therefore solvating effect, of the fluid at around this pressure. The next two features may or may not occur for any particular system, especially as the pressure range of interest may be limited. A fall, shown as C–D may occur because, at higher pressure, repulsive forces may 'squeeze' the solute out of solution. The rise shown as D–E occurs for moderately volatile solutes, if there is a critical line in the mixture phase at higher pressures.

At constant pressure and as a function of temperature, solubility has a minimum at a particular temperature, as shown, for phenanthrene as an example, in Figure 2. The initial fall occurs because, as the temperature rises, the density and solvating effect falls. However, as the temperature rises, the volatility of the solute also rises, and eventually this effect exceeds the effect of the falling solvation and the solubility rises.

The solubility of a substance is only a guide to its extractability. The solubility gives the concentration of solute in equilibrium with the pure solute. The presence of other materials in the matrix and in the fluid will effect the equilibrium concentration. The effect of the matrix will be, in general to reduce this concentration below the solubility. If the matrix can be considered

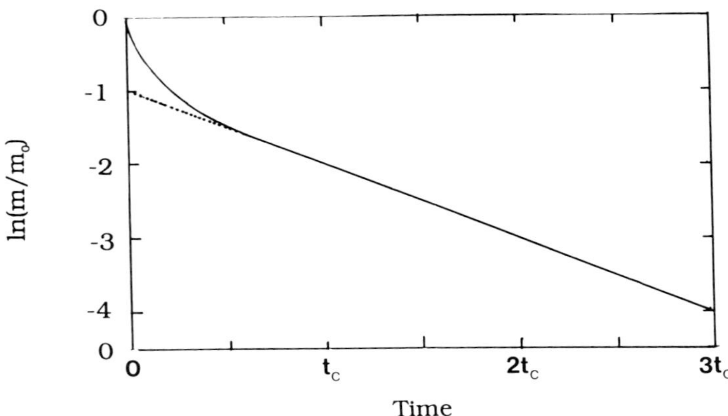

Figure 3. Plot of $\ln(m/m_0)$ versus time for SFE with little solubility limitation.

to be an ideal liquid mixture, the equilibrium concentration of each component will be equal to its solubility multiplied by its mole fraction in the matrix. Most systems will be non-ideal to an unknown extent, and the only conclusion that can be drawn is that the equilibrium concentration will fall as the concentration in the matrix falls. The presence of other solutes in the fluid, in general, enhances the equilibrium concentration [2], as the solutes usually have a stronger affinity to each other than to the solvent fluid. In addition to these effects, in any real extraction, because of the kinetic effect of diffusion out of the matrix, any solute will always be below its equilibrium concentration with the matrix, for extraction to be occurring. Nevertheless, in spite of the caveats, the extent of extraction of a solute after a given time, plotted as a function of pressure at constant temperature (see Figure 5(b) below), often follows the same shape as the corresponding solubility curve and has approximately the same threshold pressure.

We now consider the second effect, that of diffusion out of the matrix and, to illustrate this, in isolation from the solubility effect, we consider the extraction of a matrix in a flow of fluid, which is fast enough for the concentration of a particular solute to be well below its solubility limit. If the mass of solute in the matrix is m_0 initially and m after a given time, a plot of $\ln(m/m_0)$ versus time has the form given by Figure 3. It is characterised by a relatively rapid fall on to a linear portion, corresponding to an extraction 'tail' [3]. The physical explanation of the form of the curve is that the initial portion is extraction, principally out of the outer parts of the matrix particles, which establishes a smooth concentration profile across each particle, peaking at the centre and falling to zero at the surface. When this has happened, the extraction becomes an exponential decay.

The curve is characterised by two parameters: a characteristic time, t_c, and the intercept of the linear portion, $-I$. The slope of the linear portion is $-1/t_c$ and the linear portion begins at approximately $0.5t_c$. t_c is related to the effective diffusion coefficient out of the matrix, D, and a mean dimension, L (say the longest distance across), a matrix particle by the equation

$$t_c = AL^2/D \tag{1}$$

A is a contant, dependent on particle shape (which will vary) and the definition of L, and in general will not be known, but it is of the order of 0.1. For some systems, like liquid droplets, D will be a true diffusion coefficient, but often it will be an effective diffusion coefficient representing such processes as migration between adsorption sites, diffusion out of pores or passage of the fluid into and out of a porous particle. Most measurements published for D are for small molecules in relatively mobile solvents [4] and D is of the order of $10^{-5}\,cm^2\,s^{-1}$. For systems of interest to the food industry D will be of between 1 (for oils) and 4 (for solids) orders of magnitude below this value. In most cases a value will not be available, although some prediction methods are possible [5].

Equation (1) shows an inverse squared dependence on L and rationalizes the commonsense rule that for rapid extraction matrix particles must be small. This may be achieved for solids by crushing or grinding and for liquids by coating on a finely divided substrate or spraying or mechanical agitation. For solid matrix particles with L of the order of 1 mm, typical values of t_c are between 10 and 100 min.

The value of I depends on the particle shape and size distribution (for the former in particular the surface to volume ratio) and also the distribution of solute within the matrix particles (i.e. whether the solute is primarily located near the surface or in the interior of the particle). For a model system of spheres of the same size, with uniform solute concentration, it is 0.5. For real systems values of about 2 are common and prediction of the values is not really possible. Thus usually values of t_c and I can only be obtained by experiment. A small-scale dynamic extraction followed by the application of an appropriate analytical technique, as described in a later section, is therefore an important preliminary study in designing a process or a routine quantitive analytical procedure. For the former, such a study will provide information on a time scale for extraction and on the expected efficiency of extraction after a particular time, information which can then be considered in conjunction wtih solubility considerations. An experimental method for such a study is given in Section 3.

The information of Figure 3 is reproduced in Figure 4 in terms of percentage extraction versus time. As can be seen the majority is extracted in a time of $0.5t_c$; 72% for the example given, where I has been taken

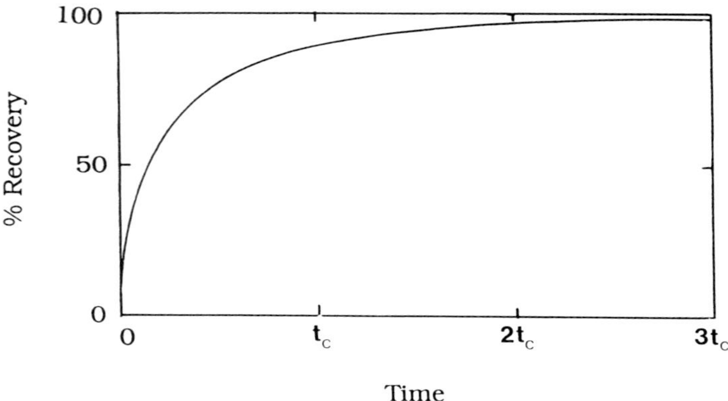

Figure 4. Plot of percentage extracted versus time for SFE with little solubility limitation.

as equal to 1. Thereafter there is a long 'tail', corresponding to slow extraction: only a further 10% is extracted in the next period of $0.5t_c$. These data represent a situation, where there is not limitation of the extraction rate by solubility considerations. In a real situation, therefore, the extraction rates will be less, depending on the nature of the extraction process used.

To illustrate the interaction between solubility and matrix-diffusion considerations, extraction of a solute from a stationary matrix in a cell by steady flow of the fluid is now discussed. Flow rates are now slow enough so that, at the beginning of the extraction, the concentration of the fluid in the solute is an appreciable fraction of its equilibrium concentration, which will vary with pressure. Figure 5(a) shows plots of the type of Figure 3, but now with solubility limitation, at a number of pressures, but constant flow rate, curve 1 being at the lowest pressure and curve 6 at the highest. The effect of solubility limitation is to reduce the dramatic high rate at the beginning of the extraction and delay the onset of the linear portion, so that it moves upwards and to the right. However, eventually solubility limitation will disappear and the slope will be that of the original curve, as in Figure 3. The extent of this effect will increase as the pressure falls and the solubility decreases. If data are taken at a particular time (this effect is most obvious at the beginning of the extraction) and plotted as a function of pressure, then Figure 5(b) is obtained, which a typical pressure-threshold curve and similar in shape to the solubility curve for the solute. A knowledge of solubility from published results, estimation by prediction, or measurement is therefore important.

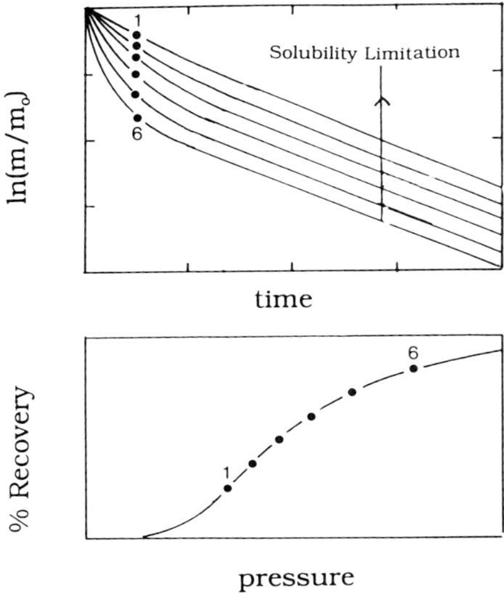

Figure 5. (a) Plots of $\ln(m/m_0)$ versus time at various pressures and constant temperatures, with the highest curve (1) at the lowest pressure and hence solubility. (b) Percentage extracted after a given time, plotted against pressure, with the points 1–6 corresponding to the points on (a).

3. EXPERIMENTAL DETERMINATION OF KINETIC EXTRACTION DATA

To obtain kinetic extraction data, suitable for plotting as in Figures 3 and 5, continuous extraction at constant flow rate, temperature and pressure is carried out. The extract is collected for a series of time intervals, e.g. by passing it through a suitable solvent, and then analysed by any appropriate method. Extraction is continued until it is complete to within 99% or better. This is done to obtain the value of m_0 to within 1% and thus values of $\ln(m/m_0)$ plotted below are fairly reliable down to a value of about -3, corresponding to 95% extraction. The method used by Hawthorne and his co-workers [6, 7], for extractions with pure or modified CO_2, is briefly described below, as an example.

The CO_2 is supplied at constant pressure using a suitably programmed syringe pump, which has a cooled pump head. Modifier, if used, is supplied at constant flow rate by a second pump. The supercritical fluid then passes to extraction cells with internal volumes of about 1 ml which have been previously filled with weighed test samples and placed in a thermostatted tube heater to maintain constant temperature. The flow rate is controlled and kept constant by a capillary restrictor, and this is monitored, in terms of liquid CO_2 volume flow, at the pump. For example, using pure CO_2 at 50° and 400 atmospheres, a flow rate of about 0.45 ml min^{-1} (measured as liquid CO_2 at the pump) is obtained with a 10 cm length of fused silica tubing of 25 μm internal diameter at the extraction cell outlet. The compounds extracted from each sample are then collected for several time intervals by purging the extraction cell effluent into about 3 ml of an appropriate solvent [7]. An internal standard is then added to each timed extract, and the concentrations of each target species determined using, for example, capillary gas chromatography. Extractionis continued until there was a very small or no further detectable extract and it was estimated that in excess of 99% has been extracted.

4. EXPERIMENTAL METHODS FOR OBTAINING SOLUBILITY

The various experimental methods used to obtain supercritical fluid solubilites can be classified in two different ways. The first relates to the way in which the saturated solution is obtained, which can be static or dynamic, i.e. in a closed cell or in a flow system. The second classification describes how this solution is analysed and the methods used can be grouped into four categories: gravimetric, spectrometric, chromatographic and miscellaneous. There is some correspondence between these two classifications. To produce enough material for accurate gravimetric analysis, a flow system is normally used. Reviews of experimental methods for supercritical solubilities have appeared recently [8, 9].

Of the techniques described, the gravimetric method is most widely used, with a few research groups, having established a particular procedure and given data in one publication, following this up with subsequent publications using essentially the same apparatus. Chromatographic analysis is the second most popular technique. Other methods tend to be limited to a single paper, describing the development of the experimental procedure and giving a body of data so obtained. A fair assessment of the present situation is therefore that the gravimentric method with its variants and to a lesser extent chromatographic analysis are the present established and reliable procedures; these two methods are described below. Also described is the simpler and faster, but less direct and sometimes qualitative, method of using chromatographic retention for obtaining solubilities.

Most of the gravimetric methods used have the same basic structure. The main motivation has been the need for solubility data for extraction processes and the philosophy of the methods reflect this interest. Briefly, these methods involve the production of a saturated solution by passing the supercritical fluid over the solute in an extraction cell, dropping the pressure to precipitate

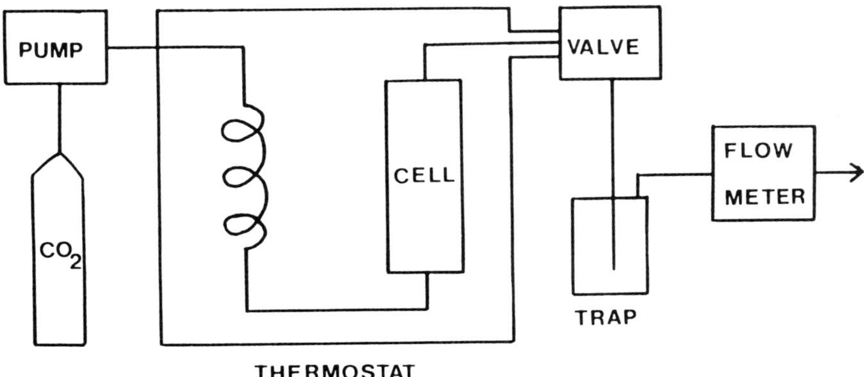

Figure 6. Schematic diagram of an apparatus for the gravimetric determination of solubility.

the solute and weighing it. The experimental procedures were developed by Eckert, Paulaitis and Reid and their co-workers in the late 1970s [10–12].

A schematic diagram of the basic system is shown in Figure 6. The CO_2 is pumped as a gas by a compressor or as a liquid by a pump with a cooled head into a thermostat, where it first passes through a preheating coil. It then passes into an extraction or equilibrium cell where a saturated solution is formed. The solute is usually dispersed in the cells, e.g. by coating it on to sand particles, and filters are positioned at the ends of the cells to prevent the entrainment of undissolved solute. The solution is then dropped to atmospheric pressure via a restrictor or valve, such as a back-regulator, which is heated to prevent the solute being lost in the valve and clogging it. The gas containing, finely divided solute precipitate, passes to a trapping system, which can vary in type (e.g. U-tubes are often used) or complexity (e.g. switching circuits between traps are sometimes described). More than one trap in series is often used and methods for ensuring complete trapping, such as cooling and/or packing with absorbent, are typically employed. Finally the CO_2 passes through some kind of flow meter. The pressure in the system, i.e. that of the experiment, and the flow rate are controlled by the pump and the valve. In the most straightforward system, a back-pressure regulator would control the pressure and the pump the flow rate.

A typical experiment is to set the flow and allow the system to reach a steady state and then switch in a weighed trap or traps for a measured period, during which the rate of flow of CO_2 is monitored. After reweighing the trap(s) and calculating the total mass of CO_2 passing in the period, the solubility is obtained directly in terms of mole fraction. Usually the errors in the experiment are quoted as 3–5% for solubilities obtained by this method.

An important consideration in the use of this method is that equilibrium has been reached in the extraction cell and different ways are used to ensure and check this including varying the flow rate [11, 12], recycling [13] and saturating at a higher temperature and then lowering the temperature [14]. The other important test is that all the precipitated solute has been collected after depressurizing. This can be done by using two or more successive traps and weighing both to show that the great majority of solute is collected in the first trap [12]. Problems can arise with precipitation inside the pressure-reducing valve, which can be overcome by washing out the valve after an experiment [15].

Most chromatographic methods used are modifications of the gravimetric methods shown in Figure 6. In one type of modification, the solute is precipitated as before in a trap, perhaps containing a solvent. The solute is then washed out of the trap, made up to volume and analysed by any suitable chromatographic method. In some cases the solute is precipitated directly on to a chromatographic plate. Gas chromatography (GC) [16], thin layer chromatography (TLC) [17], high performance liquid chromatography (HPLC) [18] including size exclusion chromatography (SEC sometimes called GPC) [19] are used. After calibration and analysis, followed by calculation of the mass of CO_2 passing during the sampling period as before, the method gives solubilities in terms of mole fraction directly. Accuracy comparable with the gravimetric method has been obtained, but some variants of the method, e.g. using direct precipitation on to a TLC plate, are intended only to give a rough indication of solubilities.

A second type of modification of the apparatus of Figure 6 for chromatographic analysis, consists of removing the trap and flow meter and inserting a sampling valve between the extraction cell and the valve or back-pressure regulator. This is commonly a multiport chromatoraphic sampling valve, with a sample loop. Using this device a fixed volume of the saturated solution is sampled and injected directly into a chromatograph (or alternatively let down to atmospheric pressure, the solute dissolved in a solvent and then injected into a chromatograph) and analysis carried out by GC [20, 21], HPLC [22] or SFC [23]. Because only small samples are needed, the extraction section of the apparatus can be small-scale and the commercially available SFE systems designed for analytical applications [20] or even a small static system [24] can be used. The apparatus is callibrated by loading a liquid solution of the same solute of known concentration into the same sample loop. The solubility results are obtained directly in terms of concentration rather than mole fraction. Careful use of this method can also give results of accuracy comparable with the gravimetric method.

A completely different way of obtaining information on solubilities, which is faster, but less direct and less accurate than the best direct experiments, is to use chromatographic retention. The degree of retention of a solute in chromatography, as measured by the capacity factor, k', is at least qualitatively inversely related to the solvating power of the mobile phase for that solute: the more soluble it is in the mobile phase, the less it will be retained. This method has been used to obtain qualitatively the dependence of solubility on pressure and to obtain 'threshold pressures' for extraction, in some cases solubility data has been given [25, 26].

Furthermore, it can be shown experimentally, that in some situations at constant temperature and for a particular column, the capacity factor for a particular solute is fairly accurately inversely proportional to its solubility in the mobile phase in terms of S [27]. Figure 7 shows $1/S$, plotted against V_m, the retention volume, which is proportional to k', for naphthalene using a particular ODS bonded-phase column at 318 K. Although both CO_2 as a supercritical fluid and methanol–water liquid mobile phases were used, and the data cover a range of 2.5 orders of magnitude, the points all lie approximately on the same line. (Because of the wide range of values, two plots are given with different scales to show the low and high values: the alternative of a log–log plot is a less obvious test of linearity. The lines drawn on both parts of Figure 7 have the same slope.)

Such plots are not obtained for all systems and it is necessary to check representative systems. However, the relationship can provide a relatively rapid way of generating large amounts of solubility data, if the proportionality constant is known. This constant may be obtained by a number of methods, including liquid solubility measurements and relatively few supercritical fluid measurements made by 'conventional' methods, and solubility data have been published obtained in this way [28].

5. SOLUBILITY MEASUREMENTS AND THEIR CORRELATION

In the context of lipids, the published solubility measurements reviewed here are restricted to long-chain fatty acids, glycerides, steroids and some pesticides. Other substances, which may be of interest in this area, may be found by reference to a recent comprehensive review of solubilities [9]. Also, this section is mainly restricted to the solubilities of pure compounds, which have been published in tabular form, rather than graphically, implying lower absolute accuracy. Details of some more qualitative solubility measurements are reported in the last section on industrial SFE. A bibliography of solubilities of interest here is reported in Table 3, which gives the compounds, temperature and pressure ranges, the experimental method used and a reference to the origianl publication. Table 3 also includes references to graphical data and mixtures. All these solubilities are in CO_2 and most of the discussion in this and the next section refers to this fluid. Examples of these data, the solubilities of triolein [24], are shown in Figure 8. These illustrate a point,

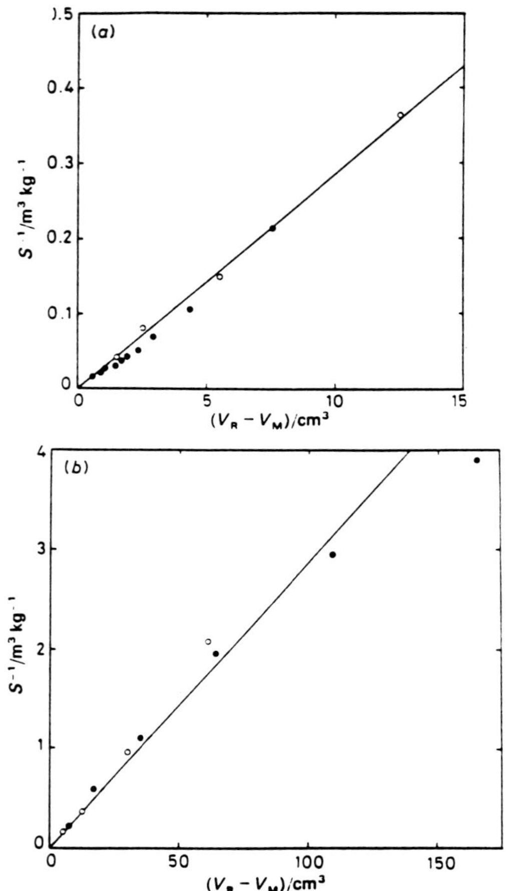

Figure 7. The reciproval solubility, $1/S$, plotted against $V_R - V_M$ for naphthalene at 318.2 K: O, liquid; ●, supercritical fluid. Lower and higher values are shown in (a) and (b), respectively, and the lines drawn in the two parts are of the same slope equal to 1/35.

important in the discussion of industrial extraction below, that at higher temperatures, where the solute is more volatile, the solubilities exhibit the feature D–E of Figure 1, i.e. are rising steeply with pressure. This feature is not present at low temperatures.

Correlation of supercritical fluid solubility data is not straightforward. All the features shown in Figure 1 can be reproduced qualitatively by any equation of state. For quantitative fitting more refined equations of state are useful in certain regions and of these the Peng–Robinson [29] has been the

Table 3. Published solubility data in CO_2 for some compounds of interest for lipids: fatty acids, glycerides and other esters, total oils, steroids and pesticides

Experimental method	Compound	Temperature range (K)	Pressure range (atm)	Reference
Spectr.	Behenic acid	313–333	30–250	[24]
Grav.	Lauric acid	313	77–248	[41]
Grav.	Myristic acid	313	82–249	[41]
Grav.		313–323	249	[52]
Grav.	Olaic acid	313–333	200–300	[52]
Grav.		308–333	84–189	[132]
Spectr.		313–333	100–250	[24]
Grav.	Palmitic acid	318–338	142–575	[133]
Grav.		313	80–248	[41]
Grav.		298–313	80–187	[13]
Grav.		308–323	200–300	[52]
Grav.		310–320	112–359	[134]
Grav.		313–333	200–300	[52]
Spectr.		313–333	100–250	[24]
Chrom.		313	270–1900	[135]
Grav.	Mono-olein	308–333	102–187	[132]
Spectr.	Tributyrin	313–333	100–250	[24]
Grav.	Trilaurin	313	90–253	[41]
Spectr.	Trilinolein	313–333	80–250	[24]
Grav.	Trimyristin	313	95–304	[41]
Grav.	Triolein	298–333	69–197	[136]
Grav.		313–333	200–300	[52]
Spectr.		313–353	80–250	[24]
Spectr.	Tripalmitin	313–353	80–250	[24]
Grav.		313	122–297	[41]
Grav.		298–313	36–182	[13]
Grav.	Tristearin	313–333	200–300	[52]
Spectr.		313–333	80–250	[24]
Spectr.	Behenyl behenate	313–333	100–250	[24]
Spectr.	Palmityl behenate	313–333	100–250	[24]
Grav.	Canola oil	298–363	100–360	[137]
Grav.	Corn germ oil	353	265	[138]
Grav.	Cottonseed oil	313–353	476–1020	[138]
Chrom.	Jojoba oil	293–353	99–2568	[47]
Grav.	Rapeseed oil	313–353	100–850	[42]
Chrom.	Soybean oil	298–353	99–2568	[139]
Chrom.		293–313	148–346	[45]
Grav.		313–343	207–689	[138]
Grav.		323–333	136–681	[48]
Chrom.	Sunflower seed oil	313	178–691	[45]
Grav.	Vegetable oil	298–328	160–480	[140]
Spectr.	Cholesterol	313–353	80–200	[24]
Chrom.	Ergosterol	313	79–198	[54]
Chrom.	Ethinylestradiol	313	79–198	[54]

Table 3. Continued

Experimental method	Compound	Temperature range (K)	Pressure range (atm)	Reference
Chrom.	Solasodin	313	79–198	[54]
Grav.	α-Tocopherol	298–313	100–183	[13]
Spectr.		313–353	100–250	[24]
Grav.	3,4-Dichloro-analine	313	197	[141]
Grav.	2,4-Dichloro-phenoxyacetic acid	313	197	[141]
Grav.	2-(4-(2,4-Dichloro-phenoxy)-phenoxy)propionic acid	313	197	[141]
Grav.	2-(4-(2,4-Dichloro-phenoxy)-phenoxy)propionic acid methyl ether	313	197	[141]
Grav.	Linuron	313	197	[141]
Grav.	Methoxychlor	313	197	[141]

most widely used. However, even this equation is not successful in fitting all the data at all pressures and temperatures. A further problem is that the parameters necessary for using the equation of state, such as the critical temperature and pressure of the solute and its vapour pressure and acentric factor, are not always available. This problem has been discussed in a recent paper by Johnston *et al.* [30]. They come to the conclusion that a cruder empirical correlation with density is the best available route for most compounds and one method of this type is outlined later.

However, if a reasonable amount of data are available for a compound of interest and the other parameters are available, then the Peng–Robinson equation of state may be used to correlate data over a limited range of pressure and temperature. This will allow interpolation and limited extrapolation of the experimental data for design purposes. An approximate version of the equations, which assumes dilute solutions and no effect of the fluid on the solute, is given in Table 4. The solute–fluid interaction parameter, d, which is a fine-tuning parameter, is not known in advance and is obtained by fitting the equations to the experimental data, a procedure which also makes up for the approximations in the equations of Table 4 [31]. The parameter, d, is in principle temperature independent and so a value can be obtained by fitting all the data; a procedure which also makes up for the approximations in the equations.However, if isotherms are fitted separately, d is sometimes found to vary with temperature. This is especially the case

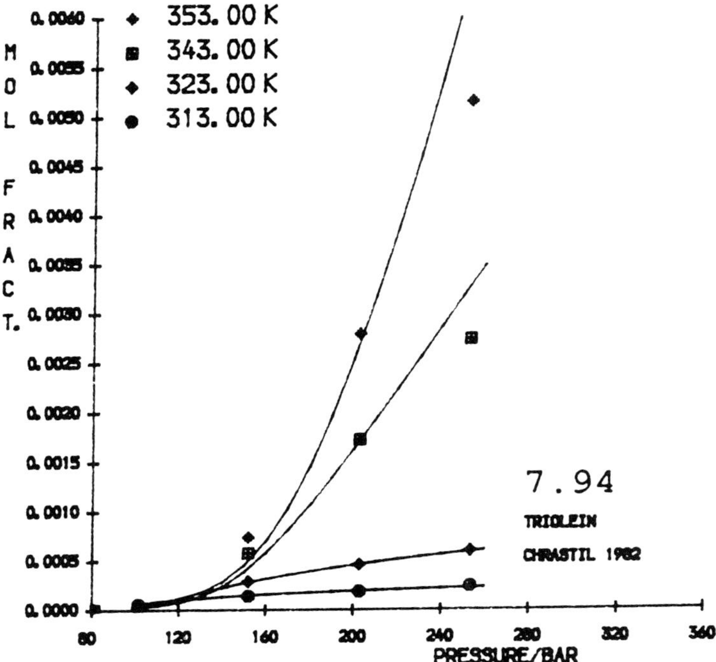

Figure 8. Published experimental values [24] for the solubility of triolein in CO_2. The lines are produced from Eq (2) and values of A' and B from Table 7.

when the solute is a solid which forms a liquid phase under certain conditions of temperature and pressure. Table 5 shows some values of the interaction parameter obtained by fitting experimental data in CO_2 for compounds of interest here to the equations of Table 4.

A much simpler correlation is to fit the solubility in terms of mole fraction, x, at constant temperature as a function of pressure to the following equation:

$$\ln(xp/p_{\text{ref}}) = A' + B(\rho - \rho_{\text{ref}}) \tag{2}$$

where p is the pressure, ρ the density of the solution (approximately the density of the pure supercritical fluid), A' and B are constants at constant temperature, and p_{ref} is a reference pressure, conveniently 1 bar. Here, ρ_{ref} is a reference density, chosen to be central to the density range of all the solubility data. A good value for pure and modified CO_2 would be $700\,\text{km m}^{-3}$ and the following discussion assumes this value. The reason for the choice of ρ_{ref} is to make the values of A' obtained much less sensitive to experimental error in the data and easier to correlate between different sets

Table 4. Simplified Peng–Robinson equations for dilute solutions, giving the mole fraction, x, at saturation of a solute in a supercritical fluid, where p_v is the vapour pressure and V_m the molar volume of the pure solute. $p_{c,i}$, $T_{c,i}$, w_i are critical pressures, temperatures and acentric factors. Subscripts: 1 = solvent (CO_2); 2 = solute. (For CO_2 $p_{c,1} = 7.383\,\text{MPa}$; $T_{c,1} = 304.2\,\text{K}$; $w_1 = 0.225$.)

$$\ln x = \ln(p_v/p) - \ln\phi + pV_m/RT$$

$$\ln\phi = (b_2/b_1)(Z-1) - \ln(Z - pb_1/RT) - (a_{11}/2\sqrt{2}b_1RT) \times (2a_{12}/a_{11} - b_2/b_1)\ln[(Z + (1+\sqrt{2})b_1p/RT)/(Z + (1-\sqrt{2})b_1p/RT)]$$

$$Z = pV/RT$$

$$b_1 = 0.0778RT_{c,1}/p_{c,1}; \quad b_2 = 0.0778RT_{c,2}/p_{c,2}$$

$$a_{12} = [0.45724(1-d)R^2T_{c,1}T_{c,2}k_1k_2]/(p_{c,1}p_{c,2})^{1/2}$$

$$a_{11} = (0.45724R^2T_{c,1}^2k_1^2)/p_{c,1}$$

$$k_1 = 1 + [1 - (T/T_{c,1})^{1/2}](0.37464 + 1.54226w_1 - 0.26992w_1^2)$$

$$k_2 = 1 + [1 - (T/T_{c,2})^{1/2}](0.37464 + 1.54226w_2 - 0.26992w_2^2)$$

of data and different temperatures. A full discussion of this correlation method has been given [9]. It is found that better correlations are obtained if data below 100 bar are not included in the fits and this has been done in the examples given below.

Values of A' and B are given in Table 6 for the solubility of naphthalene

Table 5. Peng–Robinson interaction parameters for some fatty acids, obtained from fitting experimental solubility data to the equations of Table 4 (also given are the other physical constants which were used in obtaining the parameters, with p_v given by $\log_{10}(p_v/\text{bar}) = -0.2185a/T + b_0$)

Acid T (K)	d	$T_{c,2}$ (K)	$p_{c,2}$ (bar)	w_2	a (K)	b	V_m ($cm^3\,mol^{-1}$)
Behenic		855	11.0	0.956	50 992	24.05	1312
313	0.004 7						
333	0.005 68						
Oleic		797	17.0	1.120	29 736	12.38	732
313	0.103 3						
333	0.047 5						
Palmitic		791	19.0	1.047	30 335	18.40	674
298	0.102 6						
313	0.003 3						
Stearic		810	16.5	1.085	43 518	20.84	780
313	0.071 0						
333	0.062 5						

Table 6. A' and B coefficients for naphthalene, obtained from published solubilities [9]

Temperature (K)	A'	$10^3 B$ ($m^3 kg^{-1}$)
308	−0.1394	8.00
308	−0.1121	7.82
308	−0.1517	8.50
308	−0.2659	8.86
308	−0.1700	8.45
308	−0.4714	10.16
318	0.8542	6.50
318	0.8673	6.07
323	1.2267	6.70
323	1.7216	7.50
328	1.6802	7.84
328	1.6435	7.59
328	1.7143	7.91
328	1.7125	7.41
332	1.8086	8.02
333	2.3333	8.50
338	2.5105	8.77

in CO_2, for which there is a large volume of data. The values of A' are also shown plotted against $1/T$ in Figure 9. As can be seen B does not vary greatly with temperature, whereas A' versus $1/T$ is aproximately a straight line. The slope of the line is approximately $-H_v/R$, where H_v is the enthalpy of vapourization of the solute. (In this case the slope of the line corresponds to an enthalpy change of 77 kJ mol^{-1}, compared with a value for the enthalpy of vapourization of solid naphthalene of 80 kJ mol^{-1} [32]). Hence this correlation method can also be used for the interpolation or extrapolation of experimental data to conditions of interest, as described in the next section. Table 7 shows some values of A' and B for compounds relevant to lipids.

6. SOLUBILITY PREDICTIONS

For a solute for which there is a large volume of solubility data published and for which critical parameters, vapour pressure data and hence an acentric factor is available, the best method for prediction of solubility at any pressure and temperature would be to correlate the data using an equation of state with adjustable parameter(s) in the temperature–pressure region of interest, as described using the Peng–Robinson equation in the last section.

For a solute for which there are solubility data available, but not enough information, or the will to carry out an equation-of-state correlation, less accurate estimates of the solubility may be obtained using the correlation of

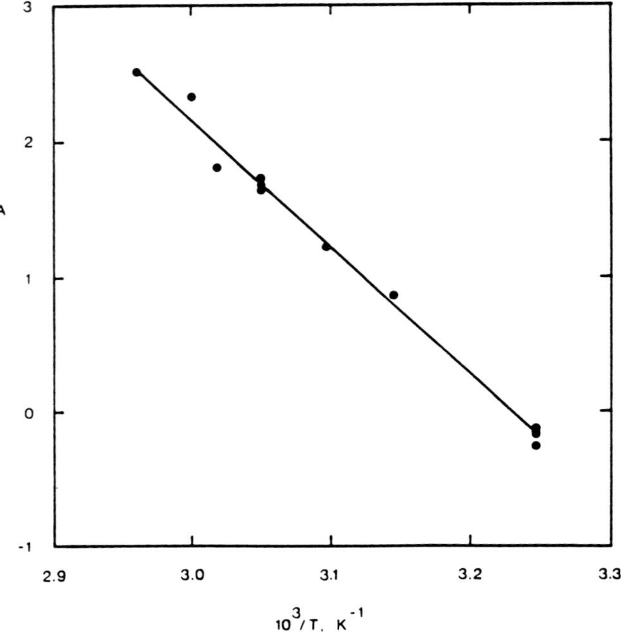

Figure 9. Plot of the values of A' in Table 6 versus $1/T$ for naphthalene.

equation (2), and the A' and B values obtained form the experimental data. The simplest method is to assume that B is constant and to correct A' for temperature using H_v. More sophisticated variants of this procedure have been described [9].

For solutes for which no experimental data are available, approximate predictions for solubilities in CO_2, which can be used as first estimates for design purposes, can be obtained using the Peng–Robinson equation, i.e. the equations of Table 4. Usually estimates of the necessary parameters and vapour pressures will have to be made by standard methods [5], and the interaction parameter, d, can be estimated using the following equation [33]:

$$d = 0.51C(w_2 - w_1)(V_{c,2}/V_{c,1})(p_{c,2}/p_{c,1})^2 \tag{3}$$

In this equation C is a constant which 0.5 for molecules which contain groups which may cause them to dimerize in the gas phase, e.g. –OH or –COOH groups, but is unity for other compounds. w_i, $V_{c,i}$ and $p_{c,i}$ are the acentric factors, critical molar volumes and pressures, respectively, of CO_2 (subscript 1) and the solute (subscript 2). Critical volumes are generally not tabulated and these have been obtained from [5]

$$V_{c,i} = (0.2918 - 0.0928w_i)RT_{c,i}/p_{c,i} \tag{4}$$

Table 7. A' and B coefficients for some compounds of interest for lipid research, obtained from published solubilities [9]

	Temperature (K)	Pressure range (bar)	A'	$10^3 B$ ($m^3\ kg^{-1}$)
Behenic acid	313	100–260	74.4180	5.16
	333	100–210	−3.7701	3.55
Lauric acid	313	100–250	−0.7981	17.64
Myristic acid	313	100–250	−1.5799	14.23
Oleic acid	308	100–180	−3.3618	11.11
	313	100–260	−3.3033	13.02
	333	100–260	−0.0550	14.57
Palmitic acid	298	100–190	−5.8572	10.77
	313	100–250	−3.4554	11.52
	318	140–360	−1.9847	14.00
	328	149–580	−0.9692	13.95
	338	140–580	−0.1586	14.16
Stearic acid	310	110–370	−3.3902	3.07
	313	100–260	−3.4297	5.00
	320	110–370	−2.7179	8.73
	333	100–260	−1.0057	3.30
Mono-olein	308	100–180	−5.3870	13.39
	333	130–190	−4.1535	11.20
Tributyrin	313	100–210	−1.3326	14.61
	333	120–260	1.0955	16.27
Trilaurin	313	100–260	−3.5034	18.11
Trimyristin	313	100–310	−4.7928	14.20
Triolein	313	100–260	−4.5760	9.43
	323	100–260	−0.6561	11.39
	343	100–260	73.1951	9.45
	353	100–260	0.5930	12.09
Tripalmitin	298	100–190	−7.5659	3.61
	313	100–260	−5.2740	6.48
	333	100–260	−4.3103	7.02
	353	100–260	−3.4841	6.41
Tristearin	313	100–210	−0.4650	15.57
	333	100–260	−6.3115	16.52
Cholesterol	313	100–210	−5.4654	18.95
	333	120–210	−3.1226	21.57
	353	150–210	−0.9417	22.73
α-Tocopherol	298	100–160	−4.0699	7.57
	313	100–190	−4.9520	13.66
	333	100–260	−1.1815	16.00
	353	100–260	2.5636	18.61

The equation was obtained by fitting data for some 20 compounds for which good solubilities and other parameters were available to the Peng–Robinson equation to obtain values of d. These values of d are plotted in Figure 10 against the group of parameters on the left-hand side of equation (4) and it

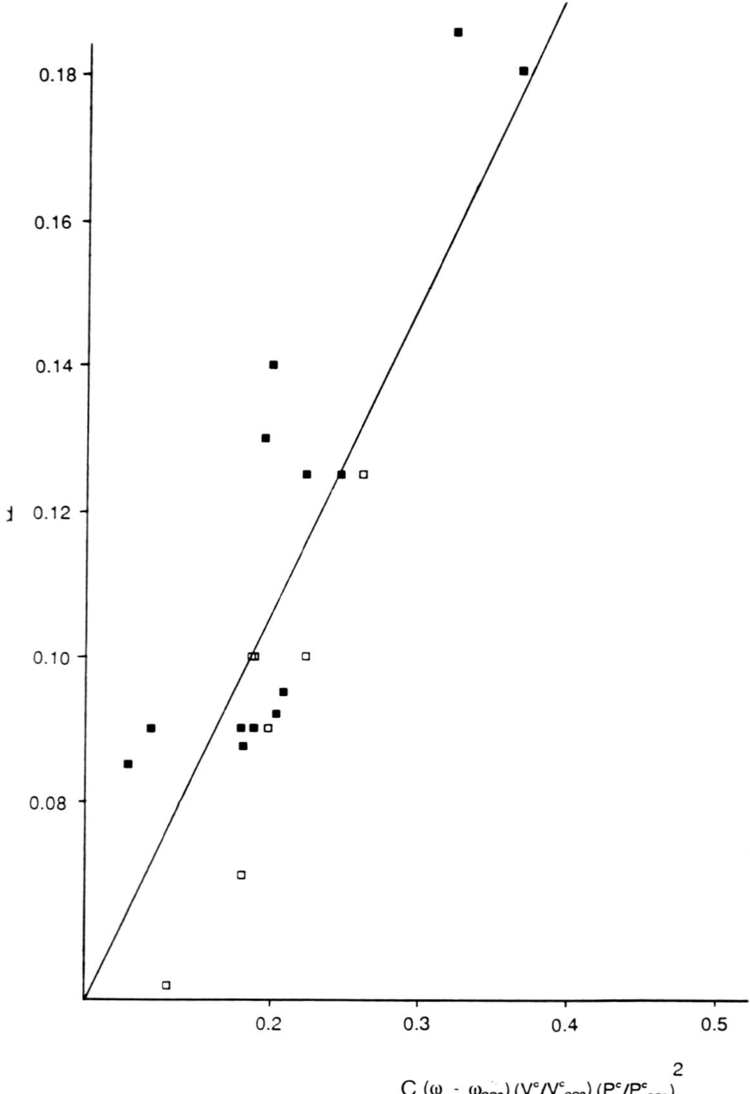

Figure 10. Plot of the Peng–Robinson interaction parameter, *d*, for 20 compounds versus a group of physical constants, to test the validity of Eq (3).

can be seen that there is a reasonable, if not perfect, correlation, with a line passing through the origin with a slope of 0.51.

7. INDUSTRIAL SUPERCRITICAL FLUID EXTRACTION

Industrial supercritical extraction processes are now well established and textbooks are available on their chemical engineering aspects [8, 34, 35]. The technique is used on a large scale, e.g. using CO_2 for coffee decaffeination and hop extraction, and on a smaller scale, for the extraction of high-value natural products, such as perfumes [36]. More than 200 US patents for supercritical extraction processes have been registered [8], and a large number outside the US mainly in West Germany. A process for the extraction and fractionation of edible oils using supercritical propane, known as the Solexol process was developed by the M.W. Kellogg Co. in 1947 [38, 39]. This process is no longer in use and at present it is CO_2 that is under consideration. Because of company confidentiality it is difficult to know to what extent industrial SFE processes for lipids are being developed to the commercial stage. It would be generally known if a process of this type was commercially sucessful and was being carried out on a very large scale, like the coffee decaffeination mentioned above, and this does not appear to be the case at the present time. However, a number of patents and scientific publications have appeared indicating considerable interest in possible processes and these were reviewed in 1986 [40]. Work is described below in five categories: lower pressure bulk extraction; bulk extraction at higher temperatures and pressures; fractionation during bulk extraction; modification of existing products; and obtaining high value components from lipid products.

In the former category a large number of studies have been made on the extraction of various lipid-bearing materials on a laboratory scale [40]. Many of these resulted in qualitative or semi-quantitative estimates of the solubilities of total oil or oil components. Recent studies, all involving pure CO_2, have been made on the solubility of pure and mixed triglycerides in supercritical CO_2 [41], on equilibria involved in the extraction of rapeseed oil [42], on the use of adsorbents in the SFE of vegetable oils [43], and on the extraction of cholesterol-free lipids from phytoplankton [44] by the same fluid. Prominent amongst the earlier studies were those of Stahl and co-workers [45–47], and much of the research has been carried out in Germany and also in the USA. Most investigations envisage the use of lower pressures and temperatures with the extract solutions obtained containing only a few percent of lipid extraction. The investigators see the advantage of CO_2 over the presently used organic solvents, such as hexane, in its non-toxicity, safety, ease of separation and cheapness. Another advantage of using SFE as opposed to hexane extraction is that the extract contains a smaller proportion of undesirable components such as phospholipids and pigments, and 'degumming' of the product can be avoided.

Many of the lipids to be extracted are sufficiently volatile at temperatures of 60–80°C for their solubility curves versus pressure to exhibit the steep rises at higher pressures illustrated schematically by the section D–E in the curve on Figure 1. This has been shown experimentally in measurements on individual components, in studies referred to in Table 3, e.g. example in the data on triolein in Figure 8. It has also been shown to be an effect for the extraction of total oil by Stahl [46, 47] and Friedrich and Pryde [48]. In 1984 a US patent (4 466 923) was granted to Friedrich which describes conditions for the replacement of hexane by supercritical CO_2 for the large-scale extraction of oilseeds, such as soy, cottonseed, sunflower, safflower peanut and linseed; cereal components, such as corn germ; and animal by-products, such as suet and offal. Although this patent was granted, the behaviour of solubility on which it is based is not unexpected and the experimental evidence was in the public domain [46] at the time of the patent aplication. The conditions suggested in the patent are to use pure CO_2, temperatures of 60–80° and pressures above 550 bar, perhaps as high as 1200 bar. By using the conditions suggested the patent author claims that the weight percent of oil in the extracted solution could be as high as 40%. The studies described in this paragraph can be summarized as advocating a high temperature and pressure route, which would have increased technological costs, but the advantage of using less solvent and a smaller volume throughput for a given yield.

Another advantage of using supercritical fluid is that separation of the product into a number of fractions can be incorporated into a bulk extraction of the type described above [49]. This can be done either by increasing the pressure in stages during a batch extraction or reducing the pressure in stages and separating the precipitated fraction following total extraction in a batch, continuous or semi-continuous process. Fatty oils, for example, can be separated into fractions which are rich in flavour and free fatty acids; rich in mono- and diglycerides; mostly triglycerides; and rich in waxes and pigments, by increasing the pressure in stages of 50–100 bar at 60°C. Similar fractionation of vegetable oils from soybean, rapeseed and sunflower seeds can be achieved [46].

These bulk processes are envisaged on a scale of millions of tons per annum, but there is also interest in smaller scale processes designed to modify the physical properties, flavour and health status of fatty materials. The former include hardening vegetable oil products to make them suitable for confectionary use, which at present is done by organic solvent extraction, and the refining of vegetable oils [50]. The latter include the removal of fatty acids from vegetable oils [51], from olive oil [52] and from cheese [53] to improve or modify their flavour and of steroids, such as cholesterol from fats, for heath reasons. This leads to the final category of SFE in relation to lipids, where natural lipid products are the source of compounds for medicinal use. This was the motivation for the study of the CO_2 solubility of a number of steroids [54]. SFE has been used in the preparation of eicosapentenoic acid derivatives [55, 56] and other medically active acids [57] from fish oils, and the separation of lecithin from soybean oil has been described [58].

8. ANALYTICAL SUPERCRITICAL FLUID EXTRACTION

The use of SFE as a first stage in an analytical is rapidly gaining importance and instrument and manufacturers have responded with the production of commercial systems in the last few years. It has advantages in speed and pressure-controlled selectivity over solvent extraction, and it can be made 'on-line' to subsequent analytical steps. There is a very recent review (1990) of this developing field [59], which gives the current situation and detailed practical advice. In an off-line system a typical procedure is that described in Section 3, with a commercial system making for easier operation. For an on-line system, for example with SFE–SFC, extraction may be carried out by flowing the fluid through a thermostatted extraction cell at a particular pressure for a given time, letting the pressure drop through a restrictor, when the extract is deposited on the beginning of the column. The extraction is then stopped and chromatographic analysis may then be carried out . a further extraction and analysis may then be carried out on the same sample to investigate remaining analytes.

Because of the solubility of most fat and oil components in CO_2, SFE using the pure gas is highly applicable to lipids stystems. Off-line SFE has been used to extract fat from meats [60], and on-line SFE–SFC to analyse fatty acids and glycerides from butter, cheese and oils [61]. Although triglycerides are very soluble under some conditions, selective extractions of organochlorine pesticides from fats have been obtained sufficiently fat free to permit their GC analysis, by extraction at 40°Cand 120 bar [62, 63].

There are still problems in using these systems for quantitative analysis, which are the result of the long extraction tail visible in Figure 4. Extraction in an initial period may extract 70% and in a subsequent extraction 7% only obtained. There may be a temptation for the analyst to conclude that a further extraction will drop in the same proportion and little extract remains. This is not the case, more than 20% remains and substantial yields in further extractions are obtained. When samples are almost constant in size, shape and composition it may be possible to extract under constant conditions and apply a factor in excess of unity (obtained from a lengthy extraction study) to the results. However, this has not been found easy or successful. No doubt these problems will find solutions soon. One possibility, where extraction is continuous, is to use the exponential tail to extrapolate the results to complete extraction and obtain quantitative analytical information in a shorter time than would be required for exhaustive extraction [3]. If extraction is carried out at least as long as the initial non-linear period to obtain an extracted mass m_1, followed by extraction over two subsequent equal time periods to obtain masses m_2 and m_3, then it can be readily shown the total mass in the sample, m_0, is given by

$$m_0 = m_1 + m_2^2/(m_2 - m_3) \tag{5}$$

Preliminary tests of the application of equation (5) to quantitation in analytical extractions is encouraging and programmes of work to establish its usefulness are being carried out. If extraction is interrupted, however, as in the SFE–SFC on-line system described above, this equation does not apply. Diffusion within the sample occurs continuously between extractions, and other mathematical solutions must be sought.

9. SUPERCRITICAL FLUID CHROMATOGRAPHY

Chromatography with a supercritical fluid as mobile phase (SFC) dates back more than 20 years, although its adoption as a viable analytical technique is recent [64, 65]. While the separating power of GC with capillary columns is unequalled, the limited volatility and thermal stability of many lipids limits its application. Many separations of lipids by HPLC have been reported [66], but there is no generally available universal detector for compounds without chromophores. SFC also offers the prospect of increased speed of analysis over HPLC.

SFC separation of high molecular mass solutes occurs at temperatures well below those at which thermal decomposition occurs. Because of the low viscosity of supercritical fluids, capillary columns may be employed, with consequent high resolution. The facility of operation of SFC with CO_2 as mobile phase means that the universal flame ionization detector (FID) as well as a range of GC and HPLC detectors may be used [67, 68]. Coupling of SFC to Fourier-transform IR (FTIR) [69] and mass spectrometric (MS) [70] detectors is also well advanced.

10. APPARATUS FOR SUPERCRITICAL FLUID CHROMATOGRAPHY

Figure 11 is a block diagram of the apparatus [64, 65] required for SFC. The mobile phase, which can be N_2O, C_5H_{12}, NH_3, etc., but which is most usually CO_2 either alone or with a small added concentration of a modifier, such as

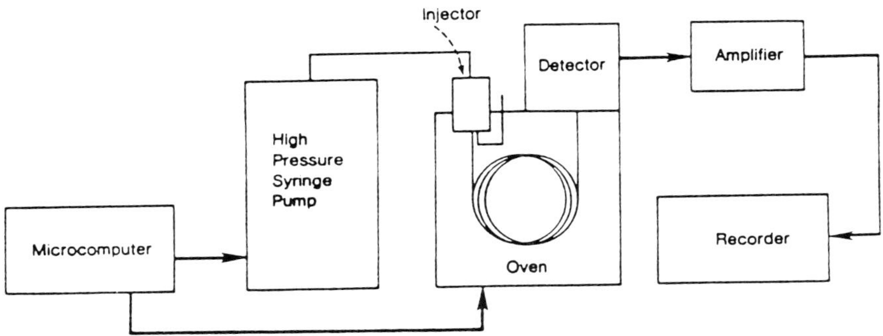

Figure 11. Schematic diagram of SFC instrumentation.

an alcohol or ether, is pressurized as liquid by a syringe pump (See also Section 3). The liquid is delivered via an injection valve to the analytical column, which can be a conventional HPLC, microbore HPLC, packed-capillary or fused-silica capillary column. The column is contained in an oven heated above the critical temperature of the mobile phase. A pressure restrictor at the end of the column ensures supercritical conditions in the column, and is installed either within the jet of an FID or in-line after a flow-cell detector such as a UV absorbance detector.

Pulse-free syringe pumps are necessary in capillary SFC because column inner diameters are restricted to 100 μm or, usually, 50 μm by the smaller diffusion coefficients in supercritical fluids compared with those in gases; the flow rates are therefore very low (of the order of $1\,\mu l\,min^{-1}$). For packed columns, reciprocating pumps are adequate. For operation with CO_2, cooling of the pump head is necessary. Since the density of the fluid controls the solvating power, and hence influences retention, the pressure is varied for density control by means of a pressure transducer in-line between pump and column, and a microcomputer programmed to generate a predetermined pressure or density profile. The oven temperature may also be concurrently varied to change solubility and hence retention, and also to increase the diffusion rates of solutes and chromatographic resolution.

To prevent loss of resolution, injection of the sample into the column as a narrow band is vitally important [71]. A conventional HPLC loop valve is adequate for packed column SFC. For capillary work the low sample capacity of 50 μm inside diameter columns necessitates first a valve with a small (200 nl) sample loop which is opened pneumatically, and second an inlet splitter, or electronically-controlled partial valve opening (so-called time-split injection) to reduce the injected volume even further. Various solvent venting procedures are available for injecting larger volumes (up to 1 μl) in trace analysis [72].

A major advantage of SFC is compatibility with both GC and HPLC

Table 8. Available detectors for SFC.

GC detectors	HPLC detectors	Coupled spectroscopic detectors
Flame ionization	UV absorption	Mass spectrometry
Thermionic ionization (N mode)	Fluorescence	FTIR
Thermionic ionization (P mode)	Light scattering	Microwave induced plasma
Flame photometric	Chemiluminescence	
Photoionization		
Electron capture		

detectors (Table 8). Flame detectors (mainly FID, but also nitrogen and phosphorus thermionic, and flame-photometric sulphur selective detectors) have been interfaced [67] to the SFC column via a capillary or frit restrictor which allows decompression of the fluid as it enters the detector. HPLC detectors [68] (UV and fluorescence) are necessary when the CO_2 contains an organic modifier; the restrictor is positioned after the flow cell. Alternative detectors for analytes without chromophores include light-scattering and ion-mobility devices.

FTIR detection after SFC may be carried out [69] by means of a flow-cell [73] in which the IR beam passes through the column effluent; mobile phase absorptions may interfere with detection of analyte, but a total absorption chromatogram may nevertheless be reconstructed in addition to the recording of spectra of individual peaks for identification. A solvent–elimination interface for FTIR detection [74] is also commonly applied; here the column effluent is sprayed onto an IR transparent plate from which the mobile phase is evaporated to leave a spot of analyte, later analysed by means of an FTIR microscope.

The interfacing of SFC to mass spectrometers has been energetically pursued [70]. The restrictor from a capillary column may be directly introduced into the ion source, although a further pumping stage before the ion source may be necessary for packed columns with higher gaseous flow rates [75].

11. OPERATING PARAMETERS IN SUPERCRITICAL FLUID CHROMATOGRAPHY AND THEIR EFFECT OF RETENTION

The retention parameter k' depends on the mobile phase density (ρ) so that members of a series may be eluted [76] by changing ρ. The variation of ρ with applied pressure is available from tabulated isotherms, usually fitted to polynomials. The analogue of temperature programming in GC or gradient elution in HPLC is SFC with density programming. Beyond the maximum

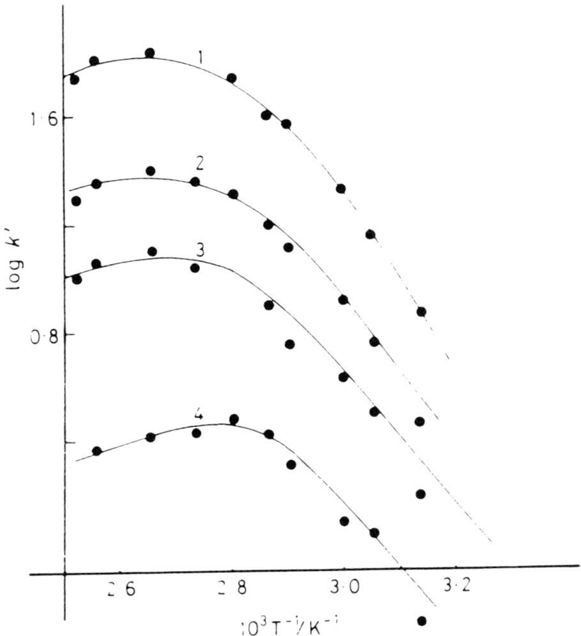

Figure 12. Graphs of log k' versus T^{-1} at constant CO_2 pressure for (1) pyrene (2) phenanthrene (3) fluorene and (4) naphthalene. Column: 25 cm ODS-modified silica. (From Ref. [79], with permission.)

available mobile phase density, migration of solutes may be achieved by adding polar modifiers [77]. Small (less than 1%) amounts of modifier give rise to large changes in k' in packed column SFC because the modifier successfully competes with the solute for active sites on the packing. For (inactive) capillary columns larger quantities of modifier are required to bring about significant changes in k', mainly by increasing solubility in the mobile phase [78].

Plots of k' against temperature show a maximum above the critical temperature in SFC [79]. This variation, is usually expressed (e.g. Figure 12) via plots of $\log k'$ versus T^{-1} at constant P. The vertical distance between these curves is fairly constant for homologues so that changing the temperature has little effect on the selectivity. The $\log k'$ versus T^{-1} curves may intersect for dissimilar compounds, however, giving rise to a reversed elution sequence and the possibility of better separation.

SFC is unique in that variation of all four significant operating parameters is possible to change retention and selectivity (Table 9). Multidimensional SFC has been demonstrated [80, 81] and SFC has also been combined with

Table 9. A comparison of chromatographic variables.

	GC	HPLC	SFC
Stationary phase	+	+	+
Mobile phase composition	−	+	+
Temperature	+	−	+
Pressure	+	−	+
High resolution	+	−	+

other separation modes (e.g. size exclusion chromatography—SEC) [82]. The possibility of carrying out GC, SFC and HPLC on the same sample within the same instrument is an attractive future possibility [83, 84].

Stationary phases for packed column SFC [85] are so far mainly based on those available for HPLC; special emphasis has been placed on the elimination of active sites—usually free hydroxyl groups on the silica and alumina support materials. Packed columns give rise to more theoretical plates per unit time than do capillary columns [86] and have higher sample capacities; they are therefore used when rapid analysis (incidentally faster with a supercritical mobile phase than for the same column in HPLC mode) [87] and larger detectable limits are required. The greater range of column packing selectivities may also be advantageous.

Wall-coated capillary columns may be operated [88], however, without the complication of stationary phase interaction with highly polar and reactive solutes. They also have the advantage of high efficiency consequent on the greater length, possible because of their permeability to the flow of supercritical fluids with low viscosities. Low volume flow rates also permit the use of toxic, inflammable or expensive fluids. Capillary columns allow greater sensitivity because of narrower peaks. Smaller pressure drops permit the maximum benefits of density programming.

Routine operation is so far only possible with minimum column diameters of 50 μm. While greater resolution is possible for smaller inside diameters, there are problems in the coating of smaller [89] diameter capillaries with a stationary phase film. On-column detection by UV, fluorescence, etc., is possible on 50 μm inside diameter capillaries.

The stationary phases for capillary SFC [88] are cross-linked to prevent solution in the mobile phase. While a number of alternative phases are available (e.g. cross-linked polyethyleneglycol or 'Carbowax'), most are based on the polysiloxane backbone with a variety of pendant groups: methyl, phenyl, octyl, biphenyl, cyanopropyl and, most recently, a smectic liquid crystal. These polymers have wide temperature ranges between glass transition and decomposition which allow particularly useful operating conditions for chromatography.

12. APPLICATIONS OF SUPERCRITICAL FLUID CHROMATOGRAPHY TO LIPIDS AND RELATED COMPOUNDS

SFC has established a niche as the most suitable technique for the separation of reactive, thermally labile and non-volatile compounds, among which groups are found many lipid related compounds. Thus, free fatty acids, sterols etc. are most usually derivatised to improve volatility for GC [90, 91], with consequent increase in analysis time. Triglycerides are commonly analysed by HPLC [66], but with difficulties consequent on variable detector response; GC can be employed, but there is always an inevitable suspicion that decomposition occurs at the high column temperatures necessary for analysis.

12.1 Packed Column Supercritical Fluid Chromatography

SFC analysis of lipids was first reported in 1984. Rawdon and Norris [92] used standard reverse-phase HPLC columns (ODS bonded silica) in the separation of oleic acid, mono-, di- and triglycerides, and soya and 'salad oil' triglycerides; the mobile phase was CO_2 modified with methanol with detection by UV absorption at 205 nm. While elution was very rapid (of the order of 2 min) resolution was inevitably limited. Perrin and Prevot [93] used adsorption mode SFC with a column packed with silica particles and a methanol-modified CO_2 mobile phase. Rapid separation (e.g. Figure 13) of mono-, di- and triglycerides was observed, with light-scattering, UV and FTIR detection (see Section 12.3); reduction of the methanol content of the mobile phase allowed a partial separation between 1,2- and 1,3-diglycerides. On an ODS column, further resolution of the individual bands (e.g. Figure 14) was possible, with separation of sunflower oil triglycerides on the basis of double bond number.

Underivatized fatty acids have been separated by SFC on a variety of packings (porous polymer [94, 95]; ODS [96], etc.); modification of the CO_2 was necessary for satisfactory chromatography on the ODS (silica based) packings, but not for the porous polymer. Modification of CO_2 with ethanol and dichloromethane gave improved separation of tocopherols with good resolution of α and β isomers on a silica column [93].

A number of attempts have been made to improve the selectivity of packed column SFC separations, particularly for glycerides with unsaturated fatty acid groups. A polar cyanopropyl silica column separated [97] saturated from unsaturated triglycerides with partial resolution of the latter, presumably by a dipole-induced dipole interaction. Similarly, partial separations were obtained for SFC on columns designed for olefin separation in HPLC with Ag^+ ions loaded onto silica [97].

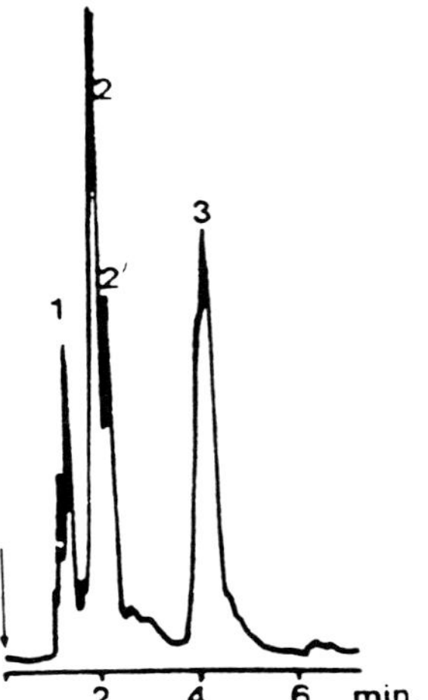

Figure 13. Separation of glycerides by SFC on a packed column. 1: triglycerides; 2: diglycerides; 3: monoglycerides. Column; 25 cm × 4 mm i.d. packed with 5-μm silica particles. Mobile phase: methanol (7.5%) modified CO_2 at 50°C. Detection: UV at 290 nm. (From Ref. [93], with permission.)

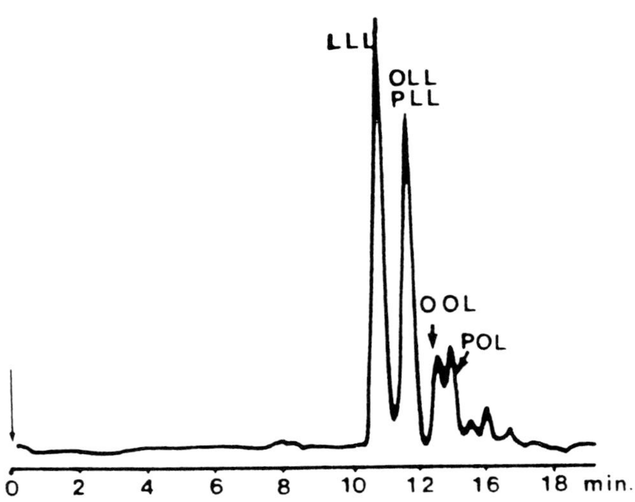

Figure 14. Separation by packed column SFC of triglycerides of sunflower oil. L: linoleic; O: oleic; P: palmitic. Column: 25 cm × 4 mm i.d. packed with 4 μm ODS modified silica particles. Mobile phase: methanol (0.7%) modified CO_2 at 40°C. Detection: UV at 290 nm. (From Ref. [93], with permission.)

12.2 Capillary Column Supercritical Fluid Chromatography

Improved SFC resolution for mixtures of lipids and related compounds has been achieved on a number of capillary columns. Early analyses by Chester [98] and White and Houck [99] demonstrated resolution similar to that available in capillary GC, but at much lower temperatures, typically below 200°C. The solubility of many lipids in CO_2 is sufficient to permit chromatography without modification of the mobile phase, although very highly polar compounds may require the addition of modifier; thus Raynor *et al.* were able to elute ecdysteroids with up to four hydroxyl substituents from a capillary column with CO_2 alone [100], but ecdysones with up to seven –OH groups were successfully chromatographed [100] (on a packed column) with methanol modified CO_2 as mobile phase (Figure 15). Kuei *et al.* preferred [101] to increase CO_2 solubility by analyte derivatization; glycosphingolipids with molecular weights between 1 and 2 $\times$ 10^3 were analysed by capillary SFC with CO_2 mobile phase after permethylation. Steroids from physiological fluids were profiled by capillary SFC with phosphorus-thermionic detection of thiophosphinic ester derivatives [102].

Choice of the appropriate stationary phase is vital in lipid analysis by capillary SFC, but effects are consistent since the column wall is inactive; this contrasts with the much less predictable situation for packed columns, with active support materials. Non-polar phases for capillary SFC, such as cross-linked methylpolysiloxanes [93, 98] allow separation of mono-, di- and triglycerides on the basis of carbon number (Figure 16). (Carbon number is the number of carbon atoms in the fatty acid moieties.) Excellent quantitation and retention time repeatability have been demonstrated [103] in such analyses (e.g. Table 10). More polar (phenylsiloxane, cyanopropylsiloxane and, particularly, Carbowax 20 M) phases yield further separation [104, 105] according to the degree of unsaturation. Thus Figure 17 illustrates the resolution of soybean oil triglycerides. It is noteworthy that pressure or density programming of the mobile phase is necessary in all such analyses if results are to be compared with those from HPLC and GC.

Huopalahti and co-workers have advocated [106] an SFC retention index system for non-polar phases, with a linear pressure programme, related to the carbon numbers of the triglycerides. Some larger shifts in the measured retention indices were attributed to asymmetric molecules containing *n*-butanoic residues. Triglyceride compositions for butter fat (Figure 18) and fish, rapeseed and soybean oil were readily obtained by this procedure (e.g. Table 11). In later work, the same group used [107] similar columns but with mass spectrometric detection (see Section 12.3). This allowed the slightly broadened single carbon-number peaks to be resolved into components from constituents with different degrees of unsaturation (Figure 18), thus account-

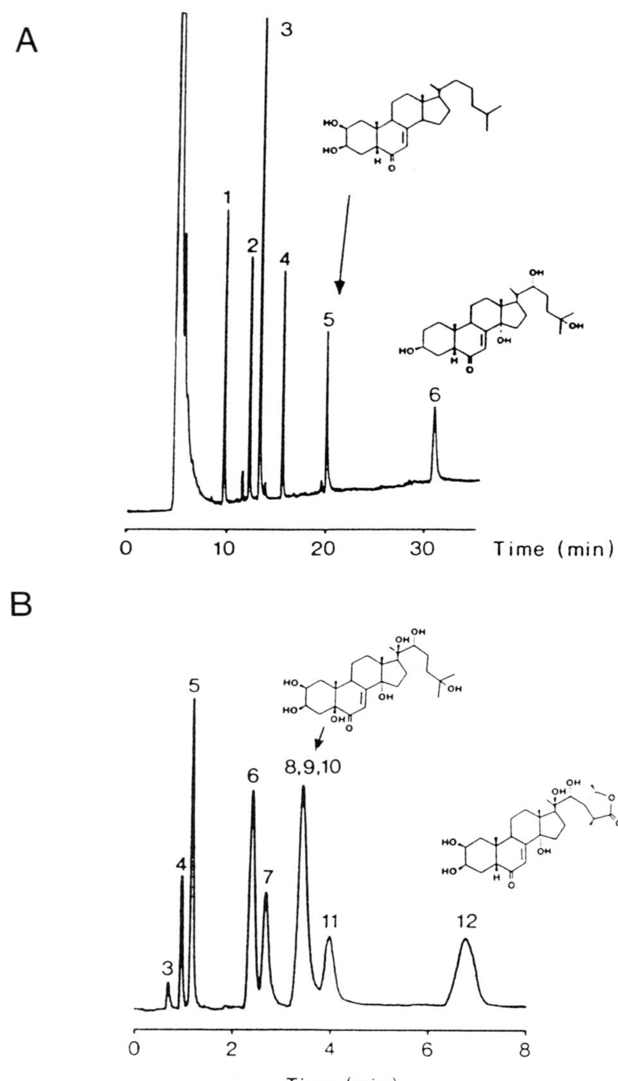

Figure 15. Separation of ecdysteroids by SFC on (A) capillary column (10 m × 50 μm) i.d. coated with 0.25 μm thick film of cross-linked cyanopropylmethylsiloxane. Mobile phase: CO_2 density programme from 0.4 to 0.71 g cm^{-3} at 0.015 g cm^{-3} min^{-1} after 5 min isoconfertic. Temperature 120°C. Detection: FID and (B) a packed column (25 cm × 4.6 mm i.d. packed with 5 μm cyanopropyl silica particles). Mobile phase: methanol (10%) modified CO_2. Detection UV at 215 nm. (From Ref. [100], with permission.)

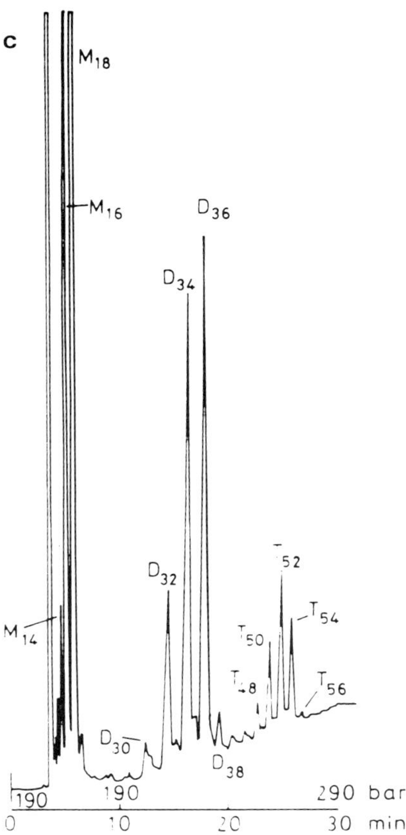

Figure 16. Separation by capillary SFC of glycerides. M: monoglycerides; D: diglycerides; T: triglycerides; subscript denotes carbon number. Column: 10 m × 100 μm i.d. coated with SE-54. Mobile phase: CO_2 pressure programme from 190 to 290 bar at 0.33 bar min^{-1}. Temperature: 170°C. Detection: FID. (From Ref. [104], with permission.)

ing for some of the differences in retention indices observed with FID detection (Table 11).

12.3 Supercritical Fluid Chromatography Detection

For CO_2 as mobile phase, the universal detection capability of the FID (identical response for chemically different solutes) makes it the detector of choice in the SFC of lipids and related materials. If a modifier is used, UV detection usually at 205 nm may be employed, but the different molar absorp-

Table 10. Repeatability of retention data and quantitation in the SFC analysis of triglycerides.

	Quantitation				Retention time (min)			
Run	$\% T_{36}$	$\% T_{42}$	$\% T_{48}$	$\% T_{54}$	t_{R36}	t_{R42}	t_{R48}	t_{R54}
1	20.76	24.89	26.03	28.32	16.41	20.84	24.44	27.55
2	20.71	24.55	26.41	28.32	16.39	20.83	24.44	27.54
3	20.90	24.84	26.36	27.89	16.41	20.86	24.44	27.54
4	20.63	24.73	26.30	28.34	16.45	20.86	24.42	27.53
X	20.75	24.75	26.27	28.22	16.41	20.85	24.43	27.54
σ	0.1134	0.1506	0.1694	0.2185	0.0251	0.015	0.010	0.008
σ_{rel} %	0.547	0.609	0.645	0.774	0.153	0.072	0.041	0.030

These data were measured on a 10 m × 100 μm SE-54 column. Column temperature: 150°C isotherm; fluid CO_2. Pressure programme: 190 bar isobaric for 10 min, then programmed from 190 to 290 bar at 5 bar min^{-1}. Compounds: trilaurin, trimyristin, tripalmitin, tristearin. (From J. High. Resolut. Chromatogr., Chromatogr. Commun., 9 (1986) 189, with permission.)

tivities of the analytes makes prior calibration necessary. Other universal detectors compatible with modified mobile phases include the light scattering detector (LSD) and the ion-mobility detector [108].

In the LSD the effluent from the restrictor of the end of the SFC column is nebulized and the resulting fine spray is evaporated as it passes down a heated drift tube to leave particles of the analyte from which light, from a continuous light source or a laser, is scattered and detected [109]. The LSD has been successfully applied [93] in the packed column separation of tocopherols and glycerides both with CO_2 alone, and with methanol-modified CO_2 as mobile phases, but no quantitative information is available.

The combination of SFC with MS detection is a powerful tool for both qualitative and quantitative analysis. A range of effective interfaces have been developed [70]; the most simple is that in which the restrictor is led directly into the ion source. The greatest sensitivity is achieved if the MS detection is in chemical ionization (CI) mode; electron impact (EI) spectra can be obtained, but detection limits are generally less good [110]. An attractive and relatively inexpensive procedure is afforded by the coupling of capillary SFC with a bench top mass spectrometer (e.g. the ion-trap detector).

Cholesterol is often used [111] to test SFC–MS systems since dehydration across the 3–4 double bond readily occurs on contact with 'active' sites in the apparatus. Kallio *et al.* used [107] EI SFC–MS in the analysis of triglycerides from butter. Monitoring of the toal-ion current to produce a universal response chromatogram gave broader chromatographic peaks because the column was overloaded to obtain the required MS sensitivity. In single-ion monitoring mode, however, the molecular ions from components of the peaks corresponding to triglycerides containing zero to three double bonds

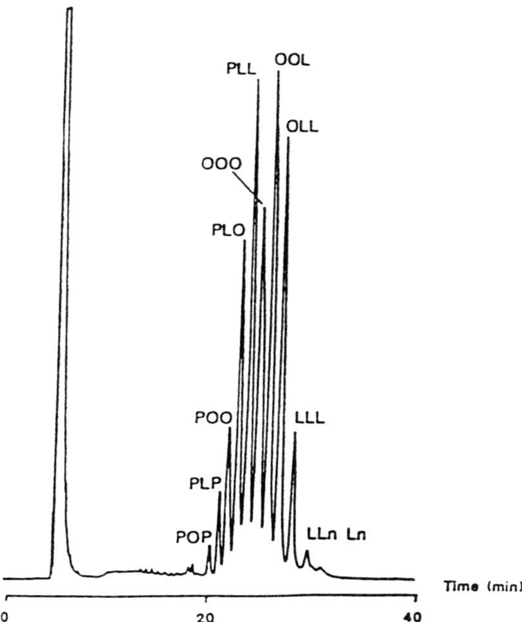

Figure 17. Separation by capillary SFC of triglycerides of soybean oil. P: palmitic; O: oleic; S: stearic; L: linoleic; Ln: linolenic. Column: 10 m × 50 μm coated with a 0.25 μm thick film of cross-linked Carbowax 20 M. Mobile phase: CO_2 density programmed from 0.25 to 0.43 g cm^{-3} at 0.005 g cm^{-3} min^{-1}. Temperature: 200°C: Detection: FID. (From Ref. [105], with permission.)

were recorded (Figure 18). CI MS spectra of ecdysteroids were obtained after SFC separation; to obtain more diagnostic fragment ions heating of the mass spectrometer jet was necessary [112].

The coupling of SFC with FTIR detection has many potential applications in the separation and positive identification of lipids. Two approaches are possible; a high pressure flowcell interface and a solvent elimination interface. Flowcells which allow universal and selective detection have been constructed for use with packed columns with volumes in the region of 8 μl. Hellgeth *et al.* [94] employed such a cell to detect free fatty acids (C_5–C_{18}) after SFC separation on a polymer column; the carbonyl absorption at 1730 cm^{-1} was used to reconstruct the chromatogram. The SFC separation of glycerides was monitored by Perrin and Prevot [93] with a flow cell and FTIR detection at 1743 cm^{-1}. Of major concern with this interface is the IR opacity of the mobile phase. At low pressures the CO_2 absorption bands obscure [69] the IR spectrum at 3500–3800 and 2200–2500 cm^{-1}. Although few organic compounds absorb in these regions, the Fermi resonance bands,

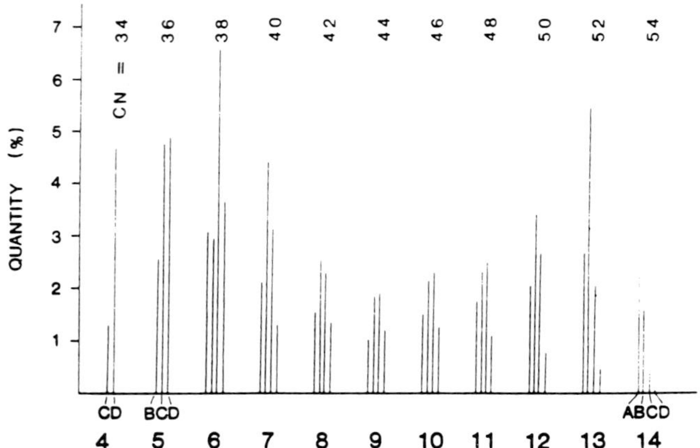

Figure 18. Distribution of triglycerides in butter fat according to carbon number (CN) and degree of unsaturated (A: 3; B: 2; C: 1; D: 0) double bonds, determined by capillary SFC with flame ionization and secondary ion mass spectrometric detection. Column: 5 m × 5 μm i.d. coated with a 0.2 μm film of DB-5 (95% dimethyl 5% diphenyl siloxane) Mobile phase: CO_2 pressure programmed from 13.8 to 27.6 MPa in 26 min and then to 34.5 MPa in 17 min after 2 min isobaric. Temperature: 150°C. (From Ref. [107], with permission.)

which increase in intensity, at 2070, 1944, 1387 and 1282 cm^{-1} reduce the quality of analyte IR spectra recorded for identification purposes. This problem is compounded when a modifier is added to the CO_2.

For capillary SFC the internal dimensions of the flowcell must be reduced: strictly, a volume of about 100 nl is required to avoid loss of chromatographic resolution but construction limitations and the requirements of IR sensitivity have led to a minimum volume [73] of 800 nl. The use of this cell with a make—up fluid accessory that reduces loss of chromatographic efficiency by halving the volume has been demonstrated; in further work, FTIR sensitivity was increased by stopping the mobile phase flow to isolate the analyte peak

Table 11. Retention indices (I_r) and quantitative analyses of the major triglycerides of rapeseed and soybean oils

I_r of peak	Rapeseed oil (%)	Soybean oil (%)
4990		2.8
5120	11.5	
5190		26.9
5370	84.3	
5400		69.3
5510	4.3	1.0

in the flowcell while the scans are recorded [113]. Capillary SFC with flow cell FTIR has been demonstrated in steroid analysis [114].

The second FTIR detection procedure involves [115] elimination of mobile phase (either pure or modified CO_2) by rapid depressurization. The non-volatile analytes separated on the capillary column are deposited as spots directly on an IR transparent support. The support is moved beneath the restrictor outlet so that chromatographed peaks are separated spatially for FTIR analysis using the beam condensing optics of a microscope accessory. This second method offers advantages in that it tolerates mobile phase additives and increases sensitivity because many spectra can be accumulated. Figure 19 illustrates application of solvent-elimination FTIR detection in the SFC analysis on steroids [69]. The complete spectrum obtained from less than 50 μg of the partially acetylated polyhydroxy sterol (Figure 19B) can be interpreted to indicate the structural features or compared with a library spectrum. This procedure was employed [115] in the analysis of the triglycerides of vegetable oil.

12.4 A Comparison of Supercritical Fluid Chromatography With Other Chromatographic Techniques In Lipid Analysis

An axiom of chromatographic analysis is that capillary GC should be employed wherever possible; the high diffusion rates in gases lead to the greatest resolution and speed of analysis. The applicability of GC in lipid analysis is somewhat limited, however, because of limited thermostability of many compounds. Thus, long-chain fatty acids are converted to methyl esters before GC analysis and steroids with often multiple hydroxyl substituents are also derivatized. SFC, which offers high resolution separations without derivatization, is clearly advantageous in the analysis of such compounds. For very high molecular weight lipids [101] such as phospholipids or glycosphingolipids with molecular weight up to 3000, SFC after methylation is the only high resolution method available.

For triglycerides, the position is less clear-cut. While various authors have claimed [116] that, particularly unsaturated, triglycerides can degrade [116] or polymerize [104] at the elevated temperatures necessary for elution in GC (350–400°C) others have shown routine application of high temperatures GC in direct triglyceride analysis. Sandra *et al.* [104] have pointed out that since a GC column can be operated much nearer its optimal linear mobile phase velocity than can an SFC column, very short analysis times with better resolution are possible in GC compared with SFC; Figure 20, for example compares the analysis of the triglycerides of milk chocolate by the two methods. Nonetheless, quantitation in the GC analysis of oils containing highly unsaturated triglycerides is likely to be more difficult, and it is probable that, in any case, repeated operation of capillary GC columns at tem-

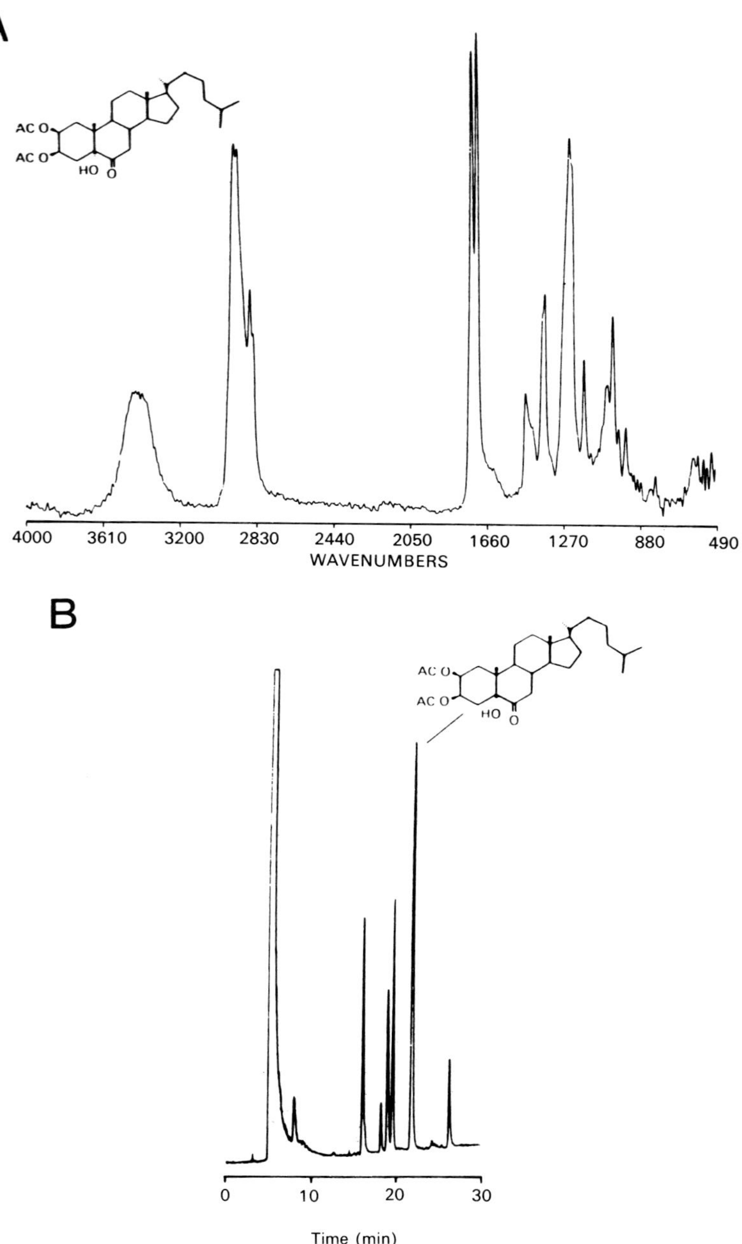

Figure 19. (A) FTIR spectrum of an acetylated steroid collected during (B) capillary SFC of a mixture. Chromatographic conditions as in Figure 15.

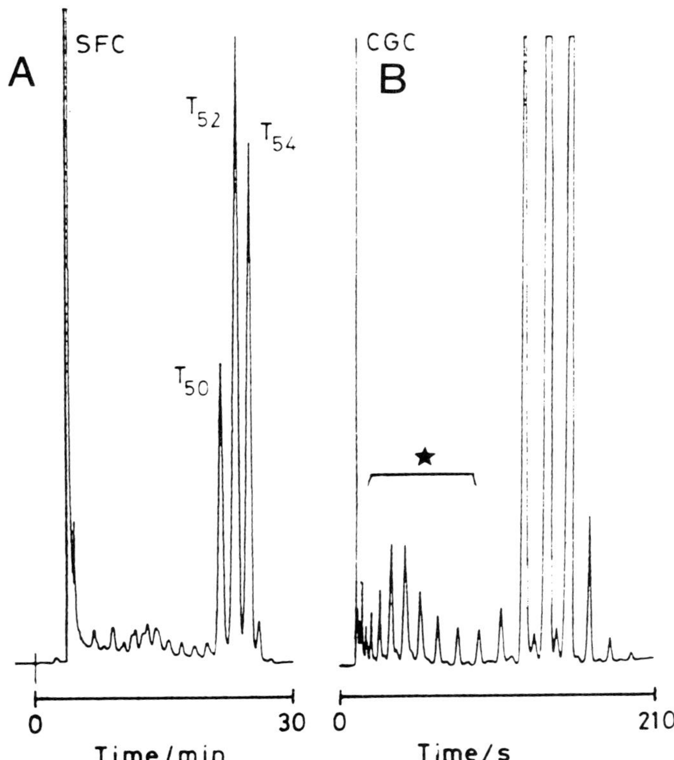

Figure 20. Analysis of triglycerides of milk chocolate by (A) capillary SFC (conditions as in Figure 16) and (B) capillary GC (column: 6 m × 25 i.d. OV-1. Temperature programme 290 to 350°C at 20°C min^{-1}; carrier gas H_2). (From Ref. [104], with permission.)

peratures up to 350°C results in impaired performance even if peak areas and retention times show small deviations [103] over consecutive runs.

Excellent selectivity can be achieved in HPLC analysis of lipids, e.g. triglycerides can be well resolved on silver ion loaded columns [117]. However, the absence of a strong chromophore in many lipid molecules means that sensitivity is low during HPLC analysis with UV detection, while response is not uniform. Refractive index detection can be used, but not for sensitive detection nor with gradient elution, and the LSD detection is gaining popularity [118].

In Figure 21 the separation [119] of a number of anabolic steroids by HPLC, and by SFC on both packed and capillary columns is compared. The most rapid analysis was achieved by packed column SFC, although with

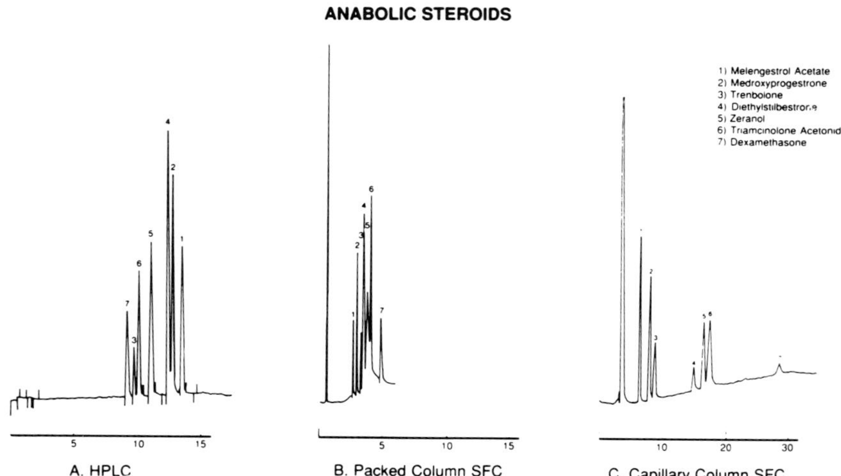

Figure 21. Analysis of a mixture of anabolic steroids by: (A) HPLC (column: octylsilane; mobile phase methanol/acetonitrile containing 20 mM ammonium formate; detection UV at 254 nm); (B) packed column SFC (column: cyanopropyl; mobile phase methanol modified CO_2 at 60°C; detection UV at 220 nm); (C) capillary DGV (column: 5 m × 50 μm cyanoprylmethylsiloxane; capillary mobile phase pressure programme CO_2 at 70°C; detection FID). (From Ref. [119], with permission.)

incomplete resolution; modification of the CO_2 with methanol was necessary to overcome the activity of column packing, and therefore detection was by UV absorption. The capillary column SFC separation took longer than that by HPLC, but resolution was complete; since unmodified CO_2 was the mobile phase, detection by FID was possible. Separation of steroid isomers by capillary SFC on a liquid crystalline stationary phase has been shown to be related to molecular shape [120].

Particular advantages of capillary SFC over HPLC in this context are thus the applicability of FID detection for ready quantitation, and the much simpler coupling to both FTIR and MS for compound identification. Application of SFC to a variety of lipids from wide-ranging origins [121–123] seems likely.

13. PREPARATIVE SUPERCRITICAL FLUID CHROMATOGRAPHY

If small quantities (milligrams) are required (for structure determination etc.), preparative scale SFC may be carried out [124] using conventional analytical packed column equipment. Separated materials may be collected,

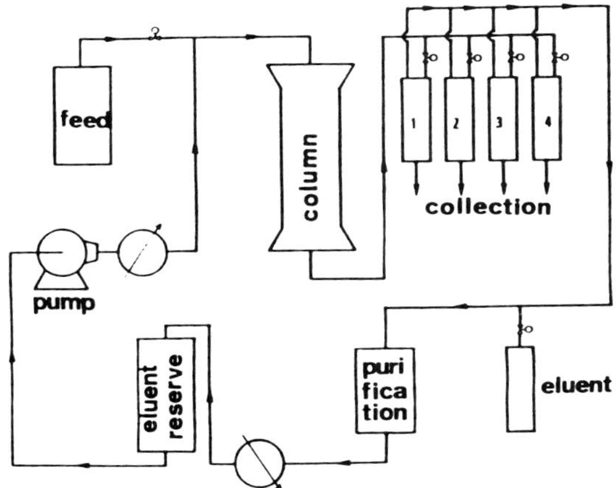

Figure 22. Schematic diagram of large-scale preparative SFC system. (From Ref. [124], with permission.)

if non-destructive detection is employed, in vessels in which depressurization of the column effluent occurs; a simple method is to immerse the restrictor tip in a suitable solvent and recover the material of interest by evaporation.

For industrial scale production (1 g to 1 kg per hour) larger scale equipment is necessary (e.g. Figure 22). Extensive discussions have appeared [124–127] on the technology required to operate columns with diameters up to 100 mm and packed by axial compression; commercial systems are on the market [126, 128]. Any mobile phase used in analytical packed column SFC can be used in preparative SFC, but most work so far reported has involved CO_2, chlorofluorocarbons, light hydrocarbons and CO_2 modified with alcohols, etc. The mobile phase is separated from the solute at the end of the column and recycled.

Particular attention has been paid to the means of collecting separated fractions so that products may be recovered free from solvent. After depressurization, the precipitated product may be separated [125] from the gaseous eluent by means of gravity or by means of a cyclone. Alternatively, the eluent may be collected while supercritical, frozen and the CO_2 allowed to sublime to leave the product.

Preparative SFC has been employed in the separation of a number of lipids and related compounds [124] including fatty acids [129]. Of particular interest is the purification [125, 130] of polyunsaturated fatty acid (PUFA) esters for medical and dietetic study in the prevention of heat disease. Fish oil methyl ester concentrates containing eicosapentaenoic acid (EPA) and docosahexaenoic acid (DHA) as methyl esters were separated on a 6 cm × 60 cm column packed with ODS modified silica with CO_2 mobile phase. Fractions containing EPA and DHA at, respectively, 56 and 78% purity were obtained at rates of 7.7 and 3.8 g h^{-1}.

Smaller dimension columns may be employed in preparative SFC if the eluent is recycled [126]. α- and β-tocopherols at, respectively, 85 and 70% purity were separated from wheat germ oil (itself obtained by supercritical fluid extraction) by recycle SFC on 10 mm silica gel columns [131].

REFERENCES

[1] Brunner, E., Hultenschmidt, G. and Schlichtharle, J. *J. Chem. Thermodynam.*, 19 (1987) 273.
[2] Kurnik, R. T. and Reid, R. C. *Fluid Phase Equil.*, 8 (1982) 93.
[3] Bartle, K. D., Clifford, A. A., Hawthorne, S. B., Langenfield, J. L., Miller, D. J. and Robinson, R. *J. Supercrit. Fluids*, 3 (1990).
[4] Tyrell, H. J. V. and Harris, K. R. *Diffusion in Liquids*. Butterworths, London, 1984.
[5] Reid, R. C., Prausnitz, J. M. and Sherwood, T. K. *The Properties of Gases and Liquids*. McGraw-Hill, New York, 1977.
[6] Hawthorne, S. B. and Miller, D. *J. Anal. Chem.*, 59 (1937) 1705.
[7] Hawthorne, S. B., Miller, D. J., Walker, D. and Wittington, D. in preparation.
[8] McHugh, M. A., Krukonis, V. J. *Supercritical Fluid Extraction—Principles and Practice*. Butterworth, Stoneham, 1986.
[9] Bartle, K. D., Clifford, A. A., Jafar, S. A. and Shilstone, G. F. *J. Phys. Chem. Ref. Data*, to be published.
[10] Johnston, K. P. and Eckert, C. A., *AICHEJ.*, 27 (1981) 773.
[11] Van Leer, R. A. and Paulaitis, M. E. *J. Chem. Eng. Data*, 25 (1980) 257.
[12] Kurnik, R. T., Holla, S. J. and Reid, R. C. *J. Chem. Eng. Data*, 26 (1981) 47.
[13] Ohgaki, K., Tsukahara, I., Semba, K. and Katayama, T. *Kagaku Kogaku Ronbunshu*, 13 (1987) 298.
[14] Tolley, K. K. and Tester, L. S. US Bureau of Mines Report of Investigations, 9216 (1989).
[15] Moradinia, I. and Teja, A. S. *Fluid Phase Equil.*, 28 (1986) 199.
[16] Kosal, E. and Holder, G. D., *J. Chem. Eng. Data*, 32 (1987) 148.
[17] Stahl, E., Schilz, W., Schutz, E. and Willing, E. *Angew. Chem. Int. Ed. Engl.*, 17 (1978) 731.
[18] Pritchard, A. M., Peakall, K. A., Smart, E. F. and Bignold, G. J. *Water Chem. Nucl. React. Syst.*, 4 (1986) 233.
[19] Kumar, S. K., Chhabrai, S. P., Reid, R. C. and Suter, U. W. *Macromolecules*, 20 (1987) 2550.
[20] Dobbs, J. M., Wong, J. M. and Johnston, K. P. *J. Chem. Eng. Data*, 31 (1986) 303.
[21] McHugh, M. A., Seckner, A. J. and Yogan, T. J. *Ind. Eng. Chem. Fundam.*, 23 (1984) 493.
[22] Schafer, K. and Baumann, W., Fresenius B. *Anal. Chem.*, 332 (1988) 122.
[23] Sako, S., Shibata, K., Ohgaki, K. and Katayama, T. *J. Supercrit. Fluids*, 2 (1989) 3.
[24] Chrastil, J. *J. Phys Chem.*, 86 (1982) 3016.
[25] Smith, R. D., Usdeth, H. R., Wright, D. W. and Yonker, C. R. *Sep. Sci. Tech.*, 22 (1987) 1065.
[26] Barker, I. K., Bartle, K. D. and Clifford, A. A. *Chem. Eng. Commun.*, 84 (1988) 4487.
[27] Bartle, K. D., Clifford, A. A. and Jafar, S. A. *J. Chem. Soc., Faraday Trans.*, 86 (1990) 855.

[28] Bartle, K. D., Clifford, A. A. and Jafar, S. A. *J. Chem. Eng. Ref. Data*, to be published.
[29] Peng, D. Y. and Robinson, D. B. *Ind. Eng. Chem. Fundam.*, 15 (1976) 59.
[30] Johnston, K. P., Peck, D. G. and Kim, S. *Ind. Eng. Chem. Res.*, 28 (1989) 1115.
[31] Bartle, K. D., Clifford, A. A. and Shilstone, G. F. *J. Supercrit. Fluids*, 2 (1989) 30.
[32] *CRC Handbook of Chemistry and Physics*, 52nd Edn. CRC Press, Florida, 1972.
[33] Bartle, K. D., Clifford, A. A. and Shilstone, G. F. To be presented at the *2nd International Conference on Supercritical Fluids*, Boston, 1991.
[34] Paulaitis, M. E., Penninger, J. M. L., Gray, R. D. and Davidson, P. *Chemical Engineering at Supercritical Fluid Conditions*. Ann Arbour Science, Ann Arbour, 1983.
[35] Charpentier, B. A. and Sevenants, M. R. (eds), *Supercritical Fluid Extraction and Chromatography* (ACS Symposium Series 366). ACS, Washington, DC, 1983.
[36] Johnston, K. P. and Penninger, J. M. L. *Supercritical Fluid Science and Technology* (ACS Symposium Series 406). ACS, Washington, DC, 1989.
[37] Moyler, D. A. In *Distilled Beverage Flavours—Recent Developments*. Piggott, J. R. and Patterson, A. (eds), Ellis Horwood, Chichester, 1989.
[38] Passino, H. *Ind. Eng. Chem.*, 41 (1949) 280.
[39] Dickinson, N. and Meyers, J. *J. Am. Oil Chem. Soc.*, 29 (1959) 235.
[40] Brunner, G. Fette, *Seifen, Anstrichm.*, 88 (1986) 464.
[41] Bamberger, T., Erickson, J. C. and Cooney, C. L. *J. Chem. Eng. Data*, 33 (1988) 327.
[42] Klein, T. and Schulz, *Ind. Eng. Chem. Res.*, 28 (1989) 1073.
[43] King, J. W., Bissler, R. L. and Friedrich, J. P. In Charpentier, B. A. and Sevenants, M. R. (eds), *Supercritical Fluid Extraction and Chromatography* (ACS Symposium Series 366). ACS, Washington, DC 1988.
[44] Polak, J. T., Balaban, M., Peplow, and Phlips, A. J. in Johnston, W. P. and Penninger, J. M. L., Carpenter, B. A. and Sevenants, M. R. (eds), *Supercritical Fluid Science and Technology* (ACS Symposium Series 106). ACS, Washington, DC, 1989.
[45] Stahl, B., Schutz, E. and Nangold, N. K. *J. Agric. Food Chem.*, 28 (1980) 1153.
[46] Stahl, E. and Quirin, K. W. presented at the *GVC Meeting*, Munster, 1982.
[47] Stahl, E., Quirin, K. W., Glatz, A., Gerard, D. and Rau, G. *Ber. Bunsenges. Phys. Chem.*, 88 (1984) 900.
[46] Friedrich, J. P. and Pryde, E. H. *J. Am. Oil Chem. Soc.*, 61 (1984) 223.
[49] Brogle, W., *Chem. Ind.*, 385 (1982).
[50] Coenen, H. and Kriegel, E. *Chem. Ing. Tech.*, MS 1162 (1983).
[51] Tiegs, C. and Peter, S. *Fette, Seifen, Anstrichm.*, 87 (1985) 231.
[52] Brunetti, L., Daghetta, A., Fedeli, E., Kikic, I. and Zanderighi, L. *J. Am. Oil Chem. Soc.*, 66 (1989) 209.
[53] Gmur, W., Bosset, J. O. and Plattner, E. *Lebensm.-Wiss u.-Technol.*, 19 (1986) 419.
[54] Stahl, E. and Glatz, A. *Fette, Seifen, Anstrichm.*, 86 (1984) 346.
[55] Willson, W. B., Stout, V. F., Gauglitz, E. J., Teeny, F. M. and Mudson, J. K. In Johnston, K. P. and Penninger, J. M. L. (eds), *Supercritical Fluid Science and Technology* (ACS Symposium Series 406). ACS, Washington, DC 1989.
[56] Eisenbach, W. *Ber. Bunsenges, Phys. Chem.*, 88 (1984) 382.
[57] Rizvi, S. S. M., Chao, R. R. and Liaw, Y. J. In Charpentier, B. A. and Sevenants, M. R. (eds), *Supercritical Fluid Extraction and Chromatography* (ACS Symposium Series 366). ACS, Washington, DC 1988.
[58] Peter, S., Schneider, M., Weidner, E. and Ziegelitz, R. *Chem. Eng. Technol.*, 10 (1987) 37.
[59] Hawthorne, S. B. *Anal. Chem.*, 62 (1990) 633A.
[60] King, J. W., Johnson, J. H. and Friedrich, J. P. *J. Agric. Food Chem.*, 37 (1989) 951.
[61] Gmur, W., Bosset, J. O. and Plattner, E. *J. Chromatogr.* 388 (1987) 335.
[62] King, J. W. *J. Chromatogr. Sci.*, 27 (1989) 355.

[63] Mam, K. S., Kapila, S., Pieczonka, G., Clevenger, T. E., Yanders, A. F., Viswanath, D. A. and Mallu, B. *Proceedings of the International Symposium on Supercritical Fluids*, French Chemical Society, Paris, France, 1988.

[64] Smith, R. M. (ed.) *Supercritical Fluid Chromatography*. Royal Society of Chemistry, London, 1988.

[65] Lee, M. L. and Markides, K. E. (eds), *Analytical Supercritical Fluid Chromatography*. Chromatography Conferences Inc. Provo, Utah, 1990.

[66] Hamilton, R. J. In *Analysis of Fats and Oils*. Hamilton, R. J. and Patterson, A. (eds), Elsevier, London, 1986, p. 243.

[67] Richter, B. E., Bornhop, D. J., Swanson, J. T., Wangsgaard, J. G. and Andersen, M. R. *J. Chromatogr. Sci.*, 27 (1989) 303.

[68] Bornhop, D. J. and Wangsgaard, J. D. *J. Chromatogr. Sci.*, 27 (1989) 293.

[69] Bartle, K. D., Raynor, M. W., Clifford, A. A., Davies, I. L., Kithinji, J. P., Shilstone, G. F., Chalmers, J. M. and Cook, B. W. *J. Chromatogr. Sci.*, 27 (1989) 283.

[70] Games, D. E., Berry, A. J., Mylchreest, I. C., Perkins, J. R. and Pleasance, S. In Smith, R. M. (ed.), *Supercritical Fluid Chromatography*. Royal Society of Chemistry, London, 1988, p. 159.

[71] Lee, M. L. and Markides, K. E. *J. High Res. Chromatogr.*, 9 (1986) 652.

[72] Moulder, R. M., Bartle, K. D., Clifford, A. A. and Davies, I. L. *Chromatographia*, 30 (1990) 618.

[73] Raynor, M. W., Clifford, A. A., Bartle, K. D., Reyner, C., Williams, A. and Cook, B. W. *J. Microcolumn. Sep.*, 1 (1989) 101.

[74] Raynor, M. W., Bartle, K. D., Davies, I. L., Williams, A., Clifford, A. A., Chalmers, J. M. and Cook, B. W. *Anal. Chem.*, 60 (1988) 427.

[75] Smith, I. D., Chapman, E. G. and Wright, B. W. *Anal. Chem.*, 57 91985) 2829.

[76] Peaden, P. A. and Lee, M. L. *J. Chromatogr.*, 269 (1983) 1.

[77] Schmidt, S., Blomberg, L. G. and Campbell, E. R., *Chromatographia*, 25 (1988) 775.

[78] Fields, S. M., Markides, K. E. and Lee, M. L. *J. Chromatogr.*, 406 (1987) 223.

[79] Bartle, K. D., Clifford, A. A., Kilhinji, J. P. and Shilstone, G. F. *J. Chem. Soc., Faraday Trans. I*, 84 (1988) 4487.

[80] Davies, I. L., Xu, B., Markides, K. E., Bartle, K. D. and Lee, M. L. *J. Microcolumn Sep.*, 1 (1989) 71.

[81] Payne, K. M., Davies, I. L., Bartle, K. D., Markides, K. E. and Lee, M. L. *J. Chromatogr.*, 447 (1989) 161.

[82] I. S. Lurie, *LC-GC*, 6 (1988) 1066.

[83] Yang, F. J. In Yang, F. J. (ed.) *Microbore Column Chromatography*. Marcel Dekker, New York, 1989, p. 1.

[84] Bartle, K. D., Davies, I. L., Raynor, M. W., Clifford, A. A. and Kithinji, J. P. *J. Microcolumn. Sep.*, 1 (1989) 63.

[85] Lee, M. L. and Markides, K. E. (eds), *Analytical Supercritical Fluid Chromatography*. Chromatography Conferences Inc., Provo, Utah, 1990, p. 64.

[86] Bartle, K. D. In Smith, R. M. (ed.) *Supercritical Fluid Chromatography*. Royal Society of Chemistry, London, 1988, p. 1.

[87] Randall, L. G. In Ahuja, S. (ed.), *Ultra High Resolution Chromatography* (ACS Symposium Series 250). ACS, Washington, DC, 1984, p. 135.

[88] Jones, B. A., Markides, K. E., Bradshaw, J. S. and Lee, M. L. *Chromatogr. Forum.*, May–June (1986) 38.

[89] Fields, S. M., Kong, R. C., Lee, M. L. and Peaden, P. A. J. High Res. *Chromatogr.*, 7 (1984) 312.

[90] Hammond, E. W. In Hamilton, R. J. and Patterson, A. (eds), *Analysis of Fats and Oils*. Elsevier, London, 1986, p. 113.

[91] Ackman, R. G. In Hamilton, R. J. and Patterson, A. (eds), *Analysis of Fats and Oils*. Elsevier, London, 1986, p. 137.
[92] Rawdon, M. G. and Norris, T. A. *Intern. Laboratory*, p. 12 (1984).
[93] Perrin, J. L. and Prevot, A. *Rev. Fr. Corps. Gras*, 35 (1988) 485.
[94] Hellgeth, J. W., Jordan, J. W., Taylor, L. T. and Khorassani, M. A. *J. Chromatogr. Sci.*, 24 (1986) 183.
[95] Liu, Y. and Yang, F. J. *J. Microcolumn Sep.*, 2 (1990) 245.
[96] Thiebaut, D., Caude, M. and Rosset, R. Analysis, 15 (1985) 528.
[97] Bartle, K. D., Clifford, A. A. and Jeffrey, B. unpublished measurements (1989).
[98] Chester, T. L. *J. Chromatogr.*, 299 (1984) 424.
[99] White, C. M. and Houck, R. K. *J. High Res. Chromatogr.*, 9 (1986) 424.
[100] Raynor, M. W., Kithinji, J. P., Barker, I. K., Bartle, K. D. and Wilson, I. D. *J. Chromatogr.*, 436 (1988) 497.
[101] Kuei, J., Her, G. R. and Reinhold, V. N., *Anal. Biochem.*, 172 (1988) 228.
[102] David, P. A. and Novotny, M. *J. Chromatogr.*, 461 (1989) 111.
[103] Proot, M., Sandra, P. and Geeraert, E. *J. High Res. Chromatogr.*, 9 (1986) 189.
[104] Sandra, P. in Smith, R. M. (ed.) *Supercritical Fluid Chromatography*. Royal Society of Chemistry, London, 1988, p. 137.
[105] Richter, B. E., Anderson, M. R., Knowles, D. E., Campbell, E. R., Porter, N. L., Nixon, L. and Later, D. W. In Charpentier, B. A. and Sevenants, M. R. (eds), *Supercritical Fluid Extraction and Chromatography* (ACS Symposium Series 366). ACS, Washington, DC, 1988, p. 179.
[106] Huopalahti, R., Laakso, P., Saaristo, J., Linko, R. R. and Kallio, H. *J. High Res. Chromatogr.*, 11 (1988) 899.
[107] Kallio, H., Laakso, P., Huopalahti, R. and Linko, R. *R. Anal. Chem.*, 61 (1989) 698.
[108] Hill, H. H. and Morrissey, M. In White, C. M. (ed.), *Modern Supercritical Fluid Chromatography*. Huthig, Heidelberg, 1988, p. 95.
[109] Hoffmann, S. and Greibrokk, T. *J. Microcolumn Sep.*, 1, 35 (1989).
[110] Voorhees, K. J., Zaugg, S. D. and DeLuca, S. J. In White, C. M. (ed.), *Modern Supercritical Fluid Chromatography*. Huthig, Heidelberg, 1988, p. 59.
[111] Owens, G. D., Burkes, L. J., Pinkston, J. D., Keough, T., Simms, J. R. and Lacey, M. P. In Charpentier, B. A. and Sevenants, M. R. (eds), *Supercritical Fluid Extraction and Chromatography* (ACS Symposium Series 366). ACS, Washington, DC, 1988), p. 191.
[112] Raynor, M. W., Kithinji, J. P., Bartle, K. D., Games, D. E., Mylchreest, I. C., Lafont, R., Morgan, E. D. and Wilson, I. D. *J. Chromatogr.*, 467 (1989) 292.
[113] Raynor, M. W., Bartle, K. D., Clifford, A. A. and Cook, B. W., *J. Microcolumn Sep.*, 2 (1990) 300.
[114] Shah, S., Ashraf-Khorassani, M. and Tyalor, L. T., *Chromatographia*, 25 (1988) 631.
[115] Pentoney, S. L., Shafer, K. H., Griffiths, P. R. and Fuoco, R. *J. High Res. Chromatogr.*, 9 (1986) 169.
[116] Mares, P., Skorepa, J., Sinkelkove, E. and Turzicka, E. *J. Chromatogr.*, 273 (1983) 172.
[117] Christie, W. W. *J. Chromatogr.*, 454 (1988) 273.
[118] Palmer, A. J. and Palmer, F. J. *J. Chromatogr.*, 465 (1989) 369.
[119] Richter, B. E. and Knowles, D. E. *Lee Scientific Applications Note Number 008*. Salt Lake City, Utah, 1990.
[120] Chang, H. C., Markides, K. E., Bradshaw, J. S. and Lee, M. L., *J. Microcolumn Sep.*, 1 (1989) 131.
[121] Holzer, G. C., Kelly, P. J. and Jones, W. J. *J. Microbiol. Methods*, 8, (1988) 161.
[122] Sakaki, K., Sako, T., Yokochi, T., Suzuki, O. and Hakuta, T. *Yukagaku*, 37 (1988) 54.
[123] Normura, A., Yamada, J., Tsunoda, K., Sakaki, K. and Yokuchi, T. *Anal. Chem.*, 61 (1989) 2076.

[124] Berger, C. and Perrut, M. *J. Chromatogr.*, 505 (1990) 37.
[125] Berger, C., Justforgues, P. and Perrut, M. *Proceedings of the International Symposium on Supercritical Fluids*, French Chemical Society, Paris, 1988, p. 397.
[126] Saito, M., Yamauchi, Y., Hondo, T. and Senda, M. *Proceedings of the International Symposium on Supercritical Fluids*, French Chemical Society, Paris 1988, p. 381.
[127] Alkio, M., Harvala, T. and Komppa, V. *Proceedings of the International Conference on Supercritical Fluids*, French Chemical Society, Paris, 1988, p. 389.
[128] *Prochrom Preparative Scale Supercritical Fluid Chromatography*, Prochrom, Champigneulles, France.
[129] Jusforgues, P. and Perrut, M. European Patent Application EP254610 A1 (1988).
[130] Alkio, M. and Komppa, V. *Kem.-Kemi*, 17 (1990) 354.
[131] Saito, M. and Yamauchi, Y. *J. Chromatogr.*, 505 (1990) 257.
[132] King, M. B., Alderson, D. A., Fallaha, F. H., Kassim, D. H. Kassim, K. M., Sheldon, J. R. and Mahmud, R. S. In Paulaitis, M. E., Penninger, J. M. L., Gray, R. D. and Davidson, P. (eds), *Chemical Engineering at Supercritical Fluid Conditions*. Ann Arbour Science, Ann Arbour, 1983.
[133] Kramer, A. and Thodos, G., *J. Chem. Eng. Data* 33 (1988) 230.
[134] Schmitt, W. J. and Reid, R. C. *Chem. Eng. Commun.*, 64 (1988) 155.
[135] Czubryt, J. J., Myers, M. N. and Giddings, J. C. *J. Phys. Chem.* 74 (1970) 4202.
[136] King, M. B., Bott, T. R., Barr, M. J., Mahmud, R. S. and Sanders, N. *Sep. Sci. Tech.*, 22 (1987) 1103.
[137] Fattori, M., Bulley, N. R. and Neisen, A. *J. Am. Oil Chem. Soc.*, 65 (1988) 968.
[138] Friedrich, J. P. US Patent 4466923 (1984).
[139] Stahl, E. and Quirin, K. W. *Fluid Phase Equil.*, 10 (1983) 269.
[140] de Pilipi, R. P. *Chem. Ind.*, 390 (1982).
[141] Schafer, K. and Baumann, W. Fresenius *Z. Anal. Chem.*, 332 (1988) 122.

INDEX

Advances in Biosensors

Edited by **A.P.F. Turner,** *Biotechnology Centre, Cranfield Institute of Technology, England*

Biosensors have captured the imagination of the world's scientific and commercial communities by combining interdisciplinary skills of biologists, physicists, chemists, and engineers to provide innovative solutions to analytical problems. Biosensors are applicable to clinical diagnostics, food analysis, cell culture monitoring, environmental control, and various military situations. Ever increasing demands for rapid and convenient analyses of a wide variety of materials in diverse locations has led to intense interest in the fusion of biology and electronics which mimics our principal concern: the effect of materials and environments on living systems.

This new series *Advances in Biosensors* will present a unique compendium of research papers, in which eminent authorities in the field of biosensors provide an up-to-date overview of their laboratory's contribution, summarizing the primary research as it has appeared, possibly scattered, in the journal and conference literature, and reflecting on their findings. The net result will be intense, yet highly readable accounts of the state of the art at this leading edge of analytical technology.

Volume 1, 1991, 296 pp.
ISBN 1-55938-240-6

CONTENTS; Preface, *A.P.F. Turner.* **Whole-Organism Based Biosensors and Microbiosensors,** *Izumi Kubo, Soka University, Tokyo, Koji Sode and Isao Karube, University of Tokyo.* **Electro-chemical Biosensors: Application to Some Real Problems,** *Marco Mascini, University of Florence and Danilla Mosconee, Universita Tor Vergata, Rome.* **Enzymatic Amplification and Elimination in Biosensors,** *Florian Schubert, Ulla Wollenebrger, Dorothea Pfeiffer and Frieder W. Scheler, Academy of Sciences of the GDR, Berlin-Buch.* **Biosensors Based on Modified Electrodes,** *J.J. Kulys, Lithuanian Academy of Sciences.* **Mediated Electro-chemistry: A Practical Approach to Biosensing,** *Marco Cardosi and Anthony P.F. Turner, Cranfield Institute of Technology.* **Eclectic Immunoassay: An Electrochemical Approach,** *Sarah H. Jenkins, H. Brian Halsall and William R. Heinemann, University of Cincinnati, Ohio.* **Optical Immunosensors: An Overview,** *G.A. Robinson, Serono Diagnostics Ltd, Woking.* **Electrochemical, Piezoelectric and Fibre-Optic Biosensors,** *George G. Guilbault, University of New Orleans and Rolf D. Schmid, GBF, Braunschweig.*

Volume 2, In preparation, Summer 1992
ISBN 1-55938-270-8

Organized Assemblies in Chemical Analysis

Edited by **Willie L. Hinze,** *Department of Chemistry Wake Forest University*

Volume 1, Reversed Micelles
In preparation, Fall 1991
ISBN 1-55938-336-4

CONTENTS: Reversed Micelles and Water-in-Oil Microemulsions: Formation and Some Relevant Properties, *Omar A. El Seoud, Universidade de Sao Paulo, Brazil.* **Overview of the Utilization of Reversed Micellar Media in Analytical Spectroscopy,** *Willie L. Hinze, Wake Forest University.* **Thermal Lens in Reversed Micelles and Related Media,** *Chieu D. Tran, Marquette University.* **Salt and Surfacrant Effects on Protein Solubilization in ACT-Isooctane Reversed Micelles,** *Brian D. Kelly, Reza S. Rahaman, and T. Alan Hatton, Massachusetts of Technology.* **Membrane-Based Protein Separations with Reversed Micelles,** *Weiyong Li and Daniel W. Armstrong, University of Missouri-Rolla.* **Reverse Micelles and Microorganization in Supercritical Fluids,** *Susan V. Olesik, Ohio State University.*

Biology of Carbohydrates

Edited by **Victor Ginsberg,** *National Institute of Health, Bethseda* and **Phillips W. Robbins,** *Massachusetts Institute of Health.*

In a continuation of their comprehensive, but incomplete survey of the biological roles played by complex carbohydrates, Ginsberg and Robbins draw together a distinguished set of contributions from leading researchers on some of the most important recent advances in carbohydrate biology and biochemistry.

The first three chapters in Voulme 3 of this series describe the important function played by carbohydrates in hormone action, circulatory haemostasis and certain biological 'targeting' reactions. Chapter 4 summarizes the vast literature on glycosidase inhibitors and shows how these inhibitors may be used to explore questions of structure and function on biotechnology and vice versa. As with the other areas reviewed in this volume, this area is set to expand rapidly in the future. It is thus the hope that these reviews, and the continuing series, will act as guideposts to the most exciting of these future developments.

Volume 3, In preparation, Spring 1991
ISBN 1-55938-014-4

CONTENTS: Introduction, *Victor Ginsberg, NIH, and Phillips W. Robbins, Massachusetts Institute of Technology.* **Structure, Synthesis and Function of the Aspargine-Linked Oligosaccharides on Pituitary Glycoprotein Hormones,** *Jacques U. Baenziger and Eric D. Green, Washington University School of Medicine.* **Regulation of the Blood Coagulation Mechanism by Anticoagulantly Active Heparin Sulphate Proteoglycans,** *James A. Marcum, Harvard Medical School and Robert D. Rosenberg, Massachusetts Institute of Technology.* **Calcium-Dependent Carbohydrate-Recognition Domains in Animal Lectins,** *Kurt Drickamer, Columbia University.* **Inhibitors of the Addition, Modification or Processing of the Oligosaccharide Chains of the N-Linked Glycooproteins,** *Alan D. Elbein, University of Texas Health Center.* **Glycosylation of Recombinant Proteins,** *James Rasmussen, Genzyme Corp., Boston.*

JAI PRESS

JAi